T0139955

Springer Optimization and Its Applications

Volume 166

Aims and Scope

Optimization has continued to expand in all directions at an astonishing rate. New algorithmic and theoretical techniques are continually developing and the diffusion into other disciplines is proceeding at a rapid pace, with a spot light on machine learning, artificial intelligence, and quantum computing. Our knowledge of all aspects of the field has grown even more profound. At the same time, one of the most striking trends in optimization is the constantly increasing emphasis on the interdisciplinary nature of the field. Optimization has been a basic tool in areas not limited to applied mathematics, engineering, medicine, economics, computer science, operations research, and other sciences.

The series **Springer Optimization and Its Applications** (SOIA) aims to publish state-of-the-art expository works (monographs, contributed volumes, textbooks, handbooks) that focus on theory, methods, and applications of optimization. Topics covered include, but are not limited to, nonlinear optimization, combinatorial optimization, continuous optimization, stochastic optimization, Bayesian optimization, optimal control, discrete optimization, multi-objective optimization, and more. New to the series portfolio include Works at the intersection of optimization and machine learning, artificial intelligence, and quantum computing.

Volumes from this series are indexed by Web of Science, zbMATH, Mathematical Reviews, and SCOPUS.

More information about this series at http://www.springer.com/series/7393

Alexander J. Zaslavski

Turnpike Theory for the Robinson–Solow–Srinivasan Model

 Springer

Alexander J. Zaslavski
Department of Mathematics
Technion – Israel Institute of Technology
Haifa, Israel

ISSN 1931-6828   ISSN 1931-6836   (electronic)
Springer Optimization and Its Applications
ISBN 978-3-030-60309-0   ISBN 978-3-030-60307-6   (eBook)
https://doi.org/10.1007/978-3-030-60307-6

Mathematics Subject Classification: 49J24, 49J99, 49K15, 49K40, 90C31

This Springer imprint is published by the registered company Springer Nature Switzerland AG
The registered company address is: Gewerbestrasse 11, 6330 Cham, Switzerland

Preface

The growing importance of the turnpike theory and infinite horizon optimal control has been recognized in recent years. This is not only due to impressive theoretical developments but also because of numerous applications to engineering, economics, life sciences, etc. This book is devoted to the study of a class of optimal control problems arising in mathematical economics, related to the Robinson–Solow–Srinivasan model. In the 1960s this model was introduced by Robinson, Solow, and Srinivasan and was studied by Robinson, Okishio, and Stiglitz. In 2005 this model was revisited by M. Ali Khan and T. Mitra and till now is an important and interesting topic in mathematical economics. As usual, for the Robinson–Solow–Srinivasan model, the existence of optimal solutions over infinite horizon and the structure of solutions on finite intervals are under consideration. In our books [117, 121], we study a class of discrete-time optimal control problems which describe many models of economic dynamics except of the Robinson–Solow–Srinivasan model. This happens because some assumptions posed in [117, 121], which are true for many models of economic dynamics, do not hold for the Robinson–Solow–Srinivasan model. Namely, for many models of economic dynamics, the turnpike is a singleton, and a local controllability property holds in a neighborhood of the turnpike. For the Robinson–Solow–Srinivasan model, the turnpike is a singleton too, but the local controllability property does not hold. This makes the situation more difficult and less understood. Nevertheless, we show in this book that the turnpike theory presented in [117] is extended for the Robinson–Solow–Srinivasan model.

In Chapter 1 we discuss turnpike properties for some classes of discrete-time optimal control problems. In Chapter 2 we present the description of the Robinson–Solow–Srinivasan model and discuss its basic properties. In particular, we show the existence of weakly optimal programs and good programs and prove an average turnpike property. Infinite horizon optimal control problems, related to the Robinson–Solow–Srinivasan model, are studied in Chapter 3, where we establish a convergence of good programs to the golden-rule stock, show the existence of overtaking optimal programs and analyze their convergence to the golden-rule stock, and consider some properties of good programs.

Turnpike properties for the Robinson–Solow–Srinivasan model are analyzed in Chapter 4. To have these properties means that the approximate solutions of the problems are essentially independent of the choice of an interval and endpoint conditions. It is shown that these turnpike properties hold and that they are stable under perturbations of an objective function.

In Chapter 5 we study infinite horizon optimal control problems related to the Robinson–Solow–Srinivasan model with a nonconcave utility function. In particular, we establish the existence of good programs and optimal programs using different optimality criterions.

In Chapter 6 we study infinite horizon optimal control problems with nonautonomous optimality criterions. The utility functions, which determine the optimality criterion, are nonconcave. The class of models contains, as a particular case, the Robinson–Solow–Srinivasan model. We establish the existence of good programs and optimal programs.

Chapter 7 contains turnpike results for a class of discrete-time optimal control problems. These control problems arise in economic dynamics and describe the one-dimensional nonstationary Robinson–Solow–Srinivasan model. In Chapter 8 we continue to study the Robinson–Solow–Srinivasan model and compare different optimality criterions. In particular, we are interested in good programs and agreeable and weakly maximal programs.

Chapter 9 is devoted to the study of the turnpike properties for the Robinson–Solow–Srinivasan model. The utility functions, which determine the optimality criterion, are nonconcave. We show that the turnpike properties hold and that they are stable under perturbations of an objective function. Moreover, we consider a class of RSS models which is identified with a complete metric space of utility functions. Using the Baire category approach, we show that the turnpike phenomenon holds for most of the models.

In Chapter 10 we consider the one-dimensional autonomous Robinson–Solow–Srinivasan model, a multiplicity of optimal programs under certain conditions, and properties of the optimal policy correspondence. The continuous-time Robinson–Solow–Srinivasan model is studied in Chapter 11.

The author believes that this book will be useful for researches interested in the turnpike theory and infinite horizon optimal control and their applications.

Haifa, Israel   Alexander J. Zaslavski
December 28, 2019

Contents

Chapter 1
Introduction

In this chapter we discuss turnpike properties and optimality criterions over infinite horizon for three classes of discrete-time dynamic optimization problems. The first class contains convex unconstrained dynamic optimization problems, the second one is a class of constrained problems without convexity (concavity) assumptions, and the third class of problems is related to the the Robinson–Solow–Srinivasan model.

1.1 The Turnpike Phenomenon for Convex Discrete-Time Problems

The study of the existence and the structure of solutions of optimal control problems and dynamic games defined on infinite intervals and on sufficiently large intervals has been a rapidly growing area of research [2, 4, 11, 16, 19, 23–26, 36, 62, 68, 69, 71, 93, 100, 106, 117, 120, 121, 124] which has various applications in engineering [1, 58, 125, 126], in models of economic growth [3, 6, 10, 17, 28, 32–34, 37–50, 61, 67, 72, 76, 80, 82, 89, 103, 108, 110, 116], in infinite discrete models of solid-state physics related to dislocations in one-dimensional crystals [7, 90], and in the theory of thermodynamical equilibrium for materials [21, 59, 64–66]. Discrete-time optimal control problems were considered in [5, 12, 15, 22, 27, 35, 51–54, 92, 94, 95, 98, 99, 104, 105, 111, 114, 115, 122], finite-dimensional continuous-time problems were analyzed in [13, 14, 18, 20, 55, 57, 60, 63, 74, 96, 97, 101, 102, 107, 112, 113, 119], infinite-dimensional optimal control was studied in [8, 9, 30, 75, 81, 88, 123], while solutions of dynamic games were discussed in [29, 31, 56, 91, 109, 118]. In this book we are interested in optimal control problems related to the Robinson–Solow–Srinivasan model, which was introduced in the 1960s by Robinson [77], Solow [83], and Srinivasan [84] and was studied by Robinson, Okishio, and Stiglitz [73, 78, 85–87]. Recently, the Robinson–Solow–Srinivasan model was studied by

© The Author(s), under exclusive license to Springer Nature Switzerland AG 2020
A. J. Zaslavski, *Turnpike Theory for the Robinson–Solow–Srinivasan Model*, Springer Optimization and Its Applications 166, https://doi.org/10.1007/978-3-030-60307-6_1

M. Ali Khan and T. Mitra [37–45], M. Ali Khan and Piazza [46–49], M. Ali Khan
and A. J. Zaslavski [51–54], and A. J. Zaslavski [92, 94, 98–101, 103–108, 110, 111,
114–116].

In this section we discuss turnpike properties and optimality criterions over
infinite horizon for a class of convex dynamic unconstrained optimization problems.

Let R^n be the n-dimensional Euclidean space with the inner product $xy = \sum_{i=1}^{n} x_i y_i$, $x, y \in R^n$ which induces the norm

$$|x| = \left(\sum_{i=1}^{n} x_i^2 \right)^{1/2}, \ x = (x_1, \ldots, x_n) \in R^n.$$

Let $v : R^n \times R^n \rightarrow R^1$ be bounded from below function. We consider the
minimization problem

$$\sum_{i=0}^{T-1} v(x_i, x_{i+1}) \rightarrow \min, \tag{P_0}$$

such that $\{x_i\}_{i=0}^{T} \subset R^n$ and $x_0 = z$, $x_T = y$,

where T is a natural number and the points $y, z \in R^n$.

The interest in discrete-time optimal problems of type (P_0) stems from the study
of various optimization problems which can be reduced to it, e.g., continuous-time
control systems which are represented by ordinary differential equations whose cost
integrand contains a discounting factor [57], tracking problems in engineering [1,
58, 125], the study of Frenkel–Kontorova model [7, 90] and the analysis of a long
slender bar of a polymeric material under tension in [21, 59, 64–66].

In this section we suppose that the function $v : R^n \times R^n \rightarrow R^1$ is strictly convex
and differentiable and satisfies the growth condition

$$v(y, z)/(|y| + |z|) \rightarrow \infty \text{ as } |y| + |z| \rightarrow \infty. \tag{1.1}$$

We intend to study the behavior of solutions of the problem (P_0) when the points
y, z and the real number T vary and T is sufficiently large. Namely, we are interested
to study a turnpike property of solutions of (P_0) which is independent of the length
of the interval T, for all sufficiently large intervals. To have this property means,
roughly speaking, that solutions of the optimal control problems are determined
mainly by the objective function v and are essentially independent of T, y, and z.
Turnpike properties are well-known in mathematical economics. The term was first
coined by Samuelson in 1948 (see [82]) where he showed that an efficient expanding
economy would spend most of the time in the vicinity of a balanced equilibrium
path (also called a von Neumann path). This property was further investigated for
optimal trajectories of models of economic dynamics (see, e.g., [61, 67, 80] and the
references mentioned there). Many turnpike results are collected in [93, 117].

In order to meet our goal, we consider the auxiliary optimization problem

$$v(x, x) \to \min, \ x \in R^n. \tag{P_1}$$

It follows from the strict convexity of v and (1.1) that the problem (P_1) has a unique solution $\bar{x}$. Let

$$\nabla v(\bar{x}, \bar{x}) = (l_1, l_2), \tag{1.2}$$

where l_1, $l_2 \in R^n$. Since $\bar{x}$ is a solution of (P_1), it follows from (1.2) that for each $h \in R^n$,

$$l_1 h + l_2 h = (l_1, l_2)(h, h)$$
$$= \lim_{t \to 0^+} t^{-1}[v(\bar{x} + th, \bar{x} + th) - v(\bar{x}, \bar{x})] \geq 0.$$

Thus

$$(l_1 + l_2)h \geq 0 \text{ for all } h \in R^n,$$

$l_2 = -l_1$ and

$$\nabla v(\bar{x}, \bar{x}) = (l_1, -l_1), \tag{1.3}$$

For each $(y, z) \in R^n \times R^n$, set

$$L(y, z) = v(y, z) - v(\bar{x}, \bar{x}) - \nabla v(\bar{x}, \bar{x})(y - \bar{x}, z - \bar{x})$$
$$= v(y, z) - v(\bar{x}, \bar{x}) - l_1(y - z). \tag{1.4}$$

It is not difficult to verify that the function $L : R^n \times R^n \to R^1$ is differentiable and strictly convex. It follows from (1.1) and (1.4) that

$$L(y, z)/(|y| + |z|) \to \infty \text{ as } |y| + |z| \to \infty. \tag{1.5}$$

Since the functions v and L are both strictly convex, it follows from (1.4) that

$$L(y, z) \geq 0 \text{ for all } (y, z) \in R^n \times R^n \tag{1.6}$$

and

$$L(y, z) = 0 \text{ if and only if } y = \bar{x}, \ z = \bar{x} \tag{1.7}$$

[70, 79].

We claim that the function $L : R^n \times R^n \to R^1$ has the following property:

(C) If a sequence $\{(y_i, z_i)\}_{i=1}^{\infty} \subset R^n \times R^n$ satisfies the equality

$$\lim_{i \to \infty} L(y_i, z_i) = 0,$$

then

$$\lim_{i \to \infty} (y_i, z_i) = (\bar{x}, \bar{x}).$$

Assume that a sequence $\{(y_i, z_i)\}_{i=1}^{\infty} \subset R^n \times R^n$ satisfies $\lim_{i \to \infty} L(y_i, z_i) = 0$. In view of (1.5), the sequence $\{(y_i, z_i)\}_{i=1}^{\infty}$ is bounded. Let (y, z) be its limit point. Then it is easy to see that the equality

$$L(y, z) = \lim_{i \to \infty} L(y_i, z_i) = 0$$

holds and by (1.7) $(y, z) = (\bar{x}, \bar{x})$. This implies that $(\bar{x}, \bar{x}) = \lim_{i \to \infty} (y_i, z_i)$.

Thus the property (C) holds, as claimed.

Consider an auxiliary minimization problem

$$\sum_{i=0}^{T-1} L(x_i, x_{i+1}) \to \min, \qquad (P_2)$$

such that $\{x_i\}_{i=0}^{T} \subset R^n$ and $x_0 = z, \ x_T = y$,

where T is a natural number and the points $y, z \in R^n$.

It follows from (1.4) that for any integer $T \geq 1$ and any sequence $\{x_i\}_{i=0}^{T} \subset R^n$, we have

$$\sum_{i=0}^{T-1} L(x_i, x_{i+1}) = \sum_{i=0}^{T-1} v(x_i, x_{i+1}) - Tv(\bar{x}, \bar{x}) - \sum_{i=0}^{T-1} l_1(x_i - x_{i+1})$$

$$= \sum_{i=0}^{T-1} v(x_i, x_{i+1}) - Tv(\bar{x}, \bar{x}) - l_1(x_0 - x_T). \qquad (1.8)$$

Relation (1.8) implies that the problems (P_0) and (P_2) are equivalent. Namely, $\{x_i\}_{i=0}^{T} \subset R^n$ is a solution of the problem (P_0) if and only if it is a solution of the problem (P_2).

Let T be a natural number and $\Delta \geq 0$. A sequence $\{x_i\}_{i=0}^{T} \subset R^n$ is called (Δ)-optimal if for any sequence $\{x_i'\}_{i=0}^{T} \subset R^n$ satisfying $x_i = x_i', i = 0, T$, the inequality

$$\sum_{i=0}^{T-1} v(x_i, x_{i+1}) \leq \sum_{i=0}^{T-1} v(x'_i, x'_{i+1}) + \Delta$$

holds. Clearly, if a sequence $\{x_i\}_{i=0}^{T} \subset R^n$ is (0)-optimal, then it is a solution of the problems (P_0) and (P_2) with $z = x_0$ and $y = x_T$.

We prove the following existence result.

Proposition 1.1 *Let $T > 1$ be an integer and $y, z \in R^n$. Then the problem (P_0) has a solution.*

Proof It is sufficient to show that the problem (P_2) has a solution. Consider a sequence $\{x'_i\}_{i=0}^{T} \subset R^n$ such that $x'_0 = z$, $x'_T = y$. Set

$$M_1 = \sum_{i=0}^{T-1} L(x'_i, x'_{i+1})$$

and

$$M_2 = \inf \left\{ \sum_{i=0}^{T-1} L(x_i, x_{i+1}) : \{x_i\}_{i=0}^{T} \subset R^n, \ x_0 = z, \ x_T = y \right\}. \tag{1.9}$$

Clearly,

$$0 \leq M_2 \leq M_1.$$

We may assume without loss of generality that

$$M_2 < M_1. \tag{1.10}$$

There exists a sequence $\{x_i^{(k)}\}_{i=0}^{T} \subset R^n$, $k = 1, 2, \ldots$ such that for any natural number k,

$$x_0^{(k)} = z, \ x_T^{(k)} = y \tag{1.11}$$

and

$$\lim_{k \to \infty} \sum_{i=0}^{T-1} L\left(x_i^{(k)}, x_{i+1}^{(k)}\right) = M_2. \tag{1.12}$$

In view of (1.10)–(1.12), we may assume that

$$\sum_{i=0}^{T-1} L\left(x_i^{(k)}, x_{i+1}^{(k)}\right) < M_1 \text{ for all integers } k \geq 1. \tag{1.13}$$

By (1.13) and (1.5), there is $M_3 > 0$ such that

$$|x_i^{(k)}| \leq M_3 \text{ for all } i = 0, \ldots, T \text{ and all integers } k \geq 1. \tag{1.14}$$

In view of (1.14), extracting subsequences, using diagonalization process and re-indexing, if necessary, we may assume without loss of generality that for each $i \in \{0, \ldots, T\}$, there exists

$$\widehat{x}_i = \lim_{k \to \infty} x_i^{(k)}. \tag{1.15}$$

By (1.11) and (1.15),

$$\widehat{x}_0 = z, \ \widehat{x}_T = y. \tag{1.16}$$

It follows from (1.15) and (1.12) that

$$\sum_{i=0}^{T-1} L(\widehat{x}_i, \widehat{x}_{i+1}) = M_2.$$

Together with (1.16) and (1.9), this implies that $\{\widehat{x}_i\}_{i=0}^{T}$ is a solution of the problem P_2. This completes the proof of Proposition 1.1.

Denote by $\mathrm{Card}(A)$ the cardinality of a set A.

The following result establishes a turnpike property for approximate solutions of the problem (P_0).

Proposition 1.2 *Let* M_1, M_2, ϵ *be positive numbers. Then there exists a natural number* k_0 *such that for each integer* $T > 1$ *and each* (M_1)*-optimal sequence* $\{x_i\}_{i=0}^{T} \subset R^n$ *satisfying*

$$|x_0| \leq M_2, \ |x_T| \leq M_2 \tag{1.17}$$

the following inequality holds:

$$\mathrm{Card}(\{i \in \{0, \ldots, T-1\} : \ |x_i - \bar{x}| + |x_{i+1} - \bar{x}| > \epsilon\}) \leq k_0.$$

Proof By condition (C) there is $\delta > 0$ such that for each $(y, z) \in R^n \times R^n$ satisfying

$$L(y, z) \leq \delta \tag{1.18}$$

we have

$$|y - \bar{x}| + |z - \bar{x}| \leq \epsilon. \qquad (1.19)$$

Set

$$M_3 = \sup\{L(y, z) : y, z \in R^n \text{ and } |y| + |z| \leq |\bar{x}| + M_2\}. \qquad (1.20)$$

Choose a natural number

$$k_0 > \delta^{-1}(M_1 + 2M_3). \qquad (1.21)$$

Assume that an integer $T > 1$ and that an (M_1)-optimal sequence $\{x_i\}_{i=0}^T \subset R^n$ satisfies (1.17). Set

$$y_0 = x_0, \quad y_T = x_T, \quad y_i = \bar{x}, \; i = 1, \ldots, T - 1. \qquad (1.22)$$

Since the sequence $\{x_i\}_{i=0}^T$ is (M_1)-optimal, it follows from (1.22) that

$$\sum_{i=0}^{T-1} v(x_i, x_{i+1}) \leq \sum_{i=0}^{T-1} v(y_i, y_{i+1}) + M_1.$$

Together with (1.7), (1.8), and (1.22), this implies that

$$\sum_{i=0}^{T-1} L(x_i, x_{i+1}) \leq \sum_{i=0}^{T-1} L(y_i, y_{i+1}) + M_1 = L(x_0, \bar{x}) + L(\bar{x}, x_T) + M_1.$$

Combined with (1.17) and (1.20), this implies that

$$\sum_{i=0}^{T-1} L(x_i, x_{i+1}) \leq M_1 + 2M_3.$$

It follows from the choice of δ (see (1.18) and (1.19)), (1.21), and the inequality above that

$$\text{Card}\left(\{i \in \{0, \ldots, T - 1\} : |x_i - \bar{x}| + |x_{i+1} - \bar{x}| > \epsilon\}\right)$$

$$\leq \text{Card}\left(\{i \in \{0, \ldots, T - 1\} : L(x_i, x_{i+1}) > \delta\}\right)$$

$$\leq \delta^{-1} \sum_{i=0}^{T-1} L(x_i, x_{i+1}) \leq \delta^{-1}(M_1 + 2M_3) \leq k_0.$$

Proposition 1.2 is proved.

Proposition 1.2 implies the following turnpike result for exact solutions of the problem (P_0).

Proposition 1.3 *Let M, ϵ be positive numbers. Then there exists a natural number k_0 such that for each integer $T > 1$, each $y, z \in R^n$ satisfying $|y|, |z| \leq M$, and each optimal sequence $\{x_i\}_{i=0}^{T} \subset R^n$ of the problem (P_0), the following inequality holds:*

$$Card\big(\{i \in \{0, \ldots, T-1\} : |x_i - \bar{x}| + |x_{i+1} - \bar{x}| > \epsilon\}\big) \leq k_0.$$

It is easy now to see that the optimal solution $\{x_i\}_{i=0}^{T}$ of the problem (P_0) spends most of the time in an ϵ-neighborhood of $\bar{x}$. By Proposition 1.3 the number of all integers $i \in \{0, \ldots, T-1\}$ such that x_i does not belong to this ϵ-neighborhood, does not exceed the constant k_0 which depends only on M, ϵ, and does not depend on T. Following the tradition, the point $\bar{x}$ is called the turnpike. Moreover we can show that the set

$$\{i \in \{0 \ldots, T\} : |x_i - \bar{x}| > \epsilon\}$$

is contained in the union of two intervals $[0, k_1] \cup [T - k_1, T]$, where k_1 is a constant depending only on M, ϵ.

We also study the infinite horizon problem associated with the problem (P_0). By (1.1) there is $M_* > 0$ such that

$$v(y, z) > |v(\bar{x}, \bar{x})| + 1 \tag{1.23}$$

for any $(y, z) \in R^n \times R^n$ satisfying $|y| + |z| \geq M_*$.

We suppose that the sum over empty set is zero.

Proposition 1.4 *Let $M_0 > 0$. Then there exists $M_1 > 0$ such that for each integer $T \geq 1$ and each sequence $\{x_i\}_{i=0}^{T} \subset R^n$ satisfying $|x_0| \leq M_0$,*

$$\sum_{i=0}^{T-1} v(x_i, x_{i+1}) \geq T v(\bar{x}, \bar{x}) - M_1. \tag{1.24}$$

Proof Put

$$M_1 = |l_1|(M_0 + M_*).$$

Assume that an integer $T \geq 1$ and a sequence $\{x_i\}_{i=0}^{T} \subset R^n$ satisfies

$$|x_0| \leq M_0. \tag{1.25}$$

If $|x_i| > M_*, i = 1, \ldots, T$, then by (1.23)

$$\sum_{i=0}^{T-1} v(x_i, x_{i+1}) \geq Tv(\bar{x}, \bar{x})$$

and (1.24) holds. Therefore we may assume that there exists a natural number q such that

$$q \leq T, \ |x_q| \leq M_*. \tag{1.26}$$

We may assume without loss of generality that

$$|x_i| > M_* \text{ for all integers } i \text{ satisfying } q < i \leq T. \tag{1.27}$$

By (1.23) and (1.27),

$$\sum_{i=0}^{T-1}(v(x_i, x_{i+1}) - v(\bar{x}, \bar{x})) = \sum_{i=0}^{q-1}(v(x_i, x_{i+1}) - v(\bar{x}, \bar{x}))$$

$$+ \sum\{v(x_i, x_{i+1}) - v(\bar{x}, \bar{x}) : \text{ an integer } i \text{ satisfies } q \leq i < T\}$$

$$\geq \sum_{i=0}^{q-1}(v(x_i, x_{i+1}) - v(\bar{x}, \bar{x})).$$

It follows from the equation above, (1.6), (1.8), (1.25), (1.26), and the choice of M_1 that

$$\sum_{i=0}^{T-1}(v(x_i, x_{i+1}) - v(\bar{x}, \bar{x})) \geq \sum_{i=0}^{q-1}(v(x_i, x_{i+1}) - v(\bar{x}, \bar{x}))$$

$$= \sum_{i=0}^{q-1} L(x_i, x_{i+1}) + l_1(x_0 - x_q) \geq -|l_1|(|x_0| + |x_q|)$$

$$\geq -|l_1|(M_0 + M_*) = -M_1.$$

Proposition 1.4 is proved.

Fix a number $\tilde{M} > 0$ such that

$$\text{Proposition 1.4 holds with } M_0 = M_* \text{ and } M_1 = \tilde{M}. \tag{1.28}$$

Proposition 1.5 *Let $\{x_i\}_{i=0}^{\infty} \subset R^n$. Then either the sequence*

$$\left\{ \sum_{i=0}^{T-1} \left(v(x_i, x_{i+1}) - v(\bar{x}, \bar{x}) \right) \right\}_{T=1}^{\infty}$$

is bounded or

$$\lim_{T \to \infty} \sum_{i=0}^{T-1} \left(v(x_i, x_{i+1}) - v(\bar{x}, \bar{x}) \right) = \infty. \tag{1.29}$$

Proof It follows from (1.23) that if for all sufficiently large natural numbers i, $|x_i| \geq M_*$, then (1.29) holds. Therefore we may assume without loss of generality that there exists a strictly increasing sequence of natural numbers $\{t_k\}_{k=1}^{\infty}$ such that

$$|x_{t_k}| < M_* \text{ for all integers } k \geq 1. \tag{1.30}$$

By Proposition 1.4 the sequence $\{\sum_{i=0}^{T-1} (v(x_i, x_{i+1}) - v(\bar{x}, \bar{x}))\}_{T=1}^{\infty}$ is bounded from below.

Assume that this sequence is not bounded from above. In order to complete the proof, it is sufficient to show that (1.29) holds.

Let Q be any positive number. Then there exists a natural number T_0 such that

$$\sum_{i=0}^{T_0-1} \left(v(x_i, x_{i+1}) - v(\bar{x}, \bar{x}) \right) > Q + \tilde{M}. \tag{1.31}$$

Choose a natural number k such that

$$t_k > T_0 + 4. \tag{1.32}$$

Let an integer

$$T > t_k. \tag{1.33}$$

By (1.30), (1.32), and (1.33), there exists an integer S such that

$$T > S \geq T_0, \tag{1.34}$$

$$|x_S| \leq M_*, \tag{1.35}$$

$$|x_t| > M_* \text{ for all integers } t \text{ satisfying} \tag{1.36}$$

$$S > t \geq T_0.$$

It follows from (1.23), (1.28), (1.31), (1.34)–(1.36), and Proposition 1.4 that

$$\sum_{i=0}^{T-1}(v(x_i, x_{i+1}) - v(\bar{x}, \bar{x})) = \sum_{i=0}^{T_0-1}(v(x_i, x_{i+1}) - v(\bar{x}, \bar{x}))$$

$$+ \sum\{v(x_i, x_{i+1}) - v(\bar{x}, \bar{x}) : i \text{ is an integer and } T_0 \leq i < S\}$$

$$+ \sum_{i=S}^{T-1}(v(x_i, x_{i+1}) - v(\bar{x}, \bar{x}))$$

$$> Q + \tilde{M} + \sum_{i=S}^{T-1}(v(x_i, x_{i+1}) - v(\bar{x}, \bar{x})) > Q.$$

Thus for any integer $T > t_k$,

$$\sum_{i=0}^{T-1}(v(x_i, x_{i+1}) - v(\bar{x}, \bar{x})) > Q.$$

Since Q is any positive number, (1.29) holds. Proposition 1.5 is proved.

A sequence $\{x_i\}_{i=0}^{\infty} \subset R^n$ is called good [28, 93, 117] if the sequence $\{\sum_{i=0}^{T-1}(v(x_i, x_{i+1}) - v(\bar{x}, \bar{x}))\}_{T=1}^{\infty}$ is bounded.

Proposition 1.6

1. A sequence $\{x_i\}_{i=0}^{\infty} \subset R^n$ is good if and only if

$$\sum_{i=0}^{\infty} L(x_i, x_{i+1}) < \infty.$$

2. If a sequence $\{x_i\}_{i=0}^{\infty} \subset R^n$ is good, then it converges to $\bar{x}$.

Proof Assume that a sequence $\{x_i\}_{i=0}^{\infty} \subset R^n$ is good. Then there exists $M_0 > 0$ such that

$$\sum_{i=0}^{T-1}\left(v(x_i, x_{i+1}) - v(\bar{x}, \bar{x})\right) < M_0 \text{ for all integers } T \geq 1. \tag{1.37}$$

By (1.23) and (1.37), there exists a strictly increasing sequence of natural numbers $\{t_k\}_{k=1}^{\infty}$ such that

$$|x_{t_k}| < M \text{ for all natural numbers } k. \tag{1.38}$$

Let k be a natural number. By (1.8), (1.37), and (1.38),

$$M_0 > \sum_{i=0}^{t_k-1} \left(v(x_i, x_{i+1}) - v(\bar{x}, \bar{x})\right) = \sum_{i=0}^{t_k-1} L(x_i, x_{i+1}) + l_1(x_0 - x_{t_k})$$

$$\geq \sum_{i=0}^{t_k-1} L(x_i, x_{i+1}) - |l_1|(|x_0| + |x_{t_k}|)$$

$$\geq \sum_{i=0}^{t_k-1} L(x_i, x_{i+1}) - |l_1|(|x_0| + M_*)$$

and

$$\sum_{i=0}^{t_k-1} L(x_i, x_{i+1}) \leq M_0 + |l_1|(|x_0| + M_*).$$

Since the inequality above holds for all natural numbers k, we conclude that

$$\sum_{i=0}^{\infty} L(x_i, x_{i+1}) \leq M_0 + |l_1|(|x_0| + M_*).$$

In view of (C), the sequence $\{x_i\}_{i=0}^{\infty}$ converges to $\bar{x}$, and assertion 2 is proved.

Assume that

$$M_1 := \sum_{i=0}^{\infty} L(x_i, x_{i+1}) < \infty. \tag{1.39}$$

By (1.5) there is $M_2 > 0$ such that

$$|x_i| < M_2 \text{ for all integers } i \geq 0. \tag{1.40}$$

In view of (1.8), (1.39), and (1.40), for all natural numbers T,

$$\sum_{i=0}^{T-1} \left(v(x_i, x_{i+1}) - v(\bar{x}, \bar{x})\right) = \sum_{i=0}^{T-1} L(x_i, x_{i+1}) + l_1(x_0 - x_T)$$

$$\leq M_1 + 2|l_1|M_2.$$

Together with Proposition 1.5, this implies that the sequence $\{x_i\}_{i=0}^{\infty}$ is good. Proposition 1.6 is proved.

Proposition 1.7 *Let $x \in R^n$. Then there exists a sequence $\{x_i\}_{i=0}^{\infty} \subset R^n$ such that $x_0 = x$ and for each sequence $\{y_i\}_{i=0}^{\infty} \subset R^n$ satisfying $y_0 = x$, the inequality*

$$\sum_{i=0}^{\infty} L(x_i, x_{i+1}) \leq \sum_{i=0}^{\infty} L(y_i, y_{i+1})$$

holds.

Proof Set

$$M_0 = \inf\left\{\sum_{i=0}^{\infty} L(y_i, y_{i+1}) : \{y_i\}_{i=0}^{\infty} \subset R^n \text{ and } y_0 = x\right\}. \tag{1.41}$$

Clearly, M_0 is well-defined and $M_0 \geq 0$. There exists a sequence $\{x_i^{(k)}\}_{i=0}^{\infty} \subset R^n$, $k = 1, 2, \ldots$ such that

$$x_0^{(k)} = x, \ k = 1, 2, \ldots, \tag{1.42}$$

$$\lim_{k\to\infty} \sum_{i=0}^{\infty} L\left(x_i^{(k)}, x_{i+1}^{(k)}\right) = M_0. \tag{1.43}$$

By (1.5) and (1.43), there exists $M_1 > 0$ such that

$$|x_i^{(k)}| < M_1 \text{ for all integers } i \geq 0 \text{ for all integers } k \geq 1. \tag{1.44}$$

In view of (1.44) using diagonalization process, extracting subsequences, and re-indexing, we may assume without loss of generality that for any integer $i \geq 0$, there is

$$x_i = \lim_{k\to\infty} x_i^{(k)}. \tag{1.45}$$

By (1.42) and (1.45),

$$x_0 = x. \tag{1.46}$$

It follows from (1.6), (1.43), and (1.45) that for any natural number T,

$$\sum_{i=0}^{T-1} L(x_i, x_{i+1}) = \lim_{k\to\infty} \sum_{i=0}^{T-1} L\left(x_i^{(k)}, x_{i+1}^{(k)}\right) \leq \lim_{k\to\infty} \sum_{i=0}^{\infty} L\left(x_i^{(k)}, x_{i+1}^{(k)}\right) = M_0.$$

Since T is an arbitrary natural number, we conclude that

$$\sum_{i=0}^{\infty} L(x_i, x_{i+1}) \leq M_0.$$

Together with (1.41) and (1.46), this implies that

$$\sum_{i=0}^{\infty} L(x_i, x_{i+1}) = M_0.$$

This completes the proof of Proposition 1.7.

In our study we use the following optimality criterion introduced in the economic literature [6, 28, 89] and used in the optimal control [19, 93, 117].

A sequence $\{x_i\}_{i=0}^{\infty} \subset R^n$ is called overtaking optimal if

$$\limsup_{T \to \infty} \left[\sum_{i=0}^{T-1} v(x_i, x_{i+1}) - \sum_{i=0}^{T-1} v(y_i, y_{i+1}) \right] \leq 0$$

for any sequence $\{y_i\}_{i=0}^{\infty} \subset R^n$ satisfying $y_0 = x_0$.

Proposition 1.8 *Let* $\{x_i\}_{i=0}^{\infty} \subset R^n$. *Then the following assertions are equivalent:*

1. the sequence $\{x_i\}_{i=0}^{\infty}$ *is overtaking optimal;*
2.

$$\sum_{i=0}^{\infty} L(x_i, x_{i+1}) \leq \sum_{i=0}^{\infty} L(y_i, y_{i+1})$$

for any sequence $\{y_i\}_{i=0}^{\infty} \subset R^n$ *satisfying* $y_0 = x_0$.

Proof Assume that the sequence $\{x_i\}_{i=0}^{\infty}$ is overtaking optimal. Clearly, it is good. By Proposition 1.6,

$$\sum_{i=0}^{\infty} L(x_i, x_{i+1}) < \infty.$$

Let a sequence $\{y_i\}_{i=0}^{\infty} \subset R^n$ satisfies

$$y_0 = x_0. \tag{1.47}$$

We show that

$$\sum_{i=0}^{\infty} L(x_i, x_{i+1}) \leq \sum_{i=0}^{\infty} L(y_i, y_{i+1}).$$

We may assume that

$$\sum_{i=0}^{\infty} L(y_i, y_{i+1}) < \infty.$$

Then in view of (C),

$$\lim_{i \to \infty} y_i = \bar{x}, \quad \lim_{i \to \infty} x_i = \bar{x}. \tag{1.48}$$

Since the sequence $\{x_i\}_{i=0}^{\infty}$ is overtaking optimal, it follows from (1.47), (1.8), and (1.48) that

$$0 \geq \limsup_{T \to \infty} \left[\sum_{i=0}^{T-1} v(x_i, x_{i+1}) - \sum_{i=0}^{T-1} v(y_i, y_{i+1}) \right]$$

$$= \limsup_{T \to \infty} \left[\sum_{i=0}^{T-1} L(x_i, x_{i+1}) + l_1(x_0 - x_T) - \sum_{i=0}^{T-1} L(y_i, y_{i+1}) - l_1(y_0 - y_T) \right]$$

$$= \limsup_{T \to \infty} \left[\sum_{i=0}^{T-1} L(x_i, x_{i+1}) - \sum_{i=0}^{T-1} L(y_i, y_{i+1}) + l_1(y_T - x_T) \right]$$

$$= \sum_{i=0}^{\infty} L(x_i, x_{i+1}) - \sum_{i=0}^{\infty} L(y_i, y_{i+1}).$$

Thus assertion 2 holds.

Assume that assertion 2 holds. Let us show that the sequence $\{x_i\}_{i=0}^{\infty}$ is overtaking optimal. Clearly,

$$\sum_{i=0}^{\infty} L(x_i, x_{i+1}) < \infty.$$

By Proposition 1.6 the sequence $\{x_i\}_{i=0}^{\infty}$ is good and

$$\lim_{i \to \infty} x_i = \bar{x}. \tag{1.49}$$

Assume that a sequence $\{y_i\}_{i=0}^{\infty} \subset R^n$ satisfies

$$y_0 = x_0. \tag{1.50}$$

We show that

$$\limsup_{T \to \infty} \left[\sum_{i=0}^{T-1} v(x_i, x_{i+1}) - \sum_{i=0}^{T-1} v(y_i, y_{i+1}) \right] \leq 0.$$

We may assume without loss of generality that the sequence $\{y_i\}_{i=0}^{\infty}$ is good. Then by Proposition 1.6,

$$\lim_{i\to\infty} y_i = \bar{x}, \quad \sum_{i=0}^{\infty} L(y_i, y_{i+1}) < \infty. \tag{1.51}$$

It follows from (1.8), (1.49)–(1.51), and assertion 2 that

$$\limsup_{T\to\infty} \left[\sum_{i=0}^{T-1} v(x_i, x_{i+1}) - \sum_{i=0}^{T-1} v(y_i, y_{i+1}) \right]$$

$$= \limsup_{T\to\infty} \left[\sum_{i=0}^{T-1} L(x_i, x_{i+1}) + l_1(x_0 - x_T) - \sum_{i=0}^{T-1} L(y_i, y_{i+1}) - l_1(y_0 - y_T) \right]$$

$$= \sum_{i=0}^{\infty} L(x_i, x_{i+1}) - \sum_{i=0}^{\infty} L(y_i, y_{i+1}) + l_1 \left(\lim_{T\to\infty} y_T - \lim_{T\to\infty} x_T \right)$$

$$= \sum_{i=0}^{\infty} L(x_i, x_{i+1}) - \sum_{i=0}^{\infty} L(y_i, y_{i+1}) \leq 0.$$

Thus assertion 1 holds, and Proposition 1.8 is proved.

Proposition 1.7 and 1.8 imply the following existence result.

Proposition 1.9 *For any* $x \in R^n$, *there exists an overtaking optimal sequence* $\{x_i\}_{i=0}^{\infty} \subset R^n$ *such that* $x_0 = x$.

1.2   The Turnpike Phenomenon

In the previous section, we proved the turnpike result and the existence of overtaking optimal solutions for rather simple class of discrete-time problems. The problems of this class are unconstrained, and their objective functions are convex and differentiable. In [117] the turnpike property and the existence of solutions over infinite horizon were established for several classes of constrained optimal control problems without convexity (concavity) assumptions. In particular, in Chapter 2 of [117], we study the structure of approximate solutions of an autonomous discrete-time control system with a compact metric space of states X. This control system is described by a bounded upper semicontinuous function $v : X \times X \to R^1$ which determines an optimality criterion and by a nonempty closed set $\Omega \subset X \times X$ which determines a class of admissible trajectories (programs). We study the problem

$$\sum_{i=0}^{T-1} v(x_i, x_{i+1}) \to \max, \ \{(x_i, x_{i+1})\}_{i=0}^{T-1} \subset \Omega, \ x_0 = z, \ x_T = y, \qquad (P)$$

where $T \geq 1$ is an integer and the points $y, z \in X$.

In the classical turnpike theory, the objective function v possesses the turnpike property (TP) if there exists a point $\bar{x} \in X$ (a turnpike) such that the following condition holds:

For each positive number ϵ, there exists an integer $L \geq 1$ such that for each integer $T \geq 2L$ and each solution $\{x_i\}_{i=0}^{T} \subset X$ of the problem (P), the inequality $\rho(x_i, \bar{x}) \leq \epsilon$ is true for all $i = L, \ldots, T - L$.

It should be mentioned that the constant L depends neither on T nor on y, z.

The turnpike phenomenon has the following interpretation. If one wishes to reach a point A from a point B by a car in an optimal way, then one should turn to a turnpike, spend most of time on it, and then leave the turnpike to reach the required point.

In the classical turnpike theory [28, 67, 80, 89], the space X is a compact convex subset of a finite-dimensional Euclidean space, the set Ω is convex, and the function v is strictly concave. Under these assumptions the turnpike property can be established, and the turnpike $\bar{x}$ is a unique solution of the maximization problem $v(x, x) \to \max, (x, x) \in \Omega$. In this situation it is shown that for each program $\{x_t\}_{t=0}^{\infty}$, either the sequence $\{\sum_{t=0}^{T-1} v(x_t, x_{t+1}) - Tv(\bar{x}, \bar{x})\}_{T=1}^{\infty}$ is bounded (in this case the program $\{x_t\}_{t=0}^{\infty}$ is called (v)-good), or it diverges to $-\infty$. Moreover, it is also established that any (v)-good program converges to the turnpike $\bar{x}$. In the sequel this property is called as the asymptotic turnpike property.

In [117] we study the problems (P) with the constraint $\{(x_i, x_{i+1})\}_{i=0}^{T-1} \subset \Omega$ where Ω is an arbitrary nonempty closed subset of the metric space $X \times X$. Clearly, these constrained problems are more difficult and less understood than their unconstrained prototypes in the previous section. They are also more realistic from the point of view of mathematical economics. In general a turnpike is not necessarily a singleton. Nevertheless problems of the type (P) for which the turnpike is a singleton are of great importance because of the following reasons: there are many models of economic growth for which a turnpike is a singleton; if a turnpike is a singleton, then approximate solutions of (P) have very simple structure, and this is very important for applications; if a turnpike is a singleton, then it can be easily calculated as a solution of the problem $v(x, x) \to \max, (x, x) \in \Omega$.

The turnpike property is very important for applications. Suppose that our objective function v has the turnpike property and we know a finite number of "approximate" solutions of the problem (P). Then we know the turnpike $\bar{x}$, or at least its approximation, and the constant L (see the definition of TP) which is an estimate for the time period required to reach the turnpike. This information can be useful if we need to find an "approximate" solution of the problem (P) with a new time interval $[m_1, m_2]$ and the new values $z, y \in X$ at the endpoints m_1 and m_2. Namely, instead of solving this new problem on the "large" interval $[m_1, m_2]$, we can find an "approximate" solution of the problem (P) on the "small" interval

$[m_1, m_1 + L]$ with the values $z, \bar{x}$ at the endpoints and an "approximate" solution of the problem (P) on the "small" interval $[m_2 - L, m_2]$ with the values $\bar{x}, y$ at the endpoints. Then the concatenation of the first solution, the constant sequence $x_i = \bar{x}, i = m_1 + L, \ldots, m_2 - L$, and the second solution is an "approximate" solution of the problem (P) on the interval $[m_1, m_2]$ with the values z, y at the endpoints. Sometimes as an "approximate" solution of the problem (P), we can choose any admissible sequence $\{x_i\}_{i=m_1}^{m_2}$ satisfying

$$x_{m_1} = z, \; x_{m_2} = y \text{ and } x_i = \bar{x} \text{ for all } i = m_1 + L, \ldots, m_2 - L.$$

1.3   Turnpike Results for Problems in Metric Spaces

In [117] we study the turnpike phenomenon for a class of discrete-time optimal control problems which is considered below.

Let (X, ρ) be a compact metric space, Ω be a nonempty closed subset of $X \times X$, and $v : X \times X \to R^1$ be a bounded upper semicontinuous function.

A sequence $\{x_t\}_{t=0}^{\infty} \subset X$ is called an (Ω)-program (or just a program if the set Ω is understood) if $(x_t, x_{t+1}) \in \Omega$ for all nonnegative integers t. A sequence $\{x_t\}_{t=0}^{T}$ where $T \geq 1$ is an integer is called an (Ω)-program (or just a program if the set Ω is understood) if $(x_t, x_{t+1}) \in \Omega$ for all integers $t \in [0, T - 1]$.

We consider the problems

$$\sum_{i=0}^{T-1} v(x_i, x_{i+1}) \to \max, \; \{(x_i, x_{i+1})\}_{i=0}^{T-1} \subset \Omega, \; x_0 = y,$$

and

$$\sum_{i=0}^{T-1} v(x_i, x_{i+1}) \to \max, \; \{(x_i, x_{i+1})\}_{i=0}^{T-1} \subset \Omega, \; x_0 = y, \; x_T = z,$$

where $T \geq 1$ is an integer and the points $y, z \in X$.

We suppose that there exist a point $\bar{x} \in X$ and a positive number $\bar{c}$ such that the following assumptions hold:

(i) $(\bar{x}, \bar{x})$ is an interior point of Ω;
(ii) $\sum_{t=0}^{T-1} v(x_t, x_{t+1}) \leq T v(\bar{x}, \bar{x}) + \bar{c}$ for any natural number T and any program $\{x_t\}_{t=0}^{T}$.

The property (ii) implies that for each program $\{x_t\}_{t=0}^{\infty}$, either the sequence

$$\left\{ \sum_{t=0}^{T-1} v(x_t, x_{t+1}) - T v(\bar{x}, \bar{x}) \right\}_{T=1}^{\infty}$$

is bounded, or $\lim_{T\to\infty}[\sum_{t=0}^{T-1} v(x_t, x_{t+1}) - Tv(\bar{x}, \bar{x})] = -\infty$.

A program $\{x_t\}_{t=0}^{\infty}$ is called (v)-good if the sequence

$$\left\{\sum_{t=0}^{T-1} v(x_t, x_{t+1}) - Tv(\bar{x}, \bar{x})\right\}_{T=1}^{\infty}$$

is bounded.

We also suppose that the following assumption holds.

(iii) (the asymptotic turnpike property) For any (v)-good program $\{x_t\}_{t=0}^{\infty}$, $\lim_{t\to\infty} \rho(x_t, \bar{x}) = 0$.

Note that the properties (i)–(iii) hold for models of economic dynamics considered in the classical turnpike theory.

For each positive number M, denote by X_M the set of all points $x \in X$ for which there exists a program $\{x_t\}_{t=0}^{\infty}$ such that $x_0 = x$ and that for all natural numbers T the following inequality holds:

$$\sum_{t=0}^{T-1} v(x_t, x_{t+1}) - Tv(\bar{x}, \bar{x}) \geq -M.$$

It is not difficult to see that $\cup\{X_M : M \in (0, \infty)\}$ is the set of all points $x \in X$ for which there exists a (v)-good program $\{x_t\}_{t=0}^{\infty}$ satisfying $x_0 = x$.

Let $T \geq 1$ be an integer and $\Delta \geq 0$. A program $\{x_i\}_{i=0}^{T} \subset R^n$ is called (Δ)-optimal if for any program $\{x_i'\}_{i=0}^{T}$ satisfying $x_0 = x_0'$, the inequality

$$\sum_{i=0}^{T-1} v(x_i, x_{i+1}) \geq \sum_{i=0}^{T-1} v(x_i', x_{i+1}') - \Delta$$

holds.

In Chapter 2 of [117], we prove the following turnpike result for approximate solutions of our first optimization problem stated above.

Theorem 1.10 *Let ϵ, M be positive numbers. Then there exist a natural number L and a positive number δ such that for each integer $T > 2L$ and each (δ)-optimal program $\{x_t\}_{t=0}^{T}$ which satisfies $x_0 \in X_M$ there exist nonnegative integers $\tau_1, \tau_2 \leq L$ such that $\rho(x_t, \bar{x}) \leq \epsilon$ for all $t = \tau_1, \ldots, T - \tau_2$ and if $\rho(x_0, \bar{x}) \leq \delta$, then $\tau_1 = 0$.*

An analogous turnpike result for approximate solutions of our second optimization problem is also proved in Chapter 2 of [117].

A program $\{x_t\}_{t=0}^{\infty}$ is called (v)-overtaking optimal if for each program $\{y_t\}_{t=0}^{\infty}$ satisfying $y_0 = x_0$, the inequality $\limsup_{T\to\infty} \sum_{t=0}^{T-1}[v(y_t, y_{t+1}) - v(x_t, x_{t+1})] \leq 0$ holds.

In Chapter 2 of [117], we prove the following result which establishes the existence of an overtaking optimal program.

Theorem 1.11 *Assume that $x \in X$ and that there exists a (v)-good program $\{x_t\}_{t=0}^{\infty}$ such that $x_0 = x$. Then there exists a (v)-overtaking optimal program $\{x_t^*\}_{t=0}^{\infty}$ such that $x_0^* = x$.*

In [117] we also study the stability of the turnpike phenomenon and show that the turnpike property is stable under perturbations of the objective function v. Note that the stability of the turnpike property is crucial in practice. One reason is that in practice we deal with a problem which consists of a perturbation of the problem we wish to consider. Another reason is that the computations introduce numerical errors.

1.4 The Robinson–Solow–Srinivasan Model

Let R^1 (R_+^1) be the set of real (nonnegative) numbers, and let R^n be the n-dimensional Euclidean space with nonnegative orthant

$$R_+^n = \{x = (x_1, \ldots, x_n) \in R^n : x_i \geq 0, \ i = 1, \ldots, n\}.$$

For every pair of vectors $x = (x_1, \ldots, x_n)$, $y = (y_1, \ldots, y_n) \in R^n$, define their inner product by

$$xy = \sum_{i=1}^{n} x_i y_i,$$

and let $x \gg y, x > y, x \geq y$ have their usual meaning. Namely, for a given pair of vectors $x = (x_1, \ldots, x_n)$, $y = (y_1, \ldots, y_n) \in R^n$, we say that $x \geq y$, if $x_i \geq y_i$ for all $i = 1, \ldots, n$, $x > y$ if $x \geq y$ and $x \neq y$, and $x \gg y$ if $x_i > y_i$ for all $i = 1, \ldots, n$.

Let $e(i)$, $i = 1, \ldots, n$, be the ith unit vector in R^n, and e be an element of R_+^n all of whose coordinates are unity. For every $x \in R^n$, denote by $\|x\|$ its Euclidean norm in R^n.

Let $a = (a_1, \ldots, a_n) \gg 0, b = (b_1, \ldots, b_n) \gg 0, b_1 \geq b_2 \cdots \geq b_n, d \in (0, 1)$,

$$c_i = b_i/(1 + da_i), \ i = 1, \ldots, n.$$

We assume the following:

There exists $\sigma \in \{1, \ldots, n\}$ such that for all

$$i \in \{1, \ldots, n\} \setminus \{\sigma\}, \ c_\sigma > c_i.$$

A sequence $\{x(t), y(t)\}_{t=0}^{\infty}$ is called a program if for each integer $t \geq 0$

$$(x(t), y(t)) \in R_+^n \times R_+^n, \ x(t+1) \geq (1-d)x(t),$$
$$0 \leq y(t) \leq x(t), \ a(x(t+1) - (1-d)x(t)) + ey(t) \leq 1. \quad (1.52)$$

Let T_1, T_2 be integers such that $0 \leq T_1 < T_2$. A pair of sequences

$$\left(\{x(t)\}_{t=T_1}^{T_2}, \{y(t)\}_{t=T_1}^{T_2-1}\right)$$

is called a program if $x(T_2) \in R_+^n$ and for each integer t satisfying $T_1 \leq t < T_2$, relations (1.52) are valid.

Assume that $w : [0, \infty) \to R^1$ is a continuous strictly increasing concave and differentiable function which represents the preferences of the planner.

Define

$$\Omega = \{(x, x') \in R_+^n \times R_+^n : x' - (1-d)x \geq 0$$
$$\text{and } a(x' - (1-d)x) \leq 1\}$$

and a correspondence $\Lambda : \Omega \to R_+^n$ given by

$$\Lambda(x, x') = \left\{ y \in R_+^n : 0 \leq y \leq x \text{ and } ey \leq 1 - a(x' - (1-d)x) \right\}.$$

For every $(x, x') \in \Omega$, set

$$u(x, x') = \max\left\{ w(by) : y \in \Lambda(x, x') \right\}.$$

The golden-rule stock is $\widehat{x} \in R_+^n$ such that $(\widehat{x}, \widehat{x})$ is a solution to the problem: maximize $u(x, x')$ subject to
(i) $x' \geq x$; (ii) $(x, x') \in \Omega$.

By Theorem 2.3, which will be proved in Chapter 2, there exists a unique golden-rule stock

$$\widehat{x} = (1/(1 + da_\sigma))e(\sigma).$$

It is not difficult to see that $\widehat{x}$ is a solution to the problem

$$w(by) \to \max, \ y \in \Lambda(\widehat{x}, \widehat{x}).$$

A program $\{x(t), y(t)\}_{t=0}^{\infty}$ is good if there is a real number M such that

$$\sum_{t=0}^{T}(w(by(t)) - w(b\widehat{y})) \geq M \text{ for every nonnegative integer } T.$$

A program $\{x(t), y(t)\}_{t=0}^{\infty}$ is bad if

$$\lim_{T \to \infty} \sum_{t=0}^{T} (w(by(t)) - w(b\widehat{y})) = -\infty.$$

We will show in Chapter 2 that every program that is not good is bad and that for every point $x_0 \in R_+^n$, there exists a good program $\{x(t), y(t)\}_{t=0}^{\infty}$ satisfying $x(0) = x_0$.

In Chapter 3 we prove the following results obtained in [92].

Theorem 1.12 *Assume that the function w is strictly concave. Then for every good program $\{x(t), y(t)\}_{t=0}^{\infty}$,*

$$\lim_{t \to \infty} (x(t), y(t)) = (\widehat{x}, \widehat{x}).$$

Set

$$\xi_\sigma = 1 - d - (1/a_\sigma).$$

Theorem 1.13 *Assume that $\xi_\sigma \neq -1$. Then*

$$\lim_{t \to \infty} (x(t), y(t)) = (\widehat{x}, \widehat{x})$$

for every good program $\{x(t), y(t)\}_{t=0}^{\infty}$.

A program $\{x^*(t), y^*(t)\}_{t=0}^{\infty}$ is overtaking optimal if

$$\limsup_{T \to \infty} \sum_{t=0}^{T} [w(by(t)) - w(by^*(t))] \leq 0$$

for every program $\{x(t), y(t)\}_{t=0}^{\infty}$ which satisfies $x(0) = x^*(0)$.

In Chapter 3 we prove the following result obtained in [92].

Theorem 1.14 *Assume that for every good program $\{x(t), y(t)\}_{t=0}^{\infty}$,*

$$\lim_{t \to \infty} (x(t), y(t)) = (\widehat{x}, \widehat{x}).$$

Then for every point $x_0 \in R_+^n$, there is an overtaking optimal program $\{x(t), y(t)\}_{t=0}^{\infty}$ such that $x(0) = x_0$.

Corollary 1.15 *Assume that the function w is strictly concave. Then for every point $x_0 \in R_+^n$, there exists an overtaking optimal program $\{x(t), y(t)\}_{t=0}^{\infty}$ satisfying $x(0) = x_0$.*

Corollary 1.16 *Assume that $\xi_\sigma \neq -1$. Then for every point $x_0 \in R_+^n$, there is an overtaking optimal program $\{x(t), y(t)\}_{t=0}^\infty$ such that $x(0) = x_0$.*

Theorem 1.13 shows that if $\xi_\sigma \neq -1$, then the asymptotic turnpike property (iii) of Section 1.3 holds with $\widehat{x}$ being the turnpike. It will also be shown that property (ii) of Section 1.3 holds. But property (i) from Section 1.3 does not hold: $(\widehat{x}, \widehat{x})$ is not an interior point of the set Ω. This makes the situation more difficult and less understood.

The next auxiliary result (see Proposition 3.13) together with certain monotonicity properties of the model will help us to overcome these difficulties.

Proposition 1.17 *Let $\epsilon > 0$. Then there exists $\delta > 0$ such that for each $x, x' \in R_+^n$ satisfying*

$$\|x - \widehat{x}\|, \|x' - \widehat{x}\| \le \delta,$$

there exist $\bar{x} \ge x'$, $y \in R_+^n$ such that

$$(x, \bar{x}) \in \Omega, \ y \in \Lambda(x, \bar{x}),$$

$$\|y - \widehat{x}\| \le \epsilon, \ \|\bar{x} - \widehat{x}\| \le \epsilon.$$

It should be mentioned that in the book, sequences are denoted as $\{x(t)\}_{t=0}^\infty$ because this notation is used in the literature on the Robinson–Solow–Srinivasan model.

Chapter 2
The Description of the Robinson–Solow–Srinivasan Model and Its Basic Properties

In this chapter we present the description of the Robinson–Solow–Srinivasan model and discuss its basic properties. In particular, we show the existence of weakly optimal programs and good programs and prove an average turnpike property.

2.1 The Robinson–Solow–Srinivasan Model

In this book we use the following notation.

Let R^1 (R^1_+) be the set of real (nonnegative) numbers, and let R^n be the n-dimensional Euclidean space with nonnegative orthant

$$R^n_+ = \{x = (x_1, \ldots, x_n) \in R^n : x_i \geq 0, \ i = 1, \ldots, n\}.$$

For every pair of vectors $x = (x_1, \ldots, x_n)$, $y = (y_1, \ldots, y_n) \in R^n$, define their inner product by

$$xy = \sum_{i=1}^{n} x_i y_i,$$

and let $x \gg y$, $x > y$, $x \geq y$ have their usual meaning. Namely, for a given pair of vectors $x = (x_1, \ldots, x_n)$, $y = (y_1, \ldots, y_n) \in R^n$, we say that $x \geq y$, if $x_i \geq y_i$ for all $i = 1, \ldots, n$, $x > y$ if $x \geq y$ and $x \neq y$, and $x \gg y$ if $x_i > y_i$ for all $i = 1, \ldots, n$.

Let $e(i)$, $i = 1, \ldots, n$, be the ith unit vector in R^n, and e be an element of R^n_+ all of whose coordinates are unity. For every $x \in R^n$, denote by $\|x\|$ its Euclidean norm in R^n.

© The Author(s), under exclusive license to Springer Nature Switzerland AG 2020   25
A. J. Zaslavski, *Turnpike Theory for the Robinson–Solow–Srinivasan Model*,
Springer Optimization and Its Applications 166,
https://doi.org/10.1007/978-3-030-60307-6_2

Let $a = (a_1, \ldots, a_n) \gg 0$, $b = (b_1, \ldots, b_n) \gg 0$, $d \in (0, 1)$, $c_i = b_i/(1 + da_i)$, $i = 1, \ldots, n$.

These parameters define an economy capable of producing a finite number n of alternative types of machines. For every $i = 1, \ldots, n$, one unit of machine of type i requires $a_i > 0$ units of labor to construct it, and together with one unit of labor, each unit of it can produce $b_i > 0$ units of a single consumption good. Thus, the production possibilities of the economy are represented by an (labor) input-coefficients vector, $a = (a_1, \ldots, a_n) \gg 0$, and an output-coefficients vector, $b = (b_1, \ldots, b_n) \gg 0$. Without loss of generality, we assume that the types of machines are numbered such that $b_1 \geq b_2 \cdots \geq b_n$.

We assume that all machines depreciate at a rate $d \in (0, 1)$. Thus the effective labor cost of producing a unit of output on a machine of type i is given by $(1 + da_i)/b_i$: the direct labor cost of producing unit output and the indirect cost of replacing the depreciation of the machine in this production. We consider the reciprocal of the effective labor cost, the effective output that takes the depreciation into account, and denote it by c_i for the machine of type i.

In this chapter we assume that there is a unique machine type σ at which effective labor cost $(1 + da_i)/b_i$ is minimal, or at which the effective output per man $b_i/(1 + da_i)$ is maximal. Thus assume the following:

there exists $\sigma \in \{1, \ldots, n\}$ such that for all $i \in \{1, \ldots, n\} \setminus \{\sigma\}$, $c_\sigma > c_i$.　　(2.1)

For each nonnegative integer t, let $x(t) = (x_1(t), \ldots, x_n(t)) \geq 0$ denote the amounts of the n types of machines that are available in time-period t, and let $z(t + 1) = (z_1(t + 1), \ldots, z_n(t + 1)) \geq 0$ be the gross investments in the n types of machines during period $t + 1$. Hence, $z(t + 1) = (x(t + 1) - x(t)) + dx(t)$, the sum of net investment and of depreciation. Let $y(t) = (y_1(t), \ldots, y_n(t))$ be the amounts of the n types of machines used for production of the consumption good, $by(t)$, during period $t + 1$. Let the total labor force of the economy be stationary and positive. We normalize it to be unity. It is clear that gross investment, $z(t + 1)$, representing the production of new machines of the various types, requires $az(t+1)$ units of labor in period t. Also $y(t)$ representing the use of available machines for manufacture of the consumption good requires $ey(t)$ units of labor in period t. Thus, the availability of labor constrains employment in the consumption and investment sectors by $az(t+1)+ey(t) \leq 1$. Note that the flow of consumption and of investment (new machines) is in gestation during the period and available at the end of it. We now give a formal description of this technological structure.

A sequence $\{x(t), y(t)\}_{t=0}^{\infty}$ is called a program if for each integer $t \geq 0$

$$(x(t), y(t)) \in R_+^n \times R_+^n,$$

$$x(t + 1) \geq (1 - d)x(t),$$

$$0 \leq y(t) \leq x(t)$$

and

$$a(x(t+1) - (1-d)x(t)) + ey(t) \le 1.$$

We associate with every program $\{x(t), y(t)\}_{t=0}^{\infty}$ its gross investment sequence $\{z(t+1)\}_{t=0}^{\infty}$ such that

$$z(t+1) = x(t+1) - (1-d)x(t), \ t = 0, 1, \dots$$

and a consumption sequence $\{by(t)\}_{t=0}^{\infty}$.

The following result was obtained in [37].

Proposition 2.1 *For every program $\{x(t), y(t)\}_{t=0}^{\infty}$, there exists a constant $m(x(0)) > 0$, depending only on $x(0)$, such that $x(t) \le m(x(0))e$ for all nonnegative integers t.*

Proof For every integer $t > 0$, we have

$$ax(t) \le 1 + (1-d)ax(t-1)$$

$$\le \sum_{\tau=0}^{t-1} (1-d)^{\tau} + (1-d)^t ax(0).$$

The inclusion $d \in (0, 1)$ implies that

$$ax(t) \le d^{-1} + ax(0).$$

There exists $j \in \{1, \dots, n\}$ such that

$$a_j \le a_i, \ i = 1, \dots, n.$$

Using the inequality $a_i > 0, i = 1, \dots, n$, we deduce that for all $i = 1, \dots, n$,

$$x_i(t) \le a_j^{-1}(d^{-1} + ax(0)) := m(x(0)).$$

Proposition 2.1 is proved.

Let $w : [0, \infty) \to R^1$ be a continuous strictly increasing concave and differentiable function which represents the preferences of the planner.

In this chapter we use the following optimality criterion.

A program $\{x^*(t), y^*(t)\}_{t=0}^{\infty}$ is weakly optimal if

$$\liminf_{T \to \infty} \sum_{t=0}^{T} [w(by(t)) - w(by^*(t))] \le 0$$

for every program $\{x(t), y(t)\}_{t=0}^{\infty}$ satisfying $x(0) = x^*(0)$.

Set

$$\Omega = \Big\{(x, x') \in R_+^n \times R_+^n : x' - (1 - d)x \geq 0$$

$$\text{and } a(x' - (1 - d)x) \leq 1\Big\}.$$

We have a correspondence $\Lambda : \Omega \to R_+^n$ given by

$$\Lambda(x, x') = \{y \in R_+^n : 0 \leq y \leq x \text{ and } ey \leq 1 - a(x' - (1 - d)x)\}.$$

For any $(x, x') \in \Omega$, define

$$u(x, x') = \max \{w(by) : y \in \Lambda(x, x')\}.$$

2.2 A Golden-Rule Stock

A golden-rule stock is $\widehat{x} \in R_+^n$ such that $(\widehat{x}, \widehat{x})$ is a solution to the problem:
maximize $u(x, x')$ subject to
(i) $x' \geq x$; (ii) $(x, x') \in \Omega$.
For $i = 1, \ldots, n$, set

$$\widehat{q}_i = a_i b_i / (1 + d a_i), \quad \widehat{p}_i = w'\big(b_\sigma (1 + d a_\sigma)^{-1}\big)\widehat{q}_i.$$

Set

$$\widehat{y} = (1 + d a_\sigma)^{-1} e(\sigma).$$

The following useful lemma was obtained in [37].

Lemma 2.2 $w(b\widehat{y}) \geq w(by) + \widehat{p}x' - \widehat{p}x$ for any $(x, x') \in \Omega$ and for any $y \in \Lambda(x, x')$.

Proof Let

$$(x, x') \in \Omega, \ y \in \Lambda(x, x').$$

Set

$$z = x' - (1 - d)x.$$

It is not difficult to see that

$$b\widehat{y} - by - \widehat{q}(x' - x) = c_\sigma - by - \widehat{q}(x' - x)$$

$$= c_\sigma - by - \widehat{q}(x' - (1 - d)x) + d\widehat{q}x$$

$$= c_\sigma(1 - ey - az) + c_\sigma ey + c_\sigma az - by - \widehat{q}z + d\widehat{q}x$$

$$= c_\sigma(1 - ey - az) + \sum_{i=1}^{n}(c_\sigma - b_i)y_i + \sum_{i=1}^{n}(c_\sigma - c_i)a_i z_i + d\widehat{q}x$$

$$= c_\sigma(1 - ey - az) + \sum_{i=1}^{n}(c_\sigma - c_i)y_i + \sum_{i=1}^{n}(c_\sigma - c_i)a_i z_i + d\widehat{q}(x - y). \qquad (2.2)$$

Applying (2.1) we obtain that

$$by - b\widehat{y} \le \widehat{q}x - \widehat{q}x'.$$

The inequality above implies that

$$w(by) - w(b\widehat{y}) \le w'(b\widehat{y})(by - b\widehat{y})$$
$$\le w'(b\widehat{y})(\widehat{q}x - \widehat{q}x') = \widehat{p}x - \widehat{p}x'.$$

This completes the proof of Lemma 2.2.

The next result was established in [37].

Theorem 2.3 *There exists a unique golden-rule stock* $\widehat{x} = (1 + da_\sigma)^{-1}e(\sigma)$.

Proof Set

$$\widehat{x} = \widehat{y} = (1 + da_\sigma)^{-1}e(\sigma).$$

It is not difficult to see that

$$(\widehat{x}, \widehat{x}) \in \Omega \text{ and } \widehat{y} \in \Lambda(\widehat{x}, \widehat{x}).$$

Applying Lemma 2.2 we obtain that $\widehat{x}$ is a golden-rule stock.

Let us show that $\widehat{x}$ is a unique golden-rule stock. Suppose to the contrary that $(\tilde{x}, \tilde{x}')$ is another solution with a corresponding $\tilde{y} \in \Lambda(\tilde{x}, \tilde{x}')$ and

$$\tilde{z} = \tilde{x}' - (1 - d)\tilde{x}.$$

Since the function $w(\cdot)$ is strictly increasing, we have

$$b\tilde{y} = b\widehat{y} = c_\sigma.$$

On substituting $\tilde{x}$, $\tilde{y}$, and $\tilde{z}$ for x, y, and z in (2.2), we obtain that the right-hand side of (2.2) equals zero. This implies that each of its four terms is zero,

$$\tilde{y}_i = \tilde{z}_i = 0 \text{ for all } i \in \{1, \ldots, n\} \setminus \{\sigma\},$$

$$\tilde{x}_i = \tilde{y}_i \text{ for all } i \in \{1, \ldots, n\}$$

and that

$$\tilde{y}_\sigma + a_\sigma \tilde{z}_\sigma = 1.$$

Coupling the first assertion with the relation $c_\sigma = b\tilde{y}$, we obtain that

$$\tilde{y}_\sigma = (1 + da_\sigma)^{-1},$$

and therefore from the third assertion that

$$\tilde{x} = (1 + da_\sigma)^{-1} e(\sigma).$$

It follows from the last assertion that

$$\tilde{z}_\sigma = d(1 + da_\sigma)^{-1}$$

and that

$$\begin{aligned}
\tilde{x}' &= \tilde{z} + (1 - d)\tilde{x} \\
&= \left(d(1 + da_\sigma)^{-1} + (1 - d)(1 + da_\sigma)^{-1}\right)e(\sigma) \\
&= (1 + da_\sigma)^{-1} e(\sigma).
\end{aligned}$$

This completes the proof of Theorem 2.3.

2.3 Good Programs

The results of this section were obtained in [37].

We use the following notion of good programs introduced by Gale [28] and used in optimal control [19, 65, 93, 117].

A program $\{x(t), y(t)\}_{t=0}^\infty$ is called good if there exists $M \in R^1$ such that

$$\sum_{t=0}^{T} (w(by(t)) - w(b\hat{y})) \geq M \text{ for all integers } T \geq 0.$$

A program is called bad if

$$\lim_{T \to \infty} \sum_{t=0}^{T} (w(by(t)) - w(b\widehat{y})) = -\infty.$$

Proposition 2.4 *Let $x_0 \in R_+^n$. Then there exists a good program*

$$\{x(t), y(t)\}_{t=0}^{\infty}$$

which satisfies $x(0) = x_0$.

Proof For every nonnegative integer t, set

$$z(t + 1) = d\widehat{x}.$$

Set $y(0) = 0$, and for all nonnegative integer $t \geq 0$, define

$$y(t + 1) = (1 - d)y(t) + d\widehat{x}.$$

It is easy to see that that the sequence $\{y(t)\}_{t=0}^{\infty}$ is increasing and converges to $\widehat{x}$. Set

$$x(0) = x_0,$$

and for every nonnegative integer $t \geq 0$, define

$$x(t + 1) = (1 - d)x(t) + z(t + 1).$$

Clearly, $\{x(t), y(t)\}_{t=0}^{\infty}$ is a program. It is easy to see that for all integers $t > 1$,

$$by(t) - b\widehat{x} = (1 - d)^t (by(1) - b\widehat{x})$$

and that for all natural numbers t,

$$by(t) \geq db\widehat{x}.$$

This implies that for all integers $t \geq 0$,

$$w(b\widehat{x}) - w(by(t)) \leq w'(by(t))(b\widehat{x} - by(t)) \leq w'(db\widehat{x})(b\widehat{x} - by(1))(1 - d)^{t-1}.$$

Therefore the sequence $\{w(b\widehat{x}) - w(by(t))\}_{t=0}^{\infty}$ is summable, and the program $\{x(t), y(t)\}_{t=0}^{\infty}$ is good.

Proposition 2.5 *Let $\{x(t), y(t)\}_{t=0}^{\infty}$ be a program. Then there exists a constant $M(x(0)) \geq 0$ such that for every pair of nonnegative integers $t_1 \leq t_2$,*

$$\sum_{t=t_1}^{t_2} (w(by(t)) - w(b\widehat{y})) \leq M(x(0)).$$

Proof Let $m(x(0))$ be as guaranteed by Proposition 2.1. Lemma 2.2 implies that for every pair of nonnegative integers $t_1 \leq t_2$,

$$\sum_{t=t_1}^{t_2} (w(by(t)) - w(b\widehat{y}))$$

$$\leq \widehat{p}(x(t_1) - x(t_2 + 1))$$

$$\leq \widehat{p}x(t_1) \leq m(x(0)) \sum_{j=1}^{n} \widehat{p}_j.$$

In order to complete the proof, it is sufficient to define

$$M(x(0)) = m(x(0))w'\big(b_\sigma(1 + da_\sigma)^{-1}\big) \sum_{i=1}^{n} a_i b_i (1 + da_i)^{-1}.$$

Proposition 2.5 easily implies the following result.

Proposition 2.6 *Every program which is not good is bad.*

For any $(x, x') \in \Omega$ and any $y \in \Lambda(x, x')$, set

$$\delta(x, y, x') = \widehat{p}(x - x') - (w(by) - w(b\widehat{y})). \tag{2.3}$$

We say that a program $\{x(t), y(t)\}_{t=0}^{\infty}$ has the average turnpike property if

$$\lim_{T \to \infty} T^{-1} \sum_{t=0}^{T-1} (x(t), y(t)) = (\widehat{x}, \widehat{y}).$$

Proposition 2.7 *Assume that a program $\{x(t), y(t)\}_{t=0}^{\infty}$ is good. Then it has the average turnpike property.*

Proof For every natural number T, set

$$\bar{x}(T) = T^{-1} \sum_{t=0}^{T-1} x(t), \quad \bar{y}(T) = T^{-1} \sum_{t=0}^{T-1} y(t).$$

Proposition 2.1 implies that the sequence $\{\bar{x}(T), \bar{y}(T)\}_{T=1}^{\infty}$ has a convergent subsequence. Denote its limit by (x^∞, y^∞). It is not difficult to see that

$$(x^\infty, x^\infty) \in \Omega, \ y^\infty \in \Lambda(x^\infty, x^\infty).$$

Since the program $\{x(t), y(t)\}_{t=0}^\infty$ is good and the function w is concave, there exists a number G such that for all natural numbers T,

$$T^{-1}G \leq T^{-1} \sum_{t=0}^{T-1} (w(by(t)) - w(b\widehat{y})) \leq w(b\bar{y}(T)) - w(b\widehat{y}).$$

This implies that

$$w(by^\infty) \geq w(b\widehat{y})$$

and that x^∞ is a golden-rule stock. Since the golden-rule stock is unique and the function w is strictly increasing, we conclude that

$$x^\infty = \widehat{x}, \ y^\infty = \widehat{y}.$$

This completes the proof of Proposition 2.7.

The next result easily follows from Lemmas 2.2 and (2.3).

Proposition 2.8 *Assume that $\{x(t), y(t)\}_{t=0}^\infty$ is a program. Then for every integer $t \geq 0$,*

$$\delta(x(t), y(t), x(t+1)) \geq 0$$

and for every natural number T,

$$\sum_{t=0}^T (w(by(t)) - w(b\widehat{y}))$$

$$= \widehat{p}(x(0) - x(T+1)) - \sum_{t=0}^T \delta(x(t), y(t), x(t+1)).$$

Proposition 2.8 implies the following result.

Proposition 2.9 *A program $\{x(t), y(t)\}_{t=0}^\infty$ is good if and only if*

$$\sum_{t=0}^\infty \delta(x(t), y(t), x(t+1)) < \infty.$$

For every $x_0 \in R_+^n$, define

$$\Delta(x_0) = \inf\left\{ \sum_{t=0}^{\infty} \delta(x(t), y(t), x(t+1)) : \{x(t), y(t)\}_{t=0}^{\infty}\right.$$

$$\left. \text{is a program such that } x(0) = x_0 \right\}.$$

Proposition 2.10 *Let $x_0 \in R_+^n$. Then*

$$0 \leq \Delta(x_0) < \infty$$

and there exists a program $\{x'(t), y'(t)\}_{t=0}^{\infty}$ such that

$$x'(0) = x_0, \ \Delta(x_0) = \sum_{t=0}^{\infty} \delta(x'(t), y'(t), x'(t+1)).$$

Proof In view of Proposition 2.1, there exists a good program $\{x(t), y(t)\}_{t=0}^{\infty}$ such that

$$x(0) = x_0.$$

Proposition 2.8 implies that

$$\Delta(x_0) < \infty.$$

Evidently, $\Delta(x_0) \geq 0$. For every natural number i, there exists a program $\{x^{(i)}(t), y^{(i)}(t)\}_{t=0}^{\infty}$ such that

$$x^{(i)}(0) = x_0$$

and

$$\sum_{t=0}^{\infty} \delta\big(x^{(i)}(t), y^{(i)}(t), x^{(i)}(t+1)\big) \leq \Delta(x_0) + i^{-1}. \tag{2.4}$$

Proposition 2.1 implies that there exists a number $m_0 > 0$ such that

$$x^{(i)}(t) \leq m_0 e \text{ for all integers } i \geq 1 \text{ and all integers } t \geq 0.$$

Extracting subsequences and using the diagonalization process, we obtain the existence of a strictly increasing sequence of natural numbers $\{i_k\}_{k=1}^{\infty}$ such that for every nonnegative integer $t \geq 0$, there exist

$$x'(t) = \lim_{k \to \infty} x^{(i_k)}(t), \ y'(t) = \lim_{k \to \infty} y^{(i_k)}(t).$$

Clearly, for every integer $t \geq 0$,

$$\delta\big(x'(t), y'(t), x'(t+1)\big) = \lim_{k \to \infty} \delta\big(x^{(i_k)}(t), y^{(i_k)}(t), x^{(i_k)}(t+1)\big).$$

It is easy to see that $\{x'(t), y'(t)\}_{t=0}^{\infty}$ is a program such that

$$x(0) = x_0$$

and

$$\sum_{t=0}^{\infty} \delta\big(x'(t), y'(t), x'(t+1)\big) \geq \Delta(x_0).$$

It is not difficult to see that

$$\Delta(x_0) = \lim_{k \to \infty} \sum_{t=0}^{\infty} \delta\big(x^{(i)}(t), y^{(i)}(t), x^{(i)}(t+1)\big)$$

$$\geq \lim_{T \to \infty} \left(\lim_{k \to \infty} \sum_{t=0}^{T} \delta(x^{(i)}(t), y^{(i)}(t), x^{(i)}(t+1)) \right)$$

$$\geq \lim_{T \to \infty} \sum_{t=0}^{T} \delta\big(x'(t), y'(t), x'(t+1)\big) = \sum_{t=0}^{\infty} \delta\big(x'(t), y'(t), x'(t+1)\big).$$

This completes the proof of Proposition 2.10.

Proposition 2.11 *Assume that a program* $\{x(t), y(t)\}_{t=0}^{\infty}$ *satisfies*

$$\Delta(x(0)) = \sum_{t=0}^{\infty} \delta(x(t), y(t), x(t+1)).$$

Then it is weakly optimal.

Proof In view of Proposition 2.9, the program $\{x(t), y(t)\}_{t=0}^{\infty}$ is good. Assume that it is not weakly optimal. Then there exist a program $\{x'(t), y'(t)\}_{t=0}^{\infty}$ satisfying

$$x'(0) = x(0),$$

a number $\epsilon > 0$, and a natural number t_ϵ such that for all integers $T \geq t_\epsilon$,

$$\sum_{t=0}^{T} \big(w(by'(t)) - w(by(t))\big) > \epsilon.$$

Clearly, the program $\{x'(t), y'(t)\}_{t=0}^{\infty}$ is good. Proposition 2.7 implies that the programs $\{x(t), y(t)\}_{t=0}^{\infty}$ and $\{x'(t), y'(t)\}_{t=0}^{\infty}$ have the average turnpike property. Proposition 2.8 implies that for all integers $T \geq t_{\epsilon}$,

$$\epsilon < \sum_{t=0}^{T}(w(by'(t)) - w(by(t)))$$

$$= \widehat{p}(x(T+1) - x'(T+1))$$

$$+ \sum_{t=0}^{T} \delta(x(t), y(t), x(t+1)) - \sum_{t=0}^{T} \delta(x'(t), y'(t), x'(t+1)).$$

Since

$$\sum_{t=0}^{\infty} \delta(x(t), y(t), x(t+1)) - \sum_{t=0}^{\infty} \delta(x'(t), y'(t), x'(t+1))$$

$$= \Delta(x(0)) - \sum_{t=0}^{T} \delta(x'(t), y'(t), x'(t+1)) \leq 0$$

the relation above implies that there exists an integer $t'_{\epsilon} \geq t_{\epsilon}$ such that for all integer $T \geq t'_{\epsilon}$,

$$2^{-1}\epsilon < \widehat{p}\left((T+1)^{-1} \sum_{t=0}^{T} x(t) - (T+1)^{-1} \sum_{t=0}^{T} x'(t)\right)$$

$$= \widehat{p}\left((T+1)^{-1} \sum_{t=0}^{T} x(t) - \widehat{x} + \widehat{x} - (T+1)^{-1} \sum_{t=0}^{T} x'(t)\right) \to 0 \text{ as } T \to \infty.$$

The contradiction we have reached proves Proposition 2.11.

Propositions 2.10 and 2.11 and (2.3) imply the following result.

Theorem 2.12 *Let* $x_0 \in R_+^n$. *Then there exists a weakly optimal program* $\{x(t), y(t)\}_{t=0}^{\infty}$ *satisfying* $x(0) = x_0$. *If* $x_0 = \widehat{x}$, *then the program* $\{x(t), y(t)\}_{t=0}^{\infty}$ *satisfying*

$$x(t) = y(t) = \widehat{x}, \ t = 0, 1, \ldots$$

is weakly optimal.

2.4 The von Neumann Facet

The following auxiliary result, obtained in [37], plays an important role in our study of the RSS model.

Lemma 2.13 *Let*

$$\xi_\sigma = 1 - d - a_\sigma^{-1}.$$

The von Neumann facet

$$\{(x, x') \in \Omega : \text{ there exists } y \in \Lambda(x, x') \text{ such that } \delta(x, y, x') = 0\}$$

is a subset of the set

$$\{(x, x') \in \Omega : x_i' = x_i = 0, \ i \in \{1, \dots, n\} \setminus \{\sigma\}, \ x_\sigma' = a_\sigma^{-1} + \xi_\sigma x_\sigma\}$$

with the equality if the function w is linear. If the function w is strictly concave, then the face is the singleton $\{(\widehat{x}, \widehat{x})\}$.

Proof Assume that

$$(\tilde{x}, \tilde{x}') \in \Omega \text{ and } \tilde{y} \in \Lambda(\tilde{x}, \tilde{x}')$$

satisfy

$$\delta(\tilde{x}, \tilde{y}, \tilde{x}') = 0.$$

It follows from (2.3) that

$$w(b\tilde{y}) - w(b\widehat{y}) + \widehat{p}(\tilde{x}' - \tilde{x}) = 0.$$

In view of the concavity of w,

$$w(b\tilde{y}) - w(b\widehat{y}) \leq w'(b\widehat{y})(b\tilde{y} - b\widehat{y}),$$

$$b\widehat{y} - b\tilde{y} - \widehat{q}(\tilde{x}' - \tilde{x}) \leq 0. \tag{2.5}$$

It follows from (2.1), (2.2), (2.5), and the inequality $\tilde{z}_i \geq 0, \ i = 1, \dots, n$ that

$$b\widehat{y} - b\tilde{y} - \widehat{q}(\tilde{x}' - \tilde{x})$$

$$= c_\sigma(1 - e\tilde{y} - a\tilde{z}) + \sum_{i=1}^{n}(c_\sigma - c_i)\tilde{y}_i$$

$$+ \sum_{i=1}^{n}(c_\sigma - c_i)a_i\tilde{z}_i + d\widehat{q}(\tilde{x} - \tilde{y}) = 0. \tag{2.6}$$

The relation above implies that for all integers $i \in \{1, \ldots, n\} \setminus \{\sigma\}$,

$$\tilde{z}_i = 0 = \tilde{y}_i = \tilde{x}_i = \tilde{x}'_i,$$

$$\tilde{y}_\sigma = \tilde{x}_\sigma,$$

$$\tilde{y}_\sigma + a_\sigma \tilde{z}_\sigma = 1,$$

$$\tilde{x}_\sigma + a_\sigma \left(\tilde{x}'_\sigma - (1-d)\tilde{x}_\sigma\right) = 1$$

and

$$\tilde{x}'_\sigma = a_\sigma^{-1} + \xi_\sigma \tilde{x}_\sigma.$$

Assume that w is strictly concave and that $b\tilde{y} \neq b\hat{y}$. Then we have a strict inequality in (2.5). This contradicts (2.6). Therefore

$$b\tilde{y} = b\hat{y} = c_\sigma.$$

By the relations above,

$$\tilde{y}_\sigma = (1 + da_\sigma)^{-1} = \tilde{x}_\sigma$$

and

$$\tilde{x}'_\sigma = a_\sigma^{-1} + \xi_\sigma \tilde{x}_\sigma = (1 + da_\sigma)^{-1}.$$

For the revise implication in the linear case, pick $(x, x') \in \Omega$ such that

$$x'_\sigma = a_\sigma^{-1} + \xi_\sigma x_\sigma,$$

$$x'_i = x_i = 0, \ i \in \{1, \ldots, n\} \setminus \{\sigma\}$$

and

$$y_\sigma = x_\sigma.$$

On substituting these values in the left-hand side of (2.2), we see that it is equal to zero. This implies the assertion of the lemma in the linear case.

The next result was also obtained in [37]. It easily follows from Proposition 2.4.

Proposition 2.14 *Any weakly optimal program is good.*

Chapter 3
Infinite Horizon Optimization

We study infinite horizon optimal control problems related to the Robinson–Solow–Srinivasan model. We establish a convergence of good programs to the golden-rule stock, show the existence of overtaking optimal programs and analyze their convergence to the golden-rule stock, and consider some properties of good programs.

3.1 Overtaking Optimal Programs

Let R^1 (R^1_+) be the set of real (nonnegative) numbers, and let R^n be the n-dimensional Euclidean space with nonnegative orthant

$$R^n_+ = \{x = (x_1, \ldots, x_n) \in R^n : x_i \geq 0, \ i = 1, \ldots, n\}.$$

For every pair of vectors $x = (x_1, \ldots, x_n)$, $y = (y_1, \ldots, y_n) \in R^n$, define their inner product by

$$xy = \sum_{i=1}^{n} x_i y_i,$$

and let $x \gg y, x > y, x \geq y$ have their usual meaning.

Let $e(i), i = 1, \ldots, n$, be the ith unit vector in R^n, and e be an element of R^n_+ all of whose coordinates are unity. For every $x \in R^n$, denote by $\|x\|$ its Euclidean norm in R^n.

Let $a = (a_1, \ldots, a_n) \gg 0, b = (b_1, \ldots, b_n) \gg 0, b_1 \geq b_2 \cdots \geq b_n$, $d \in (0, 1)$,

A. J. Zaslavski, *Turnpike Theory for the Robinson–Solow–Srinivasan Model*,
Springer Optimization and Its Applications 166,
https://doi.org/10.1007/978-3-030-60307-6_3

$$c_i = b_i/(1 + da_i), \ i = 1, \ldots, n. \tag{3.1}$$

We assume the following:

There exists $\sigma \in \{1, \ldots, n\}$ such that for all

$$i \in \{1, \ldots, n\} \setminus \{\sigma\}, \ c_\sigma > c_i. \tag{3.2}$$

Recall that a sequence $\{x(t), y(t)\}_{t=0}^{\infty}$ is called a program if for each integer $t \geq 0$

$$(x(t), y(t)) \in R_+^n \times R_+^n, \ x(t+1) \geq (1 - d)x(t),$$
$$0 \leq y(t) \leq x(t), \ a(x(t+1) - (1 - d)x(t)) + ey(t) \leq 1. \tag{3.3}$$

Let T_1, T_2 be integers such that $0 \leq T_1 < T_2$. A pair of sequences

$$\left(\{x(t)\}_{t=T_1}^{T_2}, \{y(t)\}_{t=T_1}^{T_2-1} \right)$$

is called a program if $x(T_2) \in R_+^n$ and for each integer t satisfying $T_1 \leq t < T_2$, relation (3.3) is valid.

Assume that $w : [0, \infty) \to R^1$ is a continuous strictly increasing concave and differentiable function which represents the preferences of the planner.

Define

$$\Omega = \{(x, x') \in R_+^n \times R_+^n : x' - (1 - d)x \geq 0$$
$$\text{and } a(x' - (1 - d)x) \leq 1\} \tag{3.4}$$

and a correspondence $\Lambda : \Omega \to R_+^n$ given by

$$\Lambda(x, x') = \{y \in R_+^n : 0 \leq y \leq x \text{ and } ey \leq 1 - a(x' - (1 - d)x)\}. \tag{3.5}$$

For every $(x, x') \in \Omega$, set

$$u(x, x') = \max\{w(by) : y \in \Lambda(x, x')\}. \tag{3.6}$$

Recall that the golden-rule stock is $\widehat{x} \in R_+^n$ such that $(\widehat{x}, \widehat{x})$ is a solution to the problem:

maximize $u(x, x')$ subject to
(i) $x' \geq x$; (ii) $(x, x') \in \Omega$.

By Theorem 2.3, there exists a unique golden-rule stock

$$\widehat{x} = (1/(1 + da_\sigma))e(\sigma). \tag{3.7}$$

It is not difficult to see that $\widehat{x}$ is a solution to the problem

$$w(by) \to \max, \quad y \in \Lambda(\widehat{x}, \widehat{x}).$$

Set

$$\widehat{y} = \widehat{x}. \tag{3.8}$$

For $i = 1, \ldots, n$, set

$$\widehat{q}_i = a_i b_i / (1 + da_i), \quad \widehat{p}_i = w'(b\widehat{x})\widehat{q}_i. \tag{3.9}$$

In view of Lemma 2.2,

$$w(b\widehat{x}) \geq w(by) + \widehat{p}x' - \widehat{p}x \tag{3.10}$$

for every $(x, x') \in \Omega$ and for every $y \in \Lambda(x, x')$.

A program $\{x(t), y(t)\}_{t=0}^{\infty}$ is good if there is a real number M such that

$$\sum_{t=0}^{T}(w(by(t)) - w(b\widehat{y})) \geq M \text{ for every nonnegative integer } T.$$

A program $\{x(t), y(t)\}_{t=0}^{\infty}$ is bad if

$$\lim_{T \to \infty} \sum_{t=0}^{T}(w(by(t)) - w(b\widehat{y})) = -\infty.$$

By Proposition 2.6, every program that is not good is bad. Proposition 2.4 implies that for every point $x_0 \in R_+^n$, there exists a good program $\{x(t), y(t)\}_{t=0}^{\infty}$ satisfying $x(0) = x_0$.

In this chapter we prove the following results obtained in [92].

Theorem 3.1 *Assume that the function w is strictly concave. Then for every good program $\{x(t), y(t)\}_{t=0}^{\infty}$,*

$$\lim_{t \to \infty}(x(t), y(t)) = (\widehat{x}, \widehat{x}).$$

Set

$$\xi_\sigma = 1 - d - (1/a_\sigma). \tag{3.11}$$

Theorem 3.2 *Assume that $\xi_\sigma \neq -1$. Then*

$$\lim_{t \to \infty}(x(t), y(t)) = (\widehat{x}, \widehat{x})$$

for every good program $\{x(t), y(t)\}_{t=0}^{\infty}$.

In this book we use a notion of an overtaking optimal program introduced by Atsumi [6], Gale [28], and von Weizsacker [89] and a notion of a weakly optimal program introduced by Brock [17]. These optimality criterions are used in optimal control [19, 93, 117].

A program $\{\bar{x}(t), \bar{y}(t)\}_{t=0}^{\infty}$ is weakly optimal if for every program $\{x(t), y(t)\}_{t=0}^{\infty}$ which satisfies $x(0) = \bar{x}(0)$, the inequality

$$\liminf_{T \to \infty} \sum_{t=0}^{T} [w(by(t)) - w(b\bar{y}(t))] \le 0$$

is true.

A program $\{x^*(t), y^*(t)\}_{t=0}^{\infty}$ is overtaking optimal if

$$\limsup_{T \to \infty} \sum_{t=0}^{T} [w(by(t)) - w(by^*(t))] \le 0$$

for every program $\{x(t), y(t)\}_{t=0}^{\infty}$ which satisfies $x(0) = x^*(0)$.

In this chapter we prove the following result obtained in [92].

Theorem 3.3 *Assume that for every good program* $\{x(t), y(t)\}_{t=0}^{\infty}$,

$$\lim_{t \to \infty} (x(t), y(t)) = (\widehat{x}, \widehat{x}).$$

Then for every point $x_0 \in R_+^n$, *there is an overtaking optimal program* $\{x(t), y(t)\}_{t=0}^{\infty}$ *such that* $x(0) = x_0$.

Corollary 3.4 *Assume that the function* w *is strictly concave. Then for every point* $x_0 \in R_+^n$, *there exists an overtaking optimal program* $\{x(t), y(t)\}_{t=0}^{\infty}$ *satisfying* $x(0) = x_0$.

Corollary 3.5 *Assume that* $\xi_\sigma \ne -1$. *Then for every point* $x_0 \in R_+^n$, *there is an overtaking optimal program* $\{x(t), y(t)\}_{t=0}^{\infty}$ *such that* $x(0) = x_0$.

3.2   Auxiliary Results for Theorems 3.1–3.3

Proposition 3.6 *Let* m_0 *be a positive number. Then there exists a positive number* m_1 *such that for every integer* $T > 0$ *and every program* $(\{x(t)\}_{t=0}^{T}, \{y(t)\}_{t=0}^{T-1})$ *satisfying* $x(0) \le m_0 e$, *the inequality* $x(t) \le m_1 e$ *is valid for all integers* $t \in \{0, \ldots, T\}$.

For the proof of this result, see Proposition 2.1.

For every $(x, x') \in \Omega$ and every $y \in \Lambda(x, x')$, define (see (2.3))

$$\delta(x, y, x') = \widehat{p}(x - x') - (w(by) - w(b\widehat{y})). \qquad (3.12)$$

Let a program $\{x(t), y(t)\}_{t=0}^{\infty}$ be given. We denote by $\omega(\{x(t), y(t)\}_{t=0}^{\infty})$ the set of all points $(u, v, \eta) \in R^{3n}$ for which there exists a strictly increasing sequence of natural numbers $\{t_k\}_{k=1}^{\infty}$ for which

$$\lim_{k \to \infty} (x(t_k), y(t_k), x(t_k + 1)) = (u, v, \eta).$$

Proposition 3.6 implies that the set $\omega(\{x(t), y(t)\}_{t=0}^{\infty})$ is well-defined and it is a compact subset of R^{3n}.

Lemma 3.7 *Assume that a program* $\{x(t), y(t)\}_{t=0}^{\infty}$ *is good and that*

$$(u_0, v_0, \eta_0) \in \omega(\{x(t), y(t)\}_{t=0}^{\infty}). \qquad (3.13)$$

Then

$$(u_0, \eta_0) \in \Omega, \quad v_0 \in \Lambda(u_0, \eta_0), \qquad (3.14)$$

$$\delta(u_0, v_0, \eta_0) = 0 \qquad (3.15)$$

and there exists a sequence $\{u(t), v(t)\}_{t=-\infty}^{\infty}$ *such that*

$$(u(0), v(0), u(1)) = (u_0, v_0, \eta_0) \qquad (3.16)$$

and

$$(u(s), v(s), u(s + 1)) \in \omega(\{x(t), y(t)\}_{t=0}^{\infty}) \text{ for all integers } s. \qquad (3.17)$$

Proof Since the program $\{x(t), y(t)\}_{t=0}^{\infty}$ is good, there exists a real number $M_0 \in R^1$ for which

$$\sum_{t=0}^{T} (w(by(t)) - w(b\widehat{y})) \geq M_0 \text{ for all nonnegative integers } T. \qquad (3.18)$$

Proposition 3.6 implies that there exists a positive number M_1 such that

$$\|x(t)\|, \ \|y(t)\| \leq M_1 \text{ for all nonnegative integers } T. \qquad (3.19)$$

By (3.18), for every nonnegative integer T,

$$M_0 \leq \sum_{t=0}^{T} (w(by(t)) - w(b\widehat{y}))$$

$$= \sum_{i=0}^{T} [w(by(t)) - w(b\widehat{y}) + \widehat{p}x(t+1) - \widehat{p}x(t)]$$

$$+ \widehat{p}x(0) - \widehat{p}x(T+1).$$

Combined with (3.12) and (3.19), the relation above implies that for every nonnegative integer T,

$$\sum_{t=0}^{T} \delta(x(t), y(t), x(t+1))$$

$$= \sum_{t=0}^{T} [w(b\widehat{y}) - w(by(t)) + \widehat{p}x(t) - \widehat{p}x(t+1)]$$

$$\leq -M_0 + \widehat{p}x(0) - \widehat{p}x(T+1) \leq -M_0 + 2\|\widehat{p}\| M_1.$$

Therefore

$$\sum_{t=0}^{\infty} \delta(x(t), y(t), x(t+1)) < \infty. \tag{3.20}$$

By (3.13), there exists a strictly increasing sequence of natural numbers $\{t_k\}_{k=1}^{\infty}$ such that

$$\lim_{k \to \infty} (x(t_k), y(t_k), x(t_k + 1)) = (u_0, v_0, \eta_0). \tag{3.21}$$

Since the set

$$\{(x, y, x') : (x, x') \in \Omega, \ y \in \Lambda(x, x')\}$$

is closed and contains the sequence $\{x(t), y(t), x(t+1)\}_{t=0}^{\infty}$, it follows from (3.21) that

$$(u_0, \eta_0) \in \Omega$$

and

$$v_0 \in \Lambda(u_0, \eta_0).$$

Therefore relation (3.14) is true. By (3.20), (3.21), and the continuity of the function $\delta(\cdot, \cdot, \cdot)$, equality (3.15) holds.

For every integer $k \geq 1$, define a sequence

$$\left\{ x^{(k)}(s), y^{(k)}(s) \right\}_{s=-t_k}^{\infty} \subset R^n \times R^n$$

by

$$\left(x^{(k)}(s), y^{(k)}(s) \right) = (x(s + t_k), y(s + t_k)) \text{ for all integers } s \geq -t_k. \tag{3.22}$$

Extracting subsequences, re-indexing, and using diagonalization process, we obtain that there exists a strictly increasing sequence of natural numbers $\{k_j\}_{j=1}^{\infty}$ such that for every integer s, there exists the limit

$$(u(s), v(s)) = \lim_{j \to \infty} (x^{(k_j)}(s), y^{(k_j)}(s)). \tag{3.23}$$

In view of (3.22) and (3.23),

$$(u(s), v(s), u(s + 1)) \in \omega(\{x(t), y(t)\}_{t=0}^{\infty}) \text{ for all integers } s. \tag{3.24}$$

It follows from (3.21), (3.22), and (3.23) that

$$(u(0), v(0), u(1)) = \lim_{j \to \infty} (x^{(k_j)}(0), y^{(k_j)}(0), x^{(k_j)}(1))$$

$$= \lim_{j \to \infty} (x(t_{k_j}), y(t_{k_j}), x(t_{k_j} + 1)) = (u_0, v_0, \eta_0).$$

This completes the proof of Lemma 3.7.

3.3 Proofs of Theorems 3.1 and 3.2

Proof of Theorem 3.1 Suppose that the function w is strictly concave. Let $\{x(t), y(t)\}_{t=0}^{\infty}$ be a given good program. Assume that $(u_0, v_0, \eta_0) \in \omega(\{x(t), y(t)\}_{t=0}^{\infty})$. Lemma 3.7 implies that

$$(u_0, \eta_0) \in \Omega, \quad v_0 \in \Lambda(u_0, \eta_0), \quad \delta(u_0, v_0, \eta_0) = 0.$$

Together with Lemma 2.13, this implies that $u_0 = \eta_0 = \widehat{x}$ and $\delta(\widehat{x}, v_0, \widehat{x}) = 0$. Since $v_0 \leq \widehat{x}$, relation (3.12) and the strict monotonicity of the function w imply that $v_0 = \widehat{x}$. Hence

$$\omega(\{x(t), y(t)\}_{t=0}^{\infty}) = \{(\widehat{x}, \widehat{x}, \widehat{x})\}.$$

Theorem 3.1 is proved.

The following lemma is used in the proof of Theorem 3.2.

Lemma 3.8 *Assume that $\xi_\sigma \neq -1$ and that a sequence $\{x(t), y(t)\}_{t=-\infty}^{\infty} \subset R_+^{2n}$ satisfies the following conditions:*

There exists a positive number S_0 such that

$$\|x(t)\|, \ \|y(t)\| \leq S_0 \text{ for all integers } t, \tag{3.25}$$

$$x(t+1) \geq (1-d)x(t) \text{ and } 0 \leq y(t) \leq x(t) \text{ for all integers } t, \tag{3.26}$$

$$a(x(t+1) - (1-d)x(t)) + ey(t) \leq 1 \text{ for all integers } t, \tag{3.27}$$

$$\delta(x(t), y(t), x(t+1)) = 0 \text{ for all integers } t. \tag{3.28}$$

Then

$$x(t) = y(t) = \widehat{x} \text{ for all integers } t.$$

Proof Lemma 2.13 and (3.26)–(3.28) imply that for each integer t,

$$y_i(t) = x_i(t) = 0 \text{ for all } i \in \{1, \ldots, n\} \setminus \{\sigma\} \tag{3.29}$$

and that

$$x_\sigma(t+1) = 1/a_\sigma + \xi_\sigma x_\sigma(t), \ y_\sigma(t) \leq x_\sigma(t). \tag{3.30}$$

It follows from (3.12) and (3.28)–(3.30) imply that for every integer t,

$$\begin{aligned}
a(x(t+1) &- (1-d)x(t)) + ex(t) \\
&= a_\sigma(x_\sigma(t+1) - (1-d)x_\sigma(t)) + x_\sigma(t) \\
&= a_\sigma(1/a_\sigma + \xi_\sigma x_\sigma(t) - (1-d)x_\sigma(t)) + x_\sigma(t) \\
&= a_\sigma(1/a_\sigma - (1/a_\sigma)x_\sigma(t)) + x_\sigma(t) = 1.
\end{aligned}$$

It follows from (3.12), (3.28)–(3.30), strict monotonicity of w, and the relation above that for every integer t, we have

$$y_\sigma(t) = x_\sigma(t). \tag{3.31}$$

It follows from our assumptions that there are two cases:

$$|\xi_\sigma| < 1 \tag{3.32}$$

and

$$|\xi_\sigma| > 1. \tag{3.33}$$

Assume that (3.32) is valid. Fix an integer s. By induction, using (3.30) one can show that for every natural number t, we have

$$x_\sigma(s) = (\xi_\sigma)^t x_\sigma(s - t) + (1/a_\sigma) \sum_{i=0}^{t-1} \xi_\sigma^i. \tag{3.34}$$

It is clear that for $t = 1$ Equation (3.34) follows from relation (3.30). Assume that $\tau \geq 1$ is an integer and that Equation (3.34) is true for $t = \tau$.

In view of (3.34) with $t = \tau$ and (3.30), we have

$$x_\sigma(s) = (\xi_\sigma)^\tau x_\sigma(s - \tau) + (1/a_\sigma) \sum_{i=0}^{\tau-1} \xi_\sigma^i$$

$$= \xi_\sigma^\tau [1/a_\sigma + \xi_\sigma x_\sigma(s - \tau - 1)] + (1/a_\sigma) \sum_{i=0}^{\tau-1} \xi_\sigma^i$$

$$= \xi_\sigma^{\tau+1} x_\sigma(s - \tau - 1) + (1/a_\sigma) \sum_{i=0}^{\tau} \xi_\sigma^i.$$

Hence (3.34) is valid for $t = \tau + 1$. Therefore we have shown that Equation (3.34) is true for every natural number t. It follows from (3.11), (3.25), (3.32), and (3.34) that

$$x_\sigma(s) = \lim_{t \to \infty} \left[(\xi_\sigma)^t x_\sigma(s - t) + (1/a_\sigma) \sum_{i=0}^{t-1} \xi_\sigma^i \right] = (1/a_\sigma) \sum_{t=0}^{\infty} \xi_\sigma^i$$

$$= (1/a_\sigma)(1 - \xi_\sigma)^{-1} = (1/a_\sigma)(d + 1/a_\sigma)^{-1} = (da_\sigma + 1)^{-1} = \widehat{x}_\sigma.$$

Together with (3.29) and (3.31), this implies that

$$y(s) = x(s) = \widehat{x} \text{ for all integers } s. \tag{3.35}$$

Assume that (3.33) holds. It follows from (3.30) that for every integer t, we have

$$x_\sigma(t) = \xi_\sigma^{-1}[x_\sigma(t + 1) - 1/a_\sigma]. \tag{3.36}$$

Fix an integer s. By induction we show that for every natural number t, we have

$$x_\sigma(s) = (\xi_\sigma)^{-t} x_\sigma(s+t) - (1/a_\sigma) \sum_{i=1}^{t} \xi_\sigma^{-i}. \tag{3.37}$$

Evidently, for $t = 1$ equality (3.37) follows from relation (3.36). Assume that $\tau \geq 1$ is an integer and that equality (3.37) is valid for $t = \tau$.

In view of (3.37) with $t = \tau$ and (3.36), we have

$$x_\sigma(s) = (\xi_\sigma)^{-\tau} x_\sigma(s + \tau) - (1/a_\sigma) \sum_{i=1}^{\tau} \xi_\sigma^{-i}$$

$$= \xi_\sigma^{-\tau} [\xi_\sigma^{-1}(x_\sigma(s + \tau + 1) - 1/a_\sigma)] - (1/a_\sigma) \sum_{i=1}^{\tau} \xi_\sigma^{-i}$$

$$= \xi_\sigma^{-\tau-1} x_\sigma(s + \tau + 1) - (1/a_\sigma) \sum_{i=i}^{\tau+1} \xi_\sigma^{-i}.$$

Hence (3.37) is true for $t = \tau + 1$. Thus we have shown that (3.37) holds for every natural number t. By (3.11), (3.25), (3.33), and (3.37), we have

$$x_\sigma(s) = \lim_{t \to \infty} \left[(\xi_\sigma)^{-t} x_\sigma(s+t) - (1/a_\sigma) \sum_{i=1}^{t} \xi_\sigma^{-i} \right] = -(1/a_\sigma) \sum_{t=1}^{\infty} \xi_\sigma^{-i}$$

$$= -(1/a_\sigma)\xi_\sigma^{-1}(1 - \xi_\sigma^{-1})^{-1} = (-1/a_\sigma)(\xi_\sigma - 1)^{-1}$$

$$= (1/a_\sigma)(d + 1/a_\sigma)^{-1} = (da_\sigma + 1)^{-1} = \hat{x}_\sigma.$$

Together with (3.29) and (3.31), the relation above implies that (3.35) is valid for all integers s. Lemma 3.8 is proved.

Now Theorem 3.2 follows from Lemmas 3.7 and 3.8.

3.4   Proof of Theorem 3.3

Assume that $x_0 \in R_+^n$. Proposition 2.4 implies that there is a good program from x_0. By Proposition 2.9, a program $\{x(t), y(t)\}_{t=0}^{\infty}$ is good if and only if

$$\sum_{t=0}^{\infty} \delta(x(t), y(t), x(t+1)) < \infty.$$

Put

$$\Delta(x_0) = \inf \left\{ \sum_{t=0}^{\infty} \delta(x(t), y(t), x(t+1)) : \{x(t), y(t)\}_{t=0}^{\infty} \right.$$

$$\left. \text{is a program such that } x(0) = x_0 \right\}. \tag{3.38}$$

Evidently, $\Delta(x_0) < \infty$. Proposition 2.10 implies that there exists a program $\{x^*(t), y^*(t)\}_{t=0}^{\infty}$ such that

$$x^*(0) = x_0 \text{ and } \sum_{t=0}^{\infty} \delta(x^*(t), y^*(t), x^*(t+1)) = \Delta(x_0). \tag{3.39}$$

We claim that the program $\{x^*(t), y^*(t)\}_{t=0}^{\infty}$ is overtaking optimal.
    Let $\{x(t), y(t)\}_{t=0}^{\infty}$ be a program satisfying

$$x(0) = x_0. \tag{3.40}$$

Assume that $\{x(t), y(t)\}_{t=0}^{\infty}$ is not good. In view of Proposition 2.4, it is bad and

$$\lim_{T \to \infty} \sum_{t=0}^{T} (w(by(t)) - w(b\widehat{y})) = -\infty. \tag{3.41}$$

Since $\{x^*(t), y^*(t)\}_{t=0}^{\infty}$ is good, there exists a positive number M_0 for which

$$\sum_{t=0}^{T} (w(by^*(t)) - w(b\widehat{y})) \geq M_0 \text{ for all nonnegative integers } T. \tag{3.42}$$

By (3.41) and (3.42), for every nonnegative integer T, we have

$$\sum_{t=0}^{T} (w(by(t)) - w(by^*(t))) = \sum_{t=0}^{T} (w(by(t)) - w(b\widehat{y})) - \sum_{t=0}^{T} (w(by^*(t)) - w(b\widehat{y}))$$

$$\leq \sum_{t=0}^{T} (w(by(t)) - w(b\widehat{y})) - M_0 \to -\infty \text{ as } T \to \infty. \tag{3.43}$$

Assume that $\{x(t), y(t)\}_{t=0}^{\infty}$ is a good program. Then

$$\lim_{t \to \infty} (x(t), y(t)) = (\widehat{x}, \widehat{x}), \quad \lim_{t \to \infty} (x^*(t), y^*(t)) = (\widehat{x}, \widehat{x}). \tag{3.44}$$

By (3.12), (3.38), (3.39), (3.40), and (3.44), for every nonnegative integer T, we have

$$\sum_{t=0}^{T}[w(by(t)) - w(by^*(t))] = \sum_{t=0}^{T}[w(by(t)) - w(b\widehat{y})] - \sum_{t=0}^{T}[w(by^*(t)) - w(b\widehat{y})]$$

$$= \sum_{t=0}^{T}[w(by(t)) - w(b\widehat{y}) + \widehat{p}x(t+1) - \widehat{p}x(t)] + \widehat{p}x(0) - \widehat{p}x(T+1)$$

$$- \sum_{t=0}^{T}[w(by^*(t)) - w(b\widehat{y}) + \widehat{p}x^*(t+1) - \widehat{p}x^*(t)] - \widehat{p}x^*(0) + \widehat{p}x^*(T+1)$$

$$= -\sum_{t=0}^{T}\delta(x(t), y(t), x(t+1)) + \sum_{t=0}^{T}\delta(x^*(t), y^*(t), x^*(t+1))$$

$$-\widehat{p}[x(T+1) - x^*(T+1)] \rightarrow$$

$$-\sum_{t=0}^{\infty}\delta(x(t), y(t), x(t+1)) + \sum_{t=0}^{\infty}\delta(x^*(t), y^*(t), x^*(t+1))$$

$$= \Delta(x_0) - \sum_{t=0}^{\infty}\delta(x(t), y(t), x(t+1)) \leq 0$$

as $T \rightarrow \infty$. This completes the proof of Theorem 3.3.

3.5   Examples

Assume that

$$\xi_\sigma = -1 \tag{3.45}$$

and that the function w is linear. In this section we present an example of a good program which does not converge to the golden-rule stock. This example was constructed in [92].

Lemma 2.13 implies that

$$\{(x, x') \in \Omega : \text{ there exists } y \in \Lambda(x, x') \text{ such that } \delta(x, y, x') = 0\}$$

$$= \{(x, x') \in \Omega : x_i' = x_i = 0 \text{ for all } i \in \{1, \dots, n\} \setminus \{\sigma\},$$

$$x_\sigma' = (1/a_\sigma) + \xi_\sigma x_\sigma\}. \tag{3.46}$$

By (3.11) and (3.46), we have

$$d + 1/a_\sigma = 2, \ da_\sigma + 1 = 2a_\sigma \tag{3.47}$$

and

$$\hat{x} = \hat{y} = 1/(1 + da_\sigma)e(\sigma) = 1/(2a_\sigma)e(\sigma). \tag{3.48}$$

Let $\kappa \geq 0$, and define $(x, y, x') \in R^{3n}$ by

$$x = y = \kappa e(\sigma), \quad x' = (1/a_\sigma + \xi_\sigma \kappa)e(\sigma) = (1/a_\sigma - \kappa)e(\sigma). \tag{3.49}$$

By (3.47) and (3.49), we have

$$x' - (1 - d)x = (1/a_\sigma - \kappa - (1 - d)\kappa)e(\sigma) = (1/a_\sigma - \kappa/a_\sigma)e(\sigma). \tag{3.50}$$

In view of (3.50),

$$a(x' - (1 - d)x) = a_\sigma(1 - \kappa)a_\sigma^{-1} = 1 - \kappa. \tag{3.51}$$

By (3.47) and (3.49)–(3.51),

$$(x, x') \in \Omega \text{ if and only if } \kappa \leq 1. \tag{3.52}$$

It follows from (3.49) and (3.52)

$$1 - a(x' - (1 - d)x) - ex = 1 - (1 - \kappa) - \kappa = 0.$$

Combined with (3.52), the relation above implies that

$$(x, x') \in \Omega \text{ and } x \in \Lambda(x, x') \text{ if and only if } \kappa \leq 1. \tag{3.53}$$

Since (3.53) is true for every nonnegative number κ, it is easy to see that

$$(x', x) \in \Omega \text{ and } x' \in \Lambda(x', x) \text{ if and only if } 1/a_\sigma - \kappa \leq 1 \text{ and } \kappa \leq 1/a_\sigma. \tag{3.54}$$

Let

$$\kappa \in [1/a_\sigma - 1, 1] \tag{3.55}$$

be given. In view of (3.47) such that κ exists and it is nonnegative. For every nonnegative integer t set

$$y(2t) = x(2t) = \kappa e(\sigma), \quad y(2t + 1) = x(2t + 1) = (1/a_\sigma - \kappa)e(\sigma). \tag{3.56}$$

Relations (3.53)–(3.56) imply that

$$(x(t), x(t + 1)) \in \Omega, \quad y(t) \in \Lambda(x(t), x(t + 1))$$

for every integer t. Hence $\{x(t), y(t)\}_{t=0}^{\infty}$ is a program. By (3.46), (3.55), and (3.56), we have

$$\delta(x(t), x(t), x(t+1)) = 0 \text{ for every nonnegative integer } t. \qquad (3.57)$$

Thus $\{x(t), y(t)\}_{t=0}^{\infty}$ is a good program. It is clear that

$$\omega(\{x(t)\}_{t=0}^{\infty}, \{y(t)\}_{t=0}^{\infty}) = \{(\kappa e(\sigma), \kappa e(\sigma), (1/a_{\sigma} - \kappa)e(\sigma))$$

$$((1/a_{\sigma} - \kappa)e(\sigma), (1/a_{\sigma} - \kappa)e(\sigma), \kappa e(\sigma))\}$$

and it is not a singleton if $\kappa \neq (2a_{\sigma})^{-1}$. Therefore Theorem 3.3 does not hold if $\xi_{\sigma} = -1$ and w is linear.

Let us show the existence of a weakly optimal program which is not overtaking optimal.

Set

$$n = 1, \; w(z) = z \text{ for all } z \geq 0, \; b_1 = 1, \; d \in (0, 1), \; a_1 = (2 - d)^{-1}. \qquad (3.58)$$

By (3.11) and (3.58), we have

$$1/2 < a_1 < 1, \; \xi_1 = 1 - d - 1/a_1 = -1.$$

Assume that $\kappa = 1$, and define a program $\{x(t), y(t)\}_{t=0}^{\infty}$ by (3.56). Proposition 2.11 and (3.57) imply that the program $\{x(t), y(t)\}_{t=0}^{\infty}$ is weakly optimal. There are two cases:

- $\{x(t), y(t)\}_{t=0}^{\infty}$ is overtaking optimal;
- $\{x(t), y(t)\}_{t=0}^{\infty}$ is not overtaking optimal.

It is sufficient to consider only the first case. Set

$$\bar{x}(0) = 1, \; \bar{y}(0) = \hat{x}, \; \bar{x}(t) = \bar{y}(t) = \hat{x} \text{ for all natural numbers } t. \qquad (3.59)$$

It is easy to see that

$$\bar{x}(1) - (1-d)\bar{x}(0) = (1+da_1)^{-1} - (1-d) = (d - da_1 + d^2 a_1)(1 + da_1)^{-1} > 0,$$

$$a_1(\bar{x}(1) - (1-d)\bar{x}(0)) + \bar{y}(0)$$

$$= a_1(d - da_1 + d^2 a_1)(1 + da_1)^{-1} + (1 + da_1)^{-1}$$

$$= 1 + a_1(-da_1 + d^2 a_1)(1 + da_1)^{-1} < 1.$$

The relations above imply that the sequence $\{\bar{x}(t), \bar{y}(t)\}_{t=0}^{\infty}$ is a program.

We claim that $\{\bar{x}(t), \bar{y}(t)\}_{t=0}^{\infty}$ is a weakly optimal program.

For every integer $T \geq 1$, it follows from (3.48), (3.56), (3.58), and (3.59) that

$$\sum_{t=0}^{2T-1} y(t) = T(\kappa + (1/a_1 - \kappa))$$

$$= (T/a_1) = \sum_{t=0}^{2T-1} \bar{y}(t). \tag{3.60}$$

Assume that $\{\bar{x}(t), \bar{y}(t)\}_{t=0}^{\infty}$ is not a weakly optimal program. Then there is a program $\{x'(t), y'(t)\}_{t=0}^{\infty}$ for which

$$x'(0) = \bar{x}(0) = x(0),$$

$$\liminf_{T \to \infty} \sum_{t=0}^{T} [y'(t) - \bar{y}(t)] > 0.$$

Thus there exist a positive number ϵ and an integer $T_0 \geq 1$ such that for every natural number $T \geq T_0$, we have

$$\sum_{t=0}^{T-1} [y'(t) - \bar{y}(t)] \geq \epsilon. \tag{3.61}$$

It follows from (3.60) and (3.61) that for every integer $T \geq T_0$, we have

$$\sum_{t=0}^{2T-1} [-y(t) + y'(t)] = \sum_{t=0}^{2T-1} [y'(t) - \bar{y}(t)] \geq \epsilon$$

and

$$\limsup_{T \to \infty} \sum_{t=0}^{T} [y'(t) - y(t)] \geq \epsilon.$$

This contradicts our assumption that $\{x(t), y(t)\}_{t=0}^{\infty}$ is an overtaking optimal program. The contradiction we have reached proves that $\{\bar{x}(t), \bar{y}(t)\}_{t=0}^{\infty}$ is a weakly optimal program.

By (3.46), (3.47), (3.56), (3.59), and (3.60), for every integer $T \geq 1$,

$$\sum_{t=0}^{2T} y(t) - \sum_{t=0}^{2T} \bar{y}(t) = y(2T) - \bar{y}(2T) = 1 - \hat{x} = 1 - (2a_\sigma^{-1}) > 0.$$

Therefore

$$\limsup_{T \to \infty} \left[\sum_{t=0}^{T} y(t) - \sum_{t=0}^{T} \bar{y}(t) \right] > 0$$

and $\{\bar{x}(t), \bar{y}(t)\}_{t=0}^{\infty}$ is not an overtaking optimal program.

3.6 Convergence Results

In this section we prove the results obtained in [51].

Theorem 3.9 *Assume that for each good program* $\{u(t), v(t)\}_{t=0}^{\infty}$,

$$\lim_{t \to \infty} (u(t), v(t)) = (\widehat{x}, \widehat{x}).$$

Then for each program $\{x(t), y(t)\}_{t=0}^{\infty}$, *the following conditions are equivalent:*

(i) $\sum_{t=0}^{\infty} \delta(x(t), y(t), x(t+1)) = \Delta(x(0))$ *(see (3.38)).*
(ii) $\{x(t), y(t)\}_{t=0}^{\infty}$ *is overtaking optimal.*
(iii) $\{x(t), y(t)\}_{t=0}^{\infty}$ *is weakly optimal.*

Theorem 3.10 *Assume that at least one of the following conditions holds:*

(a) w is strictly concave.
(b) $\xi_{\sigma} \neq -1$.

Let $M_0, \epsilon > 0$. *Then there exists a natural number* T_0 *such that for each overtaking optimal program* $\{x(t), y(t)\}_{t=0}^{\infty}$ *satisfying* $x(0) \leq M_0 e$ *and each integer* $t \geq T_0$,

$$\|x(t) - \widehat{x}\|, \ \|y(t) - \widehat{x}\| \leq \epsilon.$$

Theorem 3.11 *Assume that at least one of the following conditions holds:*

(a) w is strictly concave.
(b) $\xi_{\sigma} \neq -1$.

Let $\epsilon > 0$. *Then there exists* $\delta > 0$ *such that for each overtaking optimal program* $\{x(t), y(t)\}_{t=0}^{\infty}$ *satisfying* $\|x(0) - \widehat{x}\| \leq \delta$, *the following inequality holds:*

$$\|x(t) - \widehat{x}\|, \ \|y(t) - \widehat{x}\| \leq \epsilon$$

for all integers $t \geq 0$.

3.7   Proof of Theorem 3.9

Let $\{x(t), y(t)\}_{t=0}^{\infty}$ be a program. In the proof of Theorem 3.3, it was proved that (i) implies (ii). Evidently (ii) implies (iii). Let us show that (iii) implies (i).

Assume that $\{x(t), y(t)\}_{t=0}^{\infty}$ is weakly optimal. By Proposition 2.14, the program $\{x(t), y(t)\}_{t=0}^{\infty}$ is good. Proposition 2.9 implies that

$$\sum_{t=0}^{\infty} \delta(x(t), y(t), x(t+1)) < \infty. \tag{3.62}$$

In view of Proposition 2.10, there exists a program $\{x^*(t), y^*(t)\}_{t=0}^{\infty}$ such that

$$x^*(0) = x(0), \ \sum_{t=0}^{\infty} \delta(x^*(t), y^*(t), x^*(t+1)) = \Delta(x(0)). \tag{3.63}$$

Since $\{x(t), y(t)\}_{t=0}^{\infty}$ and $\{x^*(t), y^*(t)\}_{t=0}^{\infty}$ are good programs, we have

$$\lim_{t \to \infty} (x(t), y(t)) = \lim_{t \to \infty} (x^*(t), y^*(t)) = (\widehat{x}, \widehat{x}). \tag{3.64}$$

By Proposition 2.8 and (3.63), for all integers $T \geq 0$,

$$\sum_{t=0}^{T}(w(by(t)) - w(b\widehat{y})) = \widehat{p}(x(0) - x(T+1)) - \sum_{t=0}^{\infty} \delta(x(t), y(t), x(t+1)),$$

$$\sum_{t=0}^{T}(w(by^*(t)) - w(b\widehat{y})) = \widehat{p}(x^*(0) - x^*(T+1)) - \sum_{t=0}^{T} \delta(x^*(t), y^*(t), x^*(t+1))$$

and

$$\sum_{t=0}^{T}(w(by^*(t)) - w(by(t))) = \widehat{p}(x(T+1) - x^*(T+1))$$

$$+ \sum_{t=0}^{T} \delta(x(t), y(t), x(t+1)) - \sum_{t=0}^{\infty} \delta(x^*(t), y^*(t), x^*(t+1)).$$

Since the program $\{x(t), y(t)\}_{t=0}^{\infty}$ is weakly optimal, it follows from (3.62)–(3.64) that

$$0 \geq \liminf_{T \to \infty} \sum_{t=0}^{T} [w(by^*(t)) - w(by(t))]$$

$$= \sum_{t=0}^{\infty} \delta(x(t), y(t), x(t+1)) - \sum_{t=0}^{\infty} \delta(x^*(t), y^*(t), x^*(t+1))$$

$$+ \widehat{p} \lim_{T \to \infty} (x(T+1) - x^*(T+1)) = \sum_{t=0}^{\infty} \delta(x(t), y(t), x(t+1)) - \Delta(x(0)).$$

Thus

$$\sum_{t=0}^{\infty} \delta(x(t), y(t), x(t+1)) = \Delta(x(0))$$

(see (3.38)). This completes the proof of Theorem 3.9.

3.8 Auxiliary Results for Theorems 3.10 and 3.11

Proposition 3.12 *Let $M_0 > 0$. Then*

$$\sup\{\Delta(x) : \ x \in R_+^n \ and \ x \leq M_0 e\} < \infty.$$

Proof By Proposition 3.6 there is $M_1 > 0$ such that for each program $\{x(t), y(t)\}_{t=0}^{\infty}$ satisfying $x(0) \leq M_0 e$, we have $x(t) \leq M_1 e$ for each integer $t \geq 0$. Set

$$z(0) = 0, \ z(t+1) = (1-d)z(t) + (1/a_\sigma)e(\sigma), \ t = 0, 1, \dots. \tag{3.65}$$

It is easy to see that

$$\lim_{T \to \infty} z(T) = \left[a_\sigma^{-1} \sum_{t=0}^{T-1} (1-d)^t \right] e(\sigma) = (da_\sigma)^{-1} e(\sigma). \tag{3.66}$$

Clearly there exists a natural number T_0 such that

$$z(T_0) \geq \widehat{x}. \tag{3.67}$$

Assume now that $x \in R_+^n$ satisfies $x \leq M_0 e$. Set

$$x(0) = x, \ x(t+1) = (1-d)x(t) + (1/a_\sigma)e(\sigma), \ t = 0, \dots, T_0 - 1, \tag{3.68}$$

$$y(t) = 0, t = 0, \ldots, T_0 - 1.$$

It is not difficult to see that $(\{x(t)\}_{t=0}^{T_0}, \{y(t)\}_{t=0}^{T_0-1})$ is a program and that

$$x(t) \geq z(t), \ t = 0, \ldots, T_0. \tag{3.69}$$

By (3.67) and (3.69),

$$x(T_0) \geq z(T_0) \geq \widehat{x}. \tag{3.70}$$

For all natural numbers $t \geq T_0$, set

$$y(t) = \widehat{x}, \ x(t+1) = (1-d)x(t) + d\widehat{x}. \tag{3.71}$$

It is easy to see that $\{x(t), y(t)\}_{t=0}^{\infty}$ is a program. In view of Proposition 2.8 and (3.7), for all natural numbers $T > T_0$,

$$\sum_{t=0}^{T} \delta(x(t), y(t), x(t+1)) = \widehat{p}(x(0) - x(T+1)) - \sum_{t=0}^{T}(w(by(t)) - w(b\widehat{y}))$$

$$\leq M_0\widehat{p}e - \sum_{t=0}^{T_0-1}[w(by(t)) - w(b\widehat{y})]. \tag{3.72}$$

It follows from (3.72) that for all natural numbers $T > T_0$,

$$\sum_{t=0}^{T} \delta(x(t), y(t), x(t+1)) \leq M_0\widehat{p}e + T_0w(b\widehat{y}) - T_0w(0).$$

This inequality implies that

$$\Delta(x) \leq \sum_{t=0}^{\infty} \delta(x(t), y(t), x(t+1)) \leq M_0\widehat{p}e + T_0w(b\widehat{y}) - T_0w(0).$$

This completes the proof of Proposition 3.12.

Proposition 3.13 *Let $\epsilon > 0$. Then there exists $\delta > 0$ such that for each $x, x' \in R_+^n$ satisfying*

$$\|x - \widehat{x}\|, \|x' - \widehat{x}\| \leq \delta$$

there exist $\bar{x} \geq x', y \in R_+^n$ such that

$$(x, \bar{x}) \in \Omega, \ y \in \Lambda(x, \bar{x}),$$

$$\|y - \widehat{x}\| \le \epsilon, \quad \|\bar{x} - \widehat{x}\| \le \epsilon.$$

Proof Choose a positive number δ such that

$$16\delta n + 2\delta \|a\| n < \min\{\epsilon, 8^{-1}d(1 + da_\sigma)^{-1}\}. \tag{3.73}$$

Let $x, x' \in R_+^n$ satisfy

$$\|x - \widehat{x}\|, \quad \|x' - \widehat{x}\| \le \delta. \tag{3.74}$$

Define $z \in R^n$ as follows:

$$z_i = \max\{x_i' - (1 - d)x_i, 0\}, \quad i = 1, \dots .n, \tag{3.75}$$

$$y = \min\{x_\sigma, (1 - az)\}e(\sigma), \quad \bar{x} = (1 - d)x + z. \tag{3.76}$$

By (3.75) and (3.76),

$$\bar{x} \ge (1 - d)x, \quad \bar{x} \ge x'. \tag{3.77}$$

In view of (3.76),

$$a(\bar{x} - (1 - d)x) = az. \tag{3.78}$$

It follows from (3.74) and (3.75) that for each $i \in \{1, \dots, n\} \setminus \{\sigma\}$

$$|x_i'|, |x_i| \le \delta, \quad -\delta \le x_i' - (1 - d)x_i \le \delta, \quad 0 \le z_i \le \delta. \tag{3.79}$$

By (3.74),

$$|[x_\sigma' - (1 - d)x_\sigma] - d(da_\sigma + 1)^{-1}|$$
$$\le |x_\sigma' - (1 + da_\sigma)^{-1}| + (1 - d)|x_\sigma - (da_\sigma + 1)^{-1}| \le 2\delta.$$

Together with (3.73)–(3.75), this inequality implies that

$$x_\sigma' - (1 - d)x_\sigma > 0, \quad z_\sigma = x_\sigma' - (1 - d)x_\sigma,$$
$$|z_\sigma - d(1 + da_\sigma)^{-1}| \le 2\delta. \tag{3.80}$$

In view of (3.73), (3.74), (3.76), (3.79), and (3.80),

$$\|\bar{x} - \widehat{x}\| \le \|(1 - d)x - (1 - d)\widehat{x}\| + \|z - d\widehat{x}\|$$
$$\le \delta + \|z - d\widehat{x}\| \le \delta + 2\delta n < \epsilon. \tag{3.81}$$

It follows from (3.73), (3.79), and (3.80) that

$$|az - ad\widehat{x}| \leq \|a\| \|z - d\widehat{x}\| \leq \|a\| 2\delta n < 8^{-1}(1 + da_\sigma)^{-1} \tag{3.82}$$

and

$$az \leq ad\widehat{x} + 8^{-1}(1 + da_\sigma)^{-1} = a_\sigma d(1 + da_\sigma)^{-1} + (8(1 + da_\sigma))^{-1} < 1. \tag{3.83}$$

Combined with (3.77) and (3.78), this relation implies that

$$a(\bar{x} - (1 - d)x) < 1 \text{ and } (x, \bar{x}) \in \Omega. \tag{3.84}$$

By (3.76), (3.78), (3.83), and (3.84), $y \in \Lambda(x, \bar{x})$. In view of (3.73), (3.74), (3.76), and (3.82),

$$\|y - \widehat{x}\| = \|y_\sigma - \widehat{x}_\sigma\| \leq |1 - az - (1 + da_\sigma)^{-1}| + |x_\sigma - \widehat{x}_\sigma|$$
$$\leq \delta + |da_\sigma(1 + da_\sigma)^{-1} - az| = \delta + |az - ad\widehat{x}| \leq \delta + \|a\| 2\delta n < \epsilon.$$

This completes the proof of Proposition 3.13.

Proposition 3.14 *Assume that for each good program* $\{x(t), y(t)\}_{t=0}^{\infty}$,

$$\lim_{t \to \infty} (x(t), y(t)) = (\widehat{x}, \widehat{x}).$$

Let $\epsilon > 0$. *Then there is* $\delta > 0$ *such that for each* $x \in R_+^n$ *satisfying* $\|x - \widehat{x}\| \leq \delta$, *the inequality* $\Delta(x) \leq \epsilon$ *holds.*

Proof Choose a positive number δ_1 such that

$$\delta_1 \|\widehat{p}\| < \epsilon/8 \text{ and } \delta_1 < \epsilon/8, \tag{3.85}$$

$$|w(by) - w(b\widehat{y})| \leq \epsilon/8 \text{ for each } y \in R_+^n \text{ satisfying } \|y - \widehat{y}\| \leq \delta_1. \tag{3.86}$$

By Proposition 3.13 there is $\delta \in (0, \delta_1/4)$ such that for each $x, x' \in R_+^n$ satisfying $\|x - \widehat{x}\|, \|x' - \widehat{x}\| \leq \delta$, there exist $\bar{x} \geq x', y \in R_+^n$ such that

$$(x, \bar{x}) \in \Omega, \ y \in \Lambda(x, \bar{x}), \ \|y - \widehat{x}\|, \|\bar{x} - \widehat{x}\| \leq \delta_1/4. \tag{3.87}$$

Let $x \in R_+^n$ satisfy

$$\|x - \widehat{x}\| \leq \delta. \tag{3.88}$$

In view of this inequality and the choice of δ, there are $\bar{x} \in R_+^n, y \in R_+^n$ such that

$$(x, \bar{x}) \in \Omega, \ y \in \Lambda(x, \bar{x}),$$

$$\bar{x} \geq \widehat{x}, \ \|\bar{x} - \widehat{x}\|, \ \|y - \widehat{y}\| \leq \delta_1/4. \tag{3.89}$$

Define

$$x(0) = x, \ y(0) = y, \ x(1) = \bar{x},$$

$$x(t+1) = (1-d)x(t) + d\widehat{x}, \ y(t) = \widehat{y} \tag{3.90}$$

for all integers $t \geq 1$. Relations (3.89) and (3.90) imply that $\{x(t), y(t)\}_{t=0}^{\infty}$ is a good program and

$$\lim_{t \to \infty} x(t) = \widehat{x}. \tag{3.91}$$

By Proposition 2.8, (3.85), (3.86), and (3.88)–(3.91),

$$\Delta(x) \leq \sum_{t=0}^{\infty} \delta(x(t), y(t), x(t+1))$$

$$= \lim_{T \to \infty} \left[\widehat{p}(x(0) - x(T+1)) - \sum_{t=0}^{T} (w(by(t)) - w(b\widehat{y})) \right]$$

$$= \widehat{p}(x - \widehat{x}) - (w(by(0)) - w(b\widehat{y})) \leq \|\widehat{p}\|\delta + |w(by(0)) - w(b\widehat{y})|$$

$$\leq \|\widehat{p}\|\delta + |w(by) - w(b\widehat{y})| < \epsilon/2.$$

Proposition 3.14 is proved.

3.9 Proofs of Theorems 3.10 and 3.11

Lemma 3.15 *Let* $S_0, \epsilon > 0$. *Then there exists* $\delta > 0$ *such that for each sequence* $\{x(t), y(t)\}_{t=-\infty}^{\infty} \subset R_+^{2n}$ *satisfying*

$$\|x(t)\|, \ \|y(t)\| \leq S_0 \text{ for all integers } t, \tag{3.92}$$

$$(x(t), x(t+1)) \in \Omega, \ y(t) \in \Lambda(x(t), x(t+1)) \text{ for all integers } t, \tag{3.93}$$

$$\delta(x(t), y(t), x(t+1)) \leq \delta \text{ for all integers } t, \tag{3.94}$$

the following inequality holds:

$$\|x(t) - \widehat{x}\|, \ \|y(t) - \widehat{x}\| \leq \epsilon \text{ for all integers } t. \tag{3.95}$$

Proof Let us assume the converse. Then for each natural number k, there exist a sequence $\{x^{(k)}(t), y^{(k)}(t)\}_{t=-\infty}^{\infty} \subset R_+^{2n}$ which satisfies

$$\|x^{(k)}(t)\|, \|y^{(k)}(t)\| \leq S_0 \text{ for all integers } t, \tag{3.96}$$

$$(x^{(k)}(t), x^{(k)}(t+1)) \in \Omega, \; y^{(k)}(t) \in \Lambda(x^{(k)}(t), x^{(k+1)}(t+1)) \text{ for all integers } t, \tag{3.97}$$

$$\delta(x^{(k}(t), y^{(k)}(t), x^{(k)}(t+1)) \leq 1/k \text{ for all integers } t \tag{3.98}$$

and an integer τ_k such that

$$\max\{\|x^{(k)}(\tau_k) - \widehat{x}\|, \; \|y^{(k)}(\tau_k) - \widehat{x}\|\} \geq \epsilon. \tag{3.99}$$

We may assume without loss of generality that

$$\tau_k = 0 \text{ for all natural numbers } k. \tag{3.100}$$

Extracting subsequences, re-indexing, and using diagonalization process, we obtain that there exists a strictly increasing sequence of natural numbers $\{k_j\}_{j=1}^{\infty}$ such that for each integer s, there is

$$(u(s), v(s)) = \lim_{j\to\infty} (x^{(k_j)}(s), y^{(k_j)}(s)). \tag{3.101}$$

Since the set

$$\{(x, y, x') \in R_+^{3n} : (x, x') \in \Omega, \; y \in \Lambda(x, x')\}$$

is closed, it follows from (3.100) and (3.101) that

$$(u(s), u(s+1)) \in \Omega, \; v(s) \in \Lambda(u(s), u(s+1)) \text{ for all integers } s. \tag{3.102}$$

In view of (3.96) and (3.101), for all integers s,

$$\|u(s)\|, \|v(s)\| \leq S_0. \tag{3.103}$$

Since the function $\delta(\cdot, \cdot, \cdot)$ is nonnegative and continuous, it follows from (3.98) and (3.101) that

$$\delta(u(s), v(s), u(s+1)) = 0 \text{ for all integers } s. \tag{3.104}$$

Relations (3.99)–(3.101) imply that

$$\max\{\|u(0) - \widehat{x}\|, \|v(0) - \widehat{x}\|\} \geq \epsilon. \tag{3.105}$$

If w is strictly concave, then it follows from (3.102), (3.103), and Lemma 2.13 that

$$u(s) = v(s) = \widehat{x} \text{ for all integers } s. \tag{3.106}$$

If $\xi_\sigma \neq -1$, then it follows from (3.102)–(3.104) and Lemma 3.8 that for all integers $s, u(s) = v(s) = \widehat{x}$. Thus (3.106) holds in both cases. This contradicts (3.105). The contradiction we have reached proves Lemma 3.15.

Choose

$$M_0 > 1 + (1 + da_\sigma)^{-1}. \tag{3.107}$$

Proof of Theorem 3.11 By Proposition 3.6, there is $S_0 > 0$ such that for each program $\{x(t), y(t)\}_{t=0}^\infty$ satisfying $x(0) \leq M_0 e$,

$$x(t) \leq S_0 e \text{ for all integers } t \geq 0. \tag{3.108}$$

In view of Lemma 3.15, there is $\epsilon_1 \in (0, \epsilon)$ such that for each sequence $\{x(t), y(t)\}_{t=-\infty}^\infty \subset R_+^{2n}$ satisfying

$$\|x(t)\|, \|y(t)\| \leq S_0 n \text{ for all integers } t, \tag{3.109}$$

$$(x(t), x(t+1)) \in \Omega, \ y(t) \in \Lambda(x(t), x(t+1)), \ \delta(x(t), y(t), x(t+1)) \leq \epsilon_1 \tag{3.110}$$

for all integers t, the following inequality holds:

$$\|x(t) - \widehat{x}\|, \|y(t) - \widehat{y}\| \leq \epsilon/8 \text{ for all integers } t. \tag{3.111}$$

By Proposition 3.14 and Theorems 3.1 and 3.2, there is $\delta_1 > 0$ such that for each $x \in R_+^n$ satisfying $\|x - \widehat{x}\| \leq \delta_1$,

$$\Delta(x) \leq \epsilon_1/4. \tag{3.112}$$

It follows from Proposition 3.13 and the continuity of the function $\delta(\cdot, \cdot, \cdot)$ that there is $\delta_2 > 0$ such that for each $x \in R_+^n$ satisfying $\|x - \widehat{x}\| \leq \delta_2$, there exist $\bar{x} \geq x$ and $y \in R_+^n$ which satisfy

$$(\widehat{x}, \bar{x}) \in \Omega, \ y \in \Lambda(\widehat{x}, \bar{x}), \tag{3.113}$$

$$\delta(\widehat{x}, y, \bar{x}) \leq \epsilon_1/4, \ \|\bar{x} - x\| \leq (\epsilon_1/4)(\|\widehat{p}\| + 1)^{-1}.$$

Set

$$\delta = \min\{\delta_1, \delta_2, 1\}. \tag{3.114}$$

Assume that $\{x(t), y(t)\}_{t=0}^{\infty}$ is an overtaking optimal program such that

$$\|x(0) - \widehat{x}\| \leq \delta. \tag{3.115}$$

By Theorems 3.1, 3.2, and 3.9, the choice of δ_1 (see (3.112)), (3.114), and (3.115),

$$\sum_{t=0}^{\infty} \delta(x(t), y(t), x(t+1)) = \Delta(x(0)) \leq \epsilon_1/4. \tag{3.116}$$

In view of (3.114), (3.115), and the choice of δ_2 (see (3.113)), there exist $\bar{x} \geq x(0)$ and $\bar{y} \in R_+^n$ such that

$$(\widehat{x}, \bar{x}) \in \Omega, \ \ \bar{y} \in \Lambda(\widehat{x}, \bar{x}), \ \ \delta(\widehat{x}, \bar{y}, \bar{x}) \leq \epsilon_1/4, \tag{3.117}$$

$$\|\bar{x} - x(0)\| \leq (\epsilon_1/4)(\|\widehat{p}\| + 1)^{-1}.$$

Set

$$z = \bar{x} - x(0). \tag{3.118}$$

Define a sequence $\{\bar{x}(t), \bar{y}(t)\}_{t=-\infty}^{\infty}$ by

$$\bar{x}(t) = \widehat{x} \text{ for all integers } t < 0,$$

$$\bar{y}(t) = \widehat{x} \text{ for all integers } t < -1,$$

$$\bar{y}(-1) = \bar{y}, \ \bar{x}(0) = \bar{x},$$

$$\bar{x}(t) = x(t) + (1 - d)^t z \text{ for all integers } t \geq 1,$$

$$\bar{y}(t) = y(t) \text{ for all integers } t \geq 0. \tag{3.119}$$

It follows from (3.117), (3.118), and (3.119) that

$$(\bar{x}(t), \bar{x}(t+1)) \in \Omega, \ \ \bar{y}(t) \in \Lambda(\bar{x}(t), \bar{x}(t+1)) \text{ for all integers } t. \tag{3.120}$$

By (3.107), (3.119), (3.120), and the choice of S_0 (see (3.108)),

$$\|\bar{x}(t)\|, \ \|\bar{y}(t)\| \leq S_0 n \text{ for all integers } t. \tag{3.121}$$

In view of (3.116), (3.119), and Theorems 3.1 and 3.2,

$$\lim_{i \to \infty} \bar{x}(t) = \lim_{t \to \infty} x(t) = \widehat{x}. \tag{3.122}$$

Proposition 2.8 and (3.119) imply that for any natural number T,

$$\sum_{t=0}^{T} \delta(\bar{x}(t), \bar{y}(t), \bar{x}(t+1)) = \widehat{p}(\bar{x}(0) - \bar{x}(T+1)) - \sum_{t=0}^{T} (w(b\bar{y}(t)) - w(b\widehat{y}))$$

$$= \widehat{p}(x(0) - x(T+1)) - \sum_{t=0}^{T} (w(by(t)) - w(b\widehat{y}))$$

$$+ \widehat{p}(\bar{x}(0) - x(0)) + \widehat{p}(x(T+1) - \bar{x}(T+1))$$

$$\leq \sum_{t=0}^{T} \delta(x(t), y(t), x(t+1)) + \|\widehat{p}\| \|\bar{x}(0) - x(0)\|$$

$$+ \|\widehat{p}\| \|x(T+1) - \bar{x}(T+1)\|.$$

Combined with (3.116), (3.117), and (3.122), this relation implies that

$$\sum_{t=0}^{\infty} \delta(\bar{x}(t), \bar{y}(t), \bar{x}(t+1))$$

$$\leq \sum_{t=0}^{\infty} \delta(x(t), y(t), x(t+1)) + \|\widehat{p}\| \|\bar{x}(0) - x(0)\| \leq \epsilon_1/4 + \epsilon_1/4 \quad (3.123)$$

It follows from (3.117), (3.119), and (3.123) that for all integers t,

$$\delta(\bar{x}(t), \bar{y}(t), \bar{x}(t+1)) \leq \epsilon_1. \tag{3.124}$$

In view of (3.120), (3.121), (3.124), and the choice of ϵ_1 (see (3.109)–(3.111)),

$$\|\bar{x}(t) - \widehat{x}\|, \ \|\bar{y}(t) - \widehat{x}\| \leq \epsilon/8 \tag{3.125}$$

for all integers t. Relations (3.117)–(3.119) imply that for all integers $t \geq 0$,

$$\|\bar{x}(t) - x(t)\| \leq \epsilon_1/4 < \epsilon/4. \tag{3.126}$$

By (3.119) and (3.125), for all integers $t \geq 0$,

$$\|y(t) - \widehat{x}\| \leq \epsilon/8.$$

In view of (3.125) and (3.126), for all integers $t \geq 0$,

$$\|x(t) - \widehat{x}\| \leq \|x(t) - \bar{x}(t)\| + \|\bar{x}(t) - \widehat{x}\| \leq \epsilon/4 + \epsilon/8 < \epsilon.$$

This completes the proof of Theorem 3.11.

Proof of Theorem 3.10 By Proposition 3.6 there is $M_1 > 0$ such that for each program $\{x(t), y(t)\}_{t=0}^{\infty}$ satisfying $x(0) \leq M_0 e$, the following inequality holds:

$$x(t) \leq M_1 e \text{ for all integers } t \geq 0. \tag{3.127}$$

By Theorem 3.11 there exists $\delta > 0$ such that for each overtaking optimal program $\{x(t), y(t)\}_{t=0}^{\infty}$ satisfying $\|x(0) - \widehat{x}\| \leq \delta$, the following inequality holds:

$$\|x(t) - \widehat{x}\|, \ \|y(t) - \widehat{x}\| \leq \epsilon \text{ for all integers } t \geq 0. \tag{3.128}$$

We show that there exists a natural number τ_0 such that the following property holds:

(P) For each overtaking optimal program $\{x(t), y(t)\}_{t=0}^{\infty}$ satisfying $x(0) \leq M_0 e$, there exists an integer t such that

$$0 \leq t \leq \tau_0 \text{ and } \|x(t) - \widehat{x}\| < \delta.$$

Let us assume the contrary. Then for each natural number k, there exists an overtaking optimal program $\{x^{(k)}(t), y^{(k)}(t)\}_{t=0}^{\infty}$ such that

$$x^{(k)}(0) \leq M_0 e, \ \|x^{(k)}(t) - \widehat{x}\| \geq \delta, \ t = 0, \ldots, k. \tag{3.129}$$

Proposition 3.12 implies that there is $D_0 > 0$ such that

$$\Delta(z) \leq D_0 \text{ for each } z \in R_+^n \text{ satisfying } z \leq M_0 e. \tag{3.130}$$

In view of (3.129), (3.130), and Theorems 3.1, 3.2, and 3.9 for each natural number k,

$$\sum_{t=0}^{\infty} \delta(x^{(k)}(t), y^{(k)}(t), x^{(k)}(t+1)) = \Delta(x^{(k)}(0)) \leq D_0. \tag{3.131}$$

By the choice of M_1 and (3.129),

$$x^{(k)}(t), \ y^{(k)}(t) \leq M_1 e \text{ for all integers } t \geq 0 \text{ and all natural numbers } k.$$

Extracting subsequences, re-indexing, and using diagonalization process, we obtain that there exists a strictly increasing sequence of natural numbers $\{k_j\}_{j=1}^{\infty}$ such that for each integer $s \geq 0$, there exists

$$(\tilde{x}(s), \tilde{y}(s)) = \lim_{j \to \infty} (x^{(k_j)}(s), y^{(k_j)}(s)). \tag{3.132}$$

It is not difficult to see that $\{\tilde{x}(t), \tilde{y}(t)\}_{t=0}^{\infty}$ is a program. In view of (3.132), (3.140), and the continuity of the function $\delta(\cdot, \cdot, \cdot)$,

$$\sum_{t=0}^{\infty} \delta(\tilde{x}(t), \tilde{y}(t), \tilde{x}(t+1)) \leq D_0.$$

This implies that $\{\tilde{x}(t), \tilde{y}(t)\}_{t=0}^{\infty}$ is a good program. Theorems 3.1 and 3.2 imply that

$$\lim_{t \to \infty} \tilde{x}(t) = \hat{x}.$$

On the other hand, it follows from (3.129) and (3.132) that

$$\|\tilde{x}(t) - \hat{x}\| \geq \delta \text{ for all integers } t \geq 0.$$

The contradiction we have reached proves that there is a natural number τ_0 such that property (P) holds.

Now assume that $\{x(t), y(t)\}_{t=0}^{\infty}$ is an overtaking optimal program satisfying $x(0) \leq M_0 e$. By property (P) there is an integer $t_0 \in [0, \tau_0]$ such that

$$\|x(t_0) - \hat{x}\| < \delta. \tag{3.133}$$

Clearly the program $\{x(t + t_0), y(t + t_0)\}_{t=0}^{\infty}$ is also overtaking optimal. In view of (3.133) and the choice of δ (see (3.128))

$$\|x(t + t_0) - \hat{x}\|, \ \|y(t + t_0) - \hat{x}\| \leq \epsilon \text{ for all integers } t \geq 0.$$

This completes the proof of Theorem 3.10.

3.10   The Structure of Good Programs in the RSS Model

We prove the following three theorems which were obtained in [98].

Theorem 3.16 *Let a program* $\{x(t), y(t)\}_{t=0}^{\infty}$ *be good. Then for each* $i \in \{1, \dots, n\} \setminus \{\sigma\}$,

$$\sum_{t=0}^{\infty} x_i(t) < \infty,$$

$$\sum_{t=0}^{\infty} (x_\sigma(t) - y_\sigma(t)) < \infty$$

and the sequence $\{\sum_{t=0}^{T-1} x_\sigma(t) - T(1 + da_\sigma)^{-1}\}_{T=1}^{\infty}$ *is bounded.*

Theorem 3.17 *Let the function w be linear. Then a program* $\{x(t), y(t)\}_{t=0}^{\infty}$ *is good if and only if for each* $i \in \{1, \ldots, n\} \setminus \{\sigma\}$,

$$\sum_{t=0}^{\infty} x_i(t) < \infty,$$

$$\sum_{t=0}^{\infty} (x_\sigma(t) - y_\sigma(t)) < \infty$$

and the sequence $\{\sum_{t=0}^{T-1} x_\sigma(t) - T(1 + da_\sigma)^{-1}\}_{T=1}^{\infty}$ *is bounded.*

Theorem 3.18 *Let* $w \in C^2$, $w''(b\widehat{x}) \neq 0$ *and let for every good program* $\{u(t), v(t)\}_{t=0}^{\infty}$,

$$\lim_{t \to \infty} (u(t), v(t)) = (\widehat{x}, \widehat{x}).$$

Then a program $\{x(t), y(t)\}_{t=0}^{\infty}$ *is good if and only if for each* $i \in \{1, \ldots, n\} \setminus \{\sigma\}$,

$$\sum_{t=0}^{\infty} x_i(t) < \infty,$$

$$\sum_{t=0}^{\infty} (x_\sigma(t) - y_\sigma(t)) < \infty,$$

$$\sum_{t=0}^{\infty} (y_\sigma(t) - \widehat{x}_\sigma)^2 < \infty,$$

and the sequence $\{\sum_{t=0}^{T-1} x_\sigma(t) - T(1 + da_\sigma)^{-1}\}_{T=1}^{\infty}$ *is bounded.*

3.11 Proofs of Theorems 3.16–3.18

Proof of Theorem 3.16 Since the program $\{x(t), y(t)\}_{t=0}^{\infty}$ is good, it follows from Proposition 2.9 that

$$\sum_{t=0}^{\infty} \delta(x(t), y(t), x(t+1)) < \infty. \tag{3.134}$$

For $t = 0, 1, \ldots$, set

$$z(t) = x(t+1) - (1-d)x(t). \tag{3.135}$$

Since w is concave [70, 79] for each $z \in [0, \infty)$,

$$w(z) - w(b\widehat{x}) \le w'(b\widehat{x})(z - b\widehat{x}). \tag{3.136}$$

Since $w'(b\widehat{x}) \ne 0$ and w is strictly increasing, it is easy to see that

$$w'(b\widehat{x}) > 0. \tag{3.137}$$

By (2.2) (see the proof of Lemma 2.2), (3.8), (3.9), (3.12), (3.135), and (3.136), for each integer $t \ge 0$,

$$
\begin{aligned}
\delta(x(t), y(t), x(t+1)) &= \widehat{p}(x(t) - x(t+1)) - (w(by(t)) - w(b\widehat{y})) \\
&\ge \widehat{p}(x(t) - x(t+1)) + w'(b\widehat{x})(b\widehat{x} - by(t)) \\
&= w'(b\widehat{x})[b\widehat{x} - by(t) + \widehat{q}(x(t) - x(t+1))] \\
&= w'(b\widehat{x})\left[c_\sigma(1 - ey(t) - az(t)) + \sum_{i=1}^{n}(c_\sigma - c_i)y_i(t) \right. \\
&\qquad\qquad \left. + \sum_{i=1}^{n}(c_\sigma - c_i)a_i z_i(t) + d\widehat{q}(x(t) - y(t)) \right].
\end{aligned}
\tag{3.138}
$$

Combined with (3.2), (3.3), (3.134), and (3.137), relation (3.138) implies that

$$
\begin{aligned}
\infty > \sum_{t=0}^{\infty} \delta(x(t), y(t), x(t+1)) &\ge w'(b\widehat{x}) \sum_{t=0}^{\infty}\left[c_\sigma(1 - ey(t) - az(t)) \right. \\
&\left. + \sum_{i=1}^{n}(c_\sigma - c_i)y_i(t) + \sum_{i=1}^{n}(c_\sigma - c_i)a_i z_i(t) + d\widehat{q}(x(t) - y(t)) \right].
\end{aligned}
\tag{3.139}
$$

Relations (3.2), (3.3), (3.135), (3.137) and (3.139) imply that for all $i \in \{1, \ldots, n\} \setminus \{\sigma\}$

$$\sum_{t=0}^{\infty} y_i(t) < \infty, \tag{3.140}$$

$$\sum_{t=0}^{\infty}(x_i(t) - y_i(t)) < \infty, \quad i = 1, \ldots, n. \tag{3.141}$$

In view of (3.140) and (3.141),

$$\sum_{t=0}^{\infty} x_i(t) < \infty \text{ for all } i \in \{1, \ldots, n\} \setminus \{\sigma\}. \tag{3.142}$$

Relations (3.2), (3.3), (3.137), and (3.139) imply that

$$\sum_{t=0}^{\infty} (1 - ey(t) - az(t)) < \infty.$$

This inequality, (3.3), and (3.135) imply that

$$\sum_{t=0}^{\infty} (1 - y_\sigma(t) - a_\sigma(x_\sigma(t+1) - (1-d)x_\sigma(t))) < \infty.$$

Therefore there is $M_1 > 0$ such that for each natural number T,

$$M_1 > \sum_{t=0}^{T-1} (1 - y_\sigma(t) - a_\sigma(x_\sigma(t+1) - x_\sigma(t) + dx_\sigma(t)))$$

$$= T - \sum_{t=0}^{T-1} y_\sigma(t) - a_\sigma(x_\sigma(T) - x_\sigma(0)) - da_\sigma \sum_{t=0}^{T-1} x_\sigma(t) \geq 0.$$

It follows from this inequality, (3.141), (3.142), and Proposition 3.6 that there is $M_2 > 0$ such that for each natural number T,

$$\left| T - \left(\sum_{t=0}^{T-1} x_\sigma(t) \right)(1 + da_\sigma) \right| < M_2.$$

Theorem 3.16 is proved.

Proof of Theorem 3.17 Assume that a program $\{x(t), y(t)\}_{t=0}^{\infty}$ is good. Then by Theorem 3.16,

$$\sum_{t=0}^{\infty} x_i(t) < \infty \text{ for all } i \in \{1, \ldots, n\} \setminus \{\sigma\}, \tag{3.143}$$

$$\sum_{t=0}^{\infty} (x_\sigma(t) - y_\sigma(t)) < \infty, \tag{3.144}$$

and the sequence $\{T(1 + da_\sigma)^{-1} - \sum_{t=0}^{T-1} x_\sigma(t)\}_{T=1}^{\infty}$ is bounded.

Now assume that $\{x(t), y(t)\}_{t=0}^{\infty}$ is a program, that (3.143) and (3.144) hold, and that the sequence $\{T(1 + da_\sigma)^{-1} - \sum_{t=0}^{T-1} x_\sigma(t)\}_{T=1}^{\infty}$ is bounded. We prove that the program $\{x(t), y(t)\}_{t=0}^{\infty}$ is good. For $t = 0, 1, \ldots$, set

$$z(t) = x(t+1) - (1-d)x(t). \tag{3.145}$$

Since the sequence $\{T(1 + da_\sigma)^{-1} - \sum_{t=0}^{T-1} x_\sigma(t)\}_{T=1}^{\infty}$ is bounded, it follows from Proposition 3.6 and (3.144) that the sequence

$$\left\{ T - \sum_{t=0}^{T-1} y_\sigma(t) - da_\sigma \sum_{t=0}^{T-1} x_\sigma(t) - a(x_\sigma(T) - x_\sigma(0)) \right\}_{T=1}^{\infty}$$

$$= \left\{ \sum_{t=0}^{T-1} (1 - y_\sigma(t) - a_\sigma(x_\sigma(t+1) - (1-d)x_\sigma(t))) \right\}_{T=1}^{\infty}$$

is also bounded. Together with (3.3) and (3.145), this implies that

$$\sum_{t=0}^{\infty} (1 - y_\sigma(t) - a_\sigma z_\sigma(t)) < \infty.$$

Combined with (3.3) and (3.143), this inequality implies that

$$\sum_{t=0}^{\infty} (1 - ey(t) - az(t)) < \infty. \tag{3.146}$$

It follows from (3.9), the linearlity of w, (3.8), (3.145), and (2.2) (see Lemma 2.2) that for each integer $t \geq 0$,

$$\delta(x(t), y(t), x(t+1)) = \hat{p}(x(t) - x(t+1)) - (w(by(t)) - w(b\hat{y}))$$

$$= \hat{p}(x(t) - x(t+1)) - w'(b\hat{x})(by(t) - b\hat{x})$$

$$= w'(b\hat{x})[-by(t) + b\hat{x} + \hat{q}(x(t) - x(t+1))]$$

$$= w'(b\hat{x})\left[c_\sigma(1 - ey(t) - az(t)) + \sum_{i=1}^{n}(c_\sigma - c_i)y_i(t) \right.$$

$$\left. + \sum_{i=1}^{n}(c_\sigma - c_i)a_i z_i(t) + d\hat{q}(x(t) - y(t)) \right]. \tag{3.147}$$

Clearly $w'(b\hat{x}) > 0$. By (3.2), (3.3), (3.143)–(3.147),

$$\sum_{t=0}^{\infty} \delta(x(t), y(t), x(t+1)) < \infty.$$

Together with Proposition 2.9, this inequality implies that $\{x(t), y(t)\}_{t=1}^{\infty}$ is a good program. Theorem 3.17 is proved.

Proof of Theorem 3.18 Assume that $\{x(t), y(t)\}_{t=0}^{\infty}$ is a program. Let $t \geq 0$ be an integer. Set

$$z(t) = x(t+1) - (1-d)x(t). \tag{3.148}$$

In view of (3.12),

$$\delta(x(t), y(t), x(t+1)) = \widehat{p}(x(t) - x(t+1)) - (w(by(t)) - w(b\widehat{y})). \tag{3.149}$$

Since $w \in C^2$, it follows from Taylor's theorem that there exists

$$\gamma_t \in [\min\{by(t), b\widehat{y}\}, \ \max\{by(t), b\widehat{y}\}] \tag{3.150}$$

such that

$$w(by(t)) - w(b\widehat{y}) = w'(b\widehat{y})(by(t) - b\widehat{y}) + 2^{-1}w''(\gamma_t)(by(t) - b\widehat{y})^2. \tag{3.151}$$

Relations (3.8), (3.9), (3.148), (3.149), (3.151), and (2.2) from Lemma 2.2 imply that

$$\begin{aligned}
\delta(x(t), y(t), x(t+1)) &= w'(b\widehat{x})(b\widehat{y} - by(t)) - 2^{-1}w''(\gamma_t)(by(t) - b\widehat{y})^2 \\
&\quad + w'(b\widehat{y})\widehat{q}(x(t) - x(t+1)) \\
&= -2^{-1}w''(\gamma_t)(by(t) - b\widehat{y})^2 + w'(b\widehat{y})[b\widehat{y} - by(t) + \widehat{q}(x(t) - x(t+1))] \\
&= -2^{-1}w''(\gamma_t)(by(t) - b\widehat{y})^2 + w'(b\widehat{y})\Bigg[c_\sigma(1 - ey(t) - az(t)) \\
&\quad + \sum_{i=1}^{n}(c_\sigma - c_i)y_i(t) \\
&\quad + \sum_{i=1}^{n}(c_\sigma - c_i)a_i z_i(t) + d\widehat{q}(x(t) - y(t)) \Bigg].
\end{aligned} \tag{3.152}$$

Since w is concave and increasing and $w'(b\widehat{x}) \neq 0$, we have

$$w'(b\widehat{x}) = w'(b\widehat{y}) > 0, \quad w''(\gamma_t) \leq 0. \tag{3.153}$$

Assume that $\{x(t), y(t)\}_{t=0}^{\infty}$ is good. Then

$$\lim_{t\to\infty} (x(t), y(t)) = (\widehat{x}, \widehat{x}).$$

(3.154)

By Proposition 2.9,

$$\sum_{t=0}^{\infty} \delta(x(t), y(t), x(t+1)) < \infty.$$

(3.155)

Since $w''(b\widehat{y}) \neq 0$ and w is concave, we conclude that

$$w''(b\widehat{y}) < 0.$$

(3.156)

Since $w \in C^2$, it follows from (3.154), (3.156), and (3.150) that there exists a natural number t_0 such that for each integer $t \geq t_0$,

$$w''(\gamma_t) \leq 2^{-1} w''(b\widehat{y}).$$

(3.157)

It follows from (3.152), the choice of t_0, and (3.157) that for each integer $t \geq t_0$,

$$\delta(x(t), y(t), x(t+1)) \geq -(4)^{-1} w''(b\widehat{y})(by(t) - b\widehat{y})^2$$

$$+ w'(b\widehat{y}) \left[c_\sigma (1 - ey(t) - az(t)) + \sum_{i=1}^{n} (c_\sigma - c_i) y_i(t) \right.$$

$$\left. + \sum_{i=1}^{n} (c_\sigma - c_i) a_i z_i(t) + d\widehat{q}(x(t) - y(t)) \right].$$

(3.158)

It follows from (3.2), (3.3), (3.148), (3.153), (3.155), (3.156), and (3.158) that

$$\sum_{t=0}^{\infty} (x_i(t) - y_i(t)) < \infty, \quad i = 1, \dots, n,$$

(3.159)

$$\sum_{t=0}^{\infty} y_i(t) < \infty, \quad i \in \{1, \dots, n\} \setminus \{\sigma\},$$

(3.160)

$$\sum_{t=0}^{\infty} (by(t) - b\widehat{y})^2 < \infty.$$

(3.161)

Relations (3.159) and (3.160) imply that

$$\sum_{t=0}^{\infty} x_i(t) < \infty, \quad i \in \{1, \dots, n\} \setminus \{\sigma\}.$$

(3.162)

By Theorem 3.16, the sequence $\{T(1+da_\sigma)^{-1} - \sum_{t=0}^{T-1} x_\sigma(t)\}_{T=1}^\infty$ is bounded. By (3.160) and (3.161),

$$\sum_{t=0}^\infty (y_\sigma(t) - \widehat{y}_\sigma)^2 < \infty. \tag{3.163}$$

Assume that (3.159), (3.162), and (3.163) hold and that the the sequence

$$\left\{ T(1+da_\sigma)^{-1} - \sum_{t=0}^{T-1} x_\sigma(t) \right\}_{T=1}^\infty$$

is bounded. We show that $\{x(t), y(t)\}_{t=0}^\infty$ is a good program. By (3.159), (3.162), and (3.163),

$$\lim_{t \to \infty} (x(t), y(t)) = (\widehat{x}, \widehat{x}). \tag{3.164}$$

Since $w''(b\widehat{y}) \neq 0$ and w is concave, we have

$$w''(b\widehat{y}) < 0. \tag{3.165}$$

By (3.150), (3.164), and (3.165), there exists a natural number t_0 such that for each integer $t \geq t_0$

$$w''(\gamma_t) \geq 2w''(b\widehat{y}). \tag{3.166}$$

It follows from (3.152), (3.166), and the choice of t_0 that for each integer $t \geq t_0$,

$$\delta(x(t), y(t), x(t+1)) \leq -w''(b\widehat{y})(by(t) - b\widehat{y})^2$$

$$+ w'(b\widehat{y}) \left[c_\sigma(1 - ey(t) - az(t)) + \sum_{i=1}^n (c_\sigma - c_i) y_i(t) \right.$$

$$\left. + \sum_{i=1}^n (c_\sigma - c_i) a_i z_i(t) + d\widehat{q}(x(t) - y(t)) \right]. \tag{3.167}$$

By (3.2), (3.3), (3.148), (3.153), (3.159), and (3.162),

$$w'(b\widehat{y}) \sum_{t=0}^\infty \sum_{i=1}^n (c_\sigma - c_i) y_i(t) < \infty, \quad w'(b\widehat{y}) \sum_{t=0}^\infty \sum_{i=1}^n (c_\sigma - c_i) a_i z_i(t) < \infty,$$

$$\tag{3.168}$$

$$w'(b\widehat{y}) \sum_{t=0}^\infty d\widehat{q}(x(t) - y(t)) < \infty.$$

By (3.162) and (3.163),

$$\sum_{t=0}^{\infty}(by(t) - b\widehat{y})^2 < \infty. \tag{3.169}$$

We show that

$$\sum_{t=0}^{\infty}c_\sigma(1 - ey(t) - az(t)) < \infty. \tag{3.170}$$

Clearly, it is sufficient to show that the sequence

$$\left\{\sum_{t=0}^{T-1}(1 - y_\sigma(t) - a_\sigma(x_\sigma(t+1) - (1-d)x_\sigma(t)))\right\}_{T=1}^{\infty}$$

is bounded. Since the sequence $\{T(1 + da_\sigma)^{-1} - \sum_{t=0}^{T-1}x_\sigma(t)\}_{T=1}^{\infty}$ is bounded, it follows from Proposition 3.6 and (3.159) that the sequence

$$\left\{\sum_{t=0}^{T-1}(1 - y_\sigma(t) - a_\sigma(x_\sigma(t+1) - x_\sigma(t)) + a_\sigma dx_\sigma(t))\right\}_{T=1}^{\infty}$$

$$= \left\{\sum_{t=0}^{T-1}(1 - y_\sigma(t) - a_\sigma dx_\sigma(t)) - a_\sigma(x(T) - x_\sigma(0))\right\}_{T=1}^{\infty}$$

is bounded. Thus (3.170) holds.

Relations (3.15), (3.159), (3.162), (3.165), (3.167), (3.169), and (3.170) imply that

$$\sum_{t=0}^{\infty}\delta(x(t), y(t), x(t+1)) < \infty.$$

Together with Proposition 2.9, this inequality implies that $\{x(t), y(t)\}_{t=0}^{\infty}$ is a good program. Theorem 3.18 is proved.

Chapter 4
Turnpike Results for the Robinson–Solow–Srinivasan Model

In this chapter we study the turnpike properties for the Robinson–Solow–Srinivasan model. To have these properties means that the approximate solutions of the problems are essentially independent of the choice of an interval and endpoint conditions. We show that these turnpike properties hold and that they are stable under perturbations of an objective function.

4.1 The Main Results

Let R^1 (R_+^1) be the set of real (nonnegative) numbers, and let R^n be the n-dimensional Euclidean space with nonnegative orthant

$$R_+^n = \{x = (x_1, \ldots, x_n) \in R^n : x_i \geq 0, \ i = 1, \ldots, n\}.$$

For every pair of vectors $x = (x_1, \ldots, x_n)$, $y = (y_1, \ldots, y_n) \in R^n$, define their inner product by

$$xy = \sum_{i=1}^{n} x_i y_i,$$

and let $x \gg y, x > y, x \geq y$ have their usual meaning.

Let $e(i)$, $i = 1, \ldots, n$, be the ith unit vector in R^n, and e be an element of R_+^n all of whose coordinates are unity. For every $x \in R^n$, denote by $\|x\|$ its Euclidean norm in R^n.

Let $a = (a_1, \ldots, a_n) \gg 0, b = (b_1, \ldots, b_n) \gg 0, d \in (0, 1)$,

$$c_i = b_i/(1 + da_i), \ i = 1, \ldots, n. \tag{4.1}$$

© The Author(s), under exclusive license to Springer Nature Switzerland AG 2020
A. J. Zaslavski, *Turnpike Theory for the Robinson–Solow–Srinivasan Model*,
Springer Optimization and Its Applications 166,
https://doi.org/10.1007/978-3-030-60307-6_4

We assume the following:

There exists $\sigma \in \{1, \ldots, n\}$ such that for all

$$i \in \{1, \ldots, n\} \setminus \{\sigma\}, \ c_\sigma > c_i. \tag{4.2}$$

Recall that a sequence $\{x(t), y(t)\}_{t=0}^\infty$ is called a program if for each integer $t \geq 0$,

$$(x(t), y(t)) \in R_+^n \times R_+^n, \ x(t+1) \geq (1-d)x(t),$$

$$0 \leq y(t) \leq x(t), \ a(x(t+1) - (1-d)x(t)) + ey(t) \leq 1. \tag{4.3}$$

Let T_1, T_2 be integers such that $0 \leq T_1 < T_2$. A pair of sequences

$$\left(\{x(t)\}_{t=T_1}^{T_2}, \{y(t)\}_{t=T_1}^{T_2-1} \right)$$

is called a program if $x(T_2) \in R_+^n$ and for each integer t satisfying $T_1 \leq t < T_2$ relations (4.3) are valid.

Assume that $w : [0, \infty) \to R^1$ is a continuous strictly increasing concave and differentiable function which represents the preferences of the planner.

Define

$$\Omega = \{(x, x') \in R_+^n \times R_+^n : x' - (1-d)x \geq 0$$

$$\text{and } a(x' - (1-d)x) \leq 1\}$$

and a correspondence $\Lambda : \Omega \to R_+^n$ given by

$$\Lambda(x, x') = \{y \in R_+^n : 0 \leq y \leq x \text{ and } ey \leq 1 - a(x' - (1-d)x)\}, \ (x, x') \in \Omega.$$

For every $(x, x') \in \Omega$, set

$$u(x, x') = \max\{w(by) : y \in \Lambda(x, x')\}.$$

Recall that the golden-rule stock is $\widehat{x} \in R_+^n$ such that $(\widehat{x}, \widehat{x})$ is a solution to the problem:

maximize $u(x, x')$ subject to
(i) $x' \geq x$; (ii) $(x, x') \in \Omega$.

By Theorem 2.3, there exists a unique golden-rule stock

$$\widehat{x} = (1/(1 + da_\sigma))e(\sigma). \tag{4.4}$$

It is not difficult to see that $\widehat{x}$ is a solution to the problem

$$w(by) \to \max, \ y \in \Lambda(\widehat{x}, \widehat{x}).$$

Set

$$\widehat{y} = \widehat{x}. \tag{4.5}$$

For $i = 1, \ldots, n$, set

$$\widehat{q}_i = a_i b_i / (1 + d a_i), \quad \widehat{p}_i = w'(b\widehat{x})\widehat{q}_i. \tag{4.6}$$

In view of Lemma 2.2,

$$w(b\widehat{x}) \ge w(by) + \widehat{p}x' - \widehat{p}x \tag{4.7}$$

for every $(x, x') \in \Omega$ and for every $y \in \Lambda(x, x')$.

For every $(x, x') \in \Omega$ and every $y \in \Lambda(x, x')$, define

$$\delta(x, y, x') = \widehat{p}(x - x') - (w(by) - w(b\widehat{y})). \tag{4.8}$$

By (4.7) and (4.8),

$$\delta(x, y, x') \ge 0 \text{ for every } (x, x') \in \Omega \text{ and every } y \in \Lambda(x, x'). \tag{4.9}$$

A program $\{x(t), y(t)\}_{t=0}^{\infty}$ is good if there is a real number M such that

$$\sum_{t=0}^{T} (w(by(t)) - w(b\widehat{y})) \ge M \text{ for every nonnegative integer } T.$$

A program $\{x(t), y(t)\}_{t=0}^{\infty}$ is bad if

$$\lim_{T \to \infty} \sum_{t=0}^{T} (w(by(t)) - w(b\widehat{y})) = -\infty.$$

In the next proposition, which follows from Propositions 2.4, 2.6, and 2.9, we collect the properties of good programs.

Proposition 4.1 *(i) Any program that is not good is bad. (ii) For any $x_0 \in R_+^n$ there exists a good program $\{x(t), y(t)\}_{t=0}^{\infty}$ such that $x(0) = x_0$. (iii) A program $\{x(t), y(t)\}_{t=0}^{\infty}$ is good if and only if $\sum_{t=0}^{\infty} \delta(x(t), y(t), x(t+1)) < \infty$.*

Let $x_0 \in R_+^n$. Set

$$\Delta(x_0) = \inf \left\{ \sum_{t=0}^{\infty} \delta(x(t), y(t), x(t+1)) : \{x(t), y(t)\}_{t=0}^{\infty} \text{ is a} \right.$$

$$\left. \text{program such that } x(0) = x_0 \right\} \tag{4.10}$$

Since there exists a good program from x_0, it follows from Proposition 4.1 that

$$\Delta(x_0) < \infty. \tag{4.11}$$

Recall that a program $\{x^*(t), y^*(t)\}_{t=0}^{\infty}$ is called overtaking optimal if

$$\limsup_{T \to \infty} \sum_{t=0}^{T} [w(by(t)) - w(by^*(t))] \leq 0$$

for every program $\{x(t), y(t)\}_{t=0}^{\infty}$ satisfying $x(0) = x^*(0)$.

A program $\{x^*(t), y^*(t)\}_{t=0}^{\infty}$ is called weakly optimal if for each program $\{x(t), y(t)\}_{t=0}^{\infty}$ satisfying $x(0) = x^*(0)$, the following inequality holds:

$$\liminf_{T \to \infty} \sum_{t=0}^{T} [w(by(t)) - w(by^*(t))] \leq 0.$$

Let $z \in R_+^n$ and $T \geq 1$ be a natural number. Set

$$U(z, T) = \sup \left\{ \sum_{t=0}^{T-1} w(by(t)) : (\{x(t)\}_{t=0}^{T}, \{y(t)\}_{t=0}^{T-1}) \right.$$

$$\left. \text{is a program such that } x(0) = z \right\} \tag{4.12}$$

Clearly $U(z, T)$ is a finite number. Let $x_0, x_1 \in R_+^n$, T_1, T_2 be integers, $0 \leq T_1 < T_2$. Define

$$U(x_0, x_1, T_1, T_2) = \sup \left\{ \sum_{t=T_1}^{T_2-1} w(by(t)) : \left(\{x(t)\}_{t=T_1}^{T_2}, \{y(t)\}_{t=T_1}^{T_2-1} \right) \right.$$

$$\left. \text{is a program such that } x(T_1) = x_0, \ x(T_2) \geq x_1 \right\}. \tag{4.13}$$

(Here we suppose that a supremum over empty set is $-\infty$.) Clearly

$$U(x_0, x_1, T_1, T_2) < \infty.$$

It is also clear that for any $z \in R_+^n$ and any integer $T \geq 1$, $U(z, T) = U(z, 0, 0, T)$.

In this chapter we assume that the following asymptotic turnpike property holds:

(ATP) Each good program $\{x(t), y(t)\}_{t=0}^{\infty}$ converges to the golden-rule stock $(\widehat{x}, \widehat{x})$:

$$\lim_{t \to \infty} (x(t), y(t)) = (\widehat{x}, \widehat{x}).$$

With Card(A) we denote in the sequel the cardinality of a finite set A.

In this chapter we prove the following two turnpike results obtained in [53].

Theorem 4.2 *Let* M, ϵ *be positive numbers and* $\Gamma \in (0, 1)$*. Then there exists a natural number* L *such that for each integer* $T > L$*, each* $z_0, z_1 \in R_+^n$ *satisfying* $z_0 \leq Me$ *and* $az_1 \leq \Gamma d^{-1}$*, and each program* $(\{x(t)\}_{t=0}^{T}, \{y(t)\}_{t=0}^{T-1})$ *which satisfies*

$$x(0) = z_0, \quad x(T) \geq z_1, \quad \sum_{t=0}^{T-1} w(by(t)) \geq U(z_0, z_1, 0, T) - M,$$

the following inequality holds:

$$Card\{i \in \{0, \ldots, T-1\} : \max\{\|x(t) - \widehat{x}\|, \|y(t) - \widehat{x}\|\} > \epsilon\} \leq L.$$

Theorem 4.3 *Let* M, ϵ *be positive numbers and* $\Gamma \in (0, 1)$*. Then there exist a natural number* L *and a positive number* γ *such that for each integer* $T > 2L$*, each* $z_0, z_1 \in R_+^n$ *satisfying* $z_0 \leq Me$ *and* $az_1 \leq \Gamma d^{-1}$*, and each program* $(\{x(t)\}_{t=0}^{T}, \{y(t)\}_{t=0}^{T-1})$ *which satisfies*

$$x(0) = z_0, \quad x(T) \geq z_1, \quad \sum_{t=0}^{T-1} w(by(t)) \geq U(z_0, z_1, 0, T) - \gamma,$$

there are integers τ_1, τ_2 *such that*

$$\tau_1 \in [0, L], \quad \tau_2 \in [T - L, T],$$

$$\|x(t) - \widehat{x}\|, \ \|y(t) - \widehat{x}\| \leq \epsilon \text{ for all } t = \tau_1, \ldots, \tau_2 - 1.$$

Moreover if $\|x(0) - \widehat{x}\| \leq \gamma$*, then* $\tau_1 = 0$*, and if* $\|x(T) - \widehat{x}\| \leq \gamma$*, then* $\tau_2 = T$*.*

Corollary 4.4 *Let* M, ϵ *be positive numbers and* $\Gamma \in (0, 1)$*. Then there exists a natural number* L *such that for each integer* $T > L$*, each* $z_0, z_1 \in R_+^n$ *satisfying* $z_0 \leq Me$ *and* $az_1 \leq \Gamma d^{-1}$*, and each pair of programs*

$$(\{x_a(t)\}_{t=0}^{T}, \{y_a(t)\}_{t=0}^{T-1}) \text{ and } (\{x_b(t)\}_{t=0}^{T}, \{y_b(t)\}_{t=0}^{T-1})$$

which satisfy

$$x_a(0) = x_b(0) = z_0, \ x_a(T) \geq z_1, \ x_b(T) \geq z_1,$$

$$\sum_{t=0}^{T-1} w(by_a(t)) \geq U(z_0, z_1, 0, T) - M,$$

$$\sum_{t=0}^{T-1} w(by_b(t)) \geq U(z_0, z_1, 0, T) - M,$$

the following inequality holds:

$$Card\{i \in \{0, \ldots, T-1\} :$$

$$\max\{\|x_a(t) - x_b(t)\|, \ \|y_a(t) - y_b(t)\|\} > \epsilon\} \leq L.$$

4.2   Auxiliary Results for Theorems 4.2 and 4.3

We suppose that the sum over empty set is zero.

Proposition 4.5 *Let $\Gamma \in (0, 1)$. Then there exists a natural number $k(\Gamma)$ such that for each $z_0 \in R_+^n$ and each $z_1 \in R_+^n$ satisfying $az_1 \leq \Gamma d^{-1}$, there is a program $(\{x(t)\}_{t=0}^{k(\Gamma)}, \{y(t)\}_{t=0}^{k(\Gamma)-1})$ such that $x(0) = z_0, x(k(\Gamma)) \geq z_1$.*

Proof Since $\sum_{i=0}^{\infty}(1-d)^i = 1/d$, there is a natural number $k(\Gamma)$ such that

$$\sum_{i=0}^{k(\Gamma)-1}(1-d)^i > \Gamma d^{-1}.$$

Assume that $z_0 \in R_+^n$ and $z_1 \in R_+^n$ satisfies $az_1 \leq \Gamma d^{-1}$. Put

$$z_2 = \Gamma^{-1} d z_1,$$

$x(0) = z_0, \ x(t+1) = (1-d)x(t) + z_2$ and $y(t) = 0$ for all integers $t \geq 0$.

It is easy to see that $\{x(t), y(t)\}_{t=0}^{\infty}$ is a program and that

$$x(k(\Gamma)) \geq \sum_{i=0}^{k(\Gamma)-1}(1-d)^i z_2 \geq \Gamma d^{-1} z_2 = z_1.$$

Proposition 4.5 is proved.

In the sequel with each $\Gamma \in (0, 1)$, we associate a natural number $k(\Gamma)$ for which the assertion of Proposition 4.5 holds.

Proposition 4.6 *There is $m > 0$ such that for each $z \in R_+^n$ and each natural number T,*

$$U(z, T) \geq Tw(b\widehat{x}) - m.$$

Proof Put

$$y(0) = 0, \ y(t + 1) = (1 - d)y(t) + d\widehat{x} \text{ for all integers } t \geq 0. \tag{4.14}$$

It is not difficult to see that the sequence $\{y(t)\}_{t=0}^{\infty}$ is non-decreasing and converges to $\widehat{x}$ as $t \to \infty$. Put

$$m = d^{-1}(b\widehat{x} - by(1))|w'(db\widehat{x})| + |w(by(0)) - w(b\widehat{x})| + |w(by(1)) - w(b\widehat{x})|. \tag{4.15}$$

Let $z \in R_+^n$. Set

$$x(0) = z, \tag{4.16}$$

$$x(t + 1) = (1 - d)x(t) + d\widehat{x} \text{ for all integers } t \geq 0.$$

It is not difficult to see that $\{x(t), y(t)\}_{t=0}^{\infty}$ is a program. It follows from the definition of the sequence $\{y(t)\}_{t=0}^{\infty}$ that for all natural numbers t,

$$by(t) \geq db\widehat{x}, \tag{4.17}$$

and that for all integers $t \geq 2$,

$$by(t) - b\widehat{x} = (1 - d)^{t-1}(by(1) - b\widehat{x}). \tag{4.18}$$

Since the function w is concave and strictly increasing, it follows from (4.17) and (4.18) that for all integers $t \geq 2$,

$$w(b\widehat{x}) - w(by(t)) \leq w'(by(t))(b\widehat{x} - by(t)) \leq w'(db\widehat{x})(b\widehat{x} - by(t))$$
$$\leq w'(db\widehat{x})(b\widehat{x} - by(1))(1 - d)^{t-1}.$$

By the relation above, (4.15), and (4.16), for each natural number T,

$$U(z, T) \geq \sum_{t=0}^{T-1} w(by(t)) = \sum\{w(by(t)) : t \in \{0, 1\} \text{ and } t \leq T - 1\}$$

$$+ \sum\{w(by(t)) : t \text{ is an integer and } 2 \leq t < T\}$$

$$\geq Tw(b\widehat{x}) - |w(by(0)) - w(b\widehat{x})| - |w(by(1)) - w(b\widehat{x})|$$

$$- \sum \{|w(b\widehat{x}) - w(by(t))| : t \text{ is an integer and } 2 \leq t < T\}$$

$$\geq Tw(b\widehat{x}) - |w(by(0)) - w(b\widehat{x})| - |w(by(1)) - w(b\widehat{x})|$$

$$- w'(db\widehat{x})(b\widehat{x} - by(1)) \sum_{t=2}^{T+1} (1 - d)^{t-1}$$

$$\geq Tw(b\widehat{x}) - |w(by(0)) - w(b\widehat{x})| - |w(by(1)) - w(b\widehat{x})|$$

$$- w'(db\widehat{x})(b\widehat{x} - by(1))d^{-1}$$

$$= Tw(b\widehat{x}) - m.$$

Proposition 4.6 is proved.

Proposition 4.7 *Let* $\Gamma \in (0, 1)$. *Then there exists* $m > 0$ *such that for each* $z_0 \in R_+^n$, *each* $z_1 \in R_+^n$ *satisfying* $az_1 \leq \Gamma d^{-1}$, *and each natural number* $T > k(\Gamma)$,

$$U(z_0, z_1, 0, T) \geq Tw(b\widehat{x}) - m.$$

Proof By Proposition 4.6 there is $m_0 > 0$ such that

$$U(z, T) \geq Tw(b\widehat{x}) - m_0 \text{ for each } z \in R_+^n \text{ and each natural number } T. \qquad (4.19)$$

Put

$$m = m_0 + k(\Gamma)(|w(b\widehat{x})| + |w(0)|).$$

Assume that $z_0 \in R_+^n$, $z_1 \in R_+^n$ satisfies $az_1 \leq \Gamma d^{-1}$ and a natural number $T > k(\Gamma)$. By (4.19) there is a program $(\{x(t)\}_{t=0}^{T-k(\Gamma)}, \{y(t)\}_{t=0}^{T-k(\Gamma)-1})$ such that

$$x(0) = z_0, \quad \sum_{t=0}^{T-k(\Gamma)-1} w(by(t)) \geq (T - k(\Gamma))w(b\widehat{x}) - m_0. \qquad (4.20)$$

By the choice of $k(\Gamma)$, Proposition 4.5, and the inequality $az_1 \leq \Gamma d^{-1}$, there is a program $(\{x(t)\}_{t=T-k(\Gamma)}^{T}, \{y(t)\}_{t=T-k(\Gamma)}^{T-1})$ such that $x(T) \geq z_1$. By (4.19), (4.20), and the choice of m,

$$U(z_0, z_1, 0, T) \geq \sum_{t=0}^{T-1} w(by(t)) \geq \sum_{t=0}^{T-k(\Gamma)-1} w(by(t)) - k(\Gamma))|w(0)|$$

$$\geq (T - k(\Gamma))w(b\widehat{x}) - m_0 - k(\Gamma)|w(0)|$$

$$\geq Tw(b\widehat{x}) - k(\Gamma)|w(b\widehat{x})| - m_0 - k(\Gamma)|w(0)|$$
$$\geq Tw(b\widehat{x}) - m.$$

Proposition 4.7 is proved.

Proposition 4.8 *Let $m_0 > 0$. Then there exists $m_2 > 0$ such that for each natural number T and each program $(\{x(t)\}_{t=0}^{T}, \{y(t)\}_{t=0}^{T-1})$ which satisfies $x(0) \leq m_0 e$, the following inequality holds:*

$$\sum_{t=0}^{T-1}[w(by(t)) - w(b\widehat{x})] \leq m_2.$$

Proof By Proposition 3.6 there exists $m_1 > 0$ such that for each natural number T and each program $(\{x(t)\}_{t=0}^{T}, \{y(t)\}_{t=0}^{T-1})$ which satisfies $x(0) \leq m_0 e$, we have

$$x(t) \leq m_1 e \text{ for all } t = 0, \dots, T. \tag{4.21}$$

Choose a number

$$m_2 \geq 2\|\widehat{p}\|m_1 n. \tag{4.22}$$

Assume that T is a natural number and a program $(\{x(t)\}_{t=0}^{T}, \{y(t)\}_{t=0}^{T-1})$ satisfies $x(0) \leq m_0 e$. Then (4.21) holds.

By (4.5) and (4.7), for each integer $t \in [0, T-1]$,

$$w(b\widehat{y}) \geq w(by(t)) + \widehat{p}(x(t+1)) - \widehat{p}x(t).$$

Together with (4.21) and (4.22), this implies that

$$\sum_{t=0}^{T-1}[w(by(t)) - w(b\widehat{x})] \leq \sum_{t=0}^{T-1}[\widehat{p}x(t) - \widehat{p}x(t+1)]$$

$$= \widehat{p}x(0) - \widehat{p}x(T) \leq 2\|\widehat{p}\|nm_1 \leq m_2.$$

Proposition 4.8 is proved.

It is easy to see that the following auxiliary result holds.

Proposition 4.9 *Assume that T_1, T_2 are nonnegative integers, $T_1 < T_2$,*

$$(\{x(t)\}_{t=T_1}^{T_2}, \{y(t)\}_{t=T_1}^{T_2-1})$$

is a program, and $u \in R_+^n$. Then $(\{x(t) + (1-d)^{t-T_1}u\}_{t=T_1}^{T_2}, \{y(t)\}_{t=T_1}^{T_2-1})$ is also a program.

Proposition 4.10 *Let τ be a natural number, $M > 0$, $x_0, x_1 \in R_+^n$, and let*

$$\left(\{x(t)\}_{t=0}^{\tau}, \ \{y(t)\}_{t=0}^{\tau-1} \right)$$

be a program such that

$$x(0) = x_0, \ x(\tau) \geq x_1, \ \sum_{t=0}^{\tau-1} w(by(t)) \geq U(x_0, x_1, 0, \tau) - M. \tag{4.23}$$

Then for each pair of integers S_1, S_2 satisfying

$$0 \leq S_1 < S_2 \leq \tau, \tag{4.24}$$

the following inequality holds:

$$\sum_{t=S_1}^{S_2-1} w(by(t)) \geq U(x(S_1), x(S_2), S_1, S_2) - M.$$

Proof Let us assume the contrary. Then there exists a pair of integers S_1, S_2 satisfying (4.24) such that

$$\sum_{t=S_1}^{S_2-1} w(by(t)) < U(x(S_1), x(S_2), S_1, S_2) - M. \tag{4.25}$$

By (4.13) and (4.25), there exists a program $(\{x'(t)\}_{t=S_1}^{S_2}, \{y'(t)\}_{t=S_1}^{S_2-1})$ such that

$$x'(S_1) = x(S_1), \ x'(S_2) \geq x(S_2), \tag{4.26}$$

$$\sum_{t=S_1}^{S_2-1} w(by(t)) < \sum_{t=S_1}^{S_2-1} w(by'(t)) - M.$$

Set

$$z = x'(S_2) - x(S_2). \tag{4.27}$$

Define sequences $\{\bar{x}(t)\}_{t=0}^{\tau}$, $\{\bar{y}(t)\}_{t=0}^{\tau-1}$ as follows:

$$\bar{x}(t) = x(t), \ t = 0, \ldots, S_1, \ \bar{y}(t) = y(t), \ t = 0, \ldots, S_1 - 1, \tag{4.28}$$

$$\bar{x}(t) = x'(t), \ t = S_1 + 1, \ldots, S_2, \ \bar{y}(t) = y'(t), \ t = S_1, \ldots, S_2 - 1,$$

$$\bar{x}(t) = x(t) + (1-d)^{t-S_2}z \text{ for all integers } t \text{ satisfying } S_2 < t \le \tau,$$

$$\bar{y}(t) = y(t) \text{ for all integers } t \text{ satisfying } S_2 \le t \le \tau - 1.$$

It follows from Proposition 4.9 and (4.26)–(4.28) that $(\{\bar{x}(t)\}_{t=0}^{\tau}, \{\bar{y}(t)\}_{t=0}^{\tau-1})$ is a program. In view of (4.23) and (4.26)–(4.28),

$$\bar{x}(0) = x(0) = x_0, \quad \bar{x}(\tau) \ge x(\tau) \ge x_1.$$

By (4.26) and (4.28),

$$\sum_{t=0}^{\tau-1} w(b\bar{y}(t)) - \sum_{t=0}^{\tau-1} w(by(t)) = \sum_{t=S_1}^{S_2-1} w(by'(t)) - \sum_{t=S_1}^{S_2-1} w(by(t)) > M.$$

This contradicts (4.23). The contradiction we have reached proves Proposition 4.10.

4.3 Four Lemmas

Lemma 4.11 *Let $m_0, m_1 > 0$, $\epsilon > 0$. Then there is a natural number τ such that for each program $(\{x(t)\}_{t=0}^{\tau}, \{y(t)\}_{t=0}^{\tau-1})$ satisfying*

$$x(0) \le m_0 e, \quad \sum_{t=0}^{\tau-1} w(by(t)) \ge \tau w(b\hat{x}) - m_1,$$

there is an integer $t \in [0, \tau-1]$ such that

$$\|y(t) - \hat{x}\|, \ \|x(t) - \hat{x}\| \le \epsilon.$$

Proof Let us assume the contrary. Then for each natural number k, there exists a program $(\{x_t^{(k)}\}_{t=0}^{k}, \{y_t^{(k)}\}_{t=0}^{k-1})$ such that

$$x^{(k)}(0) \le m_0 e, \quad \sum_{t=0}^{k-1} w(by^{(k)}(t)) \ge k w(b\hat{x}) - m_1, \tag{4.29}$$

$$\max\left\{\|y_t^{(k)} - \hat{x}\|, \ \|x_t^{(k)} - \hat{x}\|\right\} > \epsilon, \ t = 0, \ldots, k. \tag{4.30}$$

In view of (4.29) and Proposition 3.6, there is $m_2 > m_0$ such that for each natural number k,

$$x^{(k)}(t) \le m_2 e, \ t = 0, \ldots, k. \tag{4.31}$$

By Proposition 4.8, there is $m_3 > 0$ such that for each natural number T and each program $(\{x(t)\}_{t=0}^{T}, \{y(t)\}_{t=0}^{T-1})$ satisfying $x(0) \leq m_2 e$, the following inequality holds:

$$\sum_{t=0}^{T-1} (w(by(t)) - w(b\widehat{y})) \leq m_3. \tag{4.32}$$

Let k be a natural number, and let an integer s satisfy $0 < s < k$.

It follows from (4.31) and the choice of m_3 (see (4.32)) that

$$\sum_{t=s}^{k-1} [w(by^{(k)}(t)) - w(b\widehat{y})] \leq m_3.$$

Combined with (4.29) this implies that

$$\sum_{t=0}^{s-1} [w(by^{(k)}(t)) - w(b\widehat{y})] = \sum_{t=0}^{k-1} [w(by^{(k)}(t)) - w(b\widehat{y})]$$

$$- \sum_{t=s}^{k-1} [w(by^{(k)}(t)) - w(b\widehat{y})] \geq -m_1 - m_3.$$

Thus for each pair of natural numbers s, k satisfying $s < k$,

$$\sum_{t=0}^{s-1} [w(by^{(k)}(t)) - w(b\widehat{y})] \geq -m_1 - m_3. \tag{4.33}$$

By extracting subsequence and using (4.31) and diagonalization process, we obtain that there exist a strictly increasing sequence of natural numbers $\{k_j\}_{j=1}^{\infty}$ and sequences $\{x^*(t)\}_{t=0}^{\infty}, \{y^*(t)\}_{t=0}^{\infty} \subset R^n$ such that

$$x^{(k_j)}(t) \to x^*(t), \ y^{(k_j)}(t) \to y^*(t) \text{ as } j \to \infty \text{ for all integers } j \geq 0. \tag{4.34}$$

It is not difficult to see that $\{x^*(t), \ y^*(t)\}_{t=0}^{\infty}$ is a program. In view of (4.31) and (4.34),

$$x^*(t) \leq m_2 e \text{ for all integers } t \geq 0. \tag{4.35}$$

By (4.33) and (4.34), for all natural numbers s,

$$\sum_{t=0}^{s-1} (w(by^*(t)) - w(b\widehat{x})) \geq -m_3 - m_1.$$

This implies that $\{x^*(t), y^*(t)\}_{t=0}^{\infty}$ is a good program and by (ATP),

$$y^*(t) \to \widehat{x}, \ x^*(t) \to \widehat{x} \text{ as } t \to \infty. \tag{4.36}$$

On the other hand, it follows from (4.30) and (4.34) that for all integers $t \geq 0$,

$$\max\{\|y^*(t) - \widehat{x}\|, \ \|x^*(t) - \widehat{x}\|\} \geq \epsilon.$$

This contradicts (4.36). The contradiction we have reached proves Lemma 4.11.

Lemma 4.12 *Let $S_0 > 0$, and let a sequence $\{x(t), y(t)\}_{t=-\infty}^{\infty} \subset R_+^{2n}$ satisfy the following conditions:*

$$\|x(t)\|, \ \|y(t)\| \leq S_0 \text{ for all integers } t,$$

$$x(t+1) \geq (1-d)x(t), \ 0 \leq y(t) \leq x(t) \text{ for all integers } t,$$

$$a(x(t+1) - (1-d)x(t)) + ey(t) \leq 1 \text{ for all integers } t,$$

$$\delta(x(t), y(t), x(t+1)) = 0 \text{ for all integers } t.$$

Then $x(t) = y(t) = \widehat{x}$ for all integers t.

Proof Assume that the lemma does not hold. Then we may assume without loss of generality that

$$(x(0), y(0)) \neq (\widehat{x}, \widehat{x}).$$

Set

$$\kappa = \min\{1, \max\{\|x(0) - \widehat{x}\|, \ \|y(0) - \widehat{x}\|\}\}. \tag{4.37}$$

By Proposition 3.13, there exists a strictly increasing sequence $\{\delta_k\}_{k=1}^{\infty} \subset (0, \kappa)$ such that for each natural number k, the following property holds:
(P1) If $x, x' \in R_+^n$, $\|x - \widehat{x}\|, \|x' - \widehat{x}\| \leq 4\delta_k$, then there exist $\bar{x} \geq x', y \in R^n$ such that

$$(x, \bar{x}) \in \Omega, \ y \in \Lambda(x, \bar{x}), \ \|y - \widehat{x}\| \leq 8^{-k}\kappa, \ \|\bar{x} - \widehat{x}\| \leq 8^{-k}\kappa, \ \delta(x, y, \bar{x}) \leq 8^{-k}\kappa.$$

By Lemma 4.11, for each natural number k, there exist natural numbers $T_k, S_k > 4$ such that

$$(1-d)^{\min\{T_k, S_k\}} < \min\{\delta_k, \ 8^{-k}\kappa\}, \tag{4.38}$$

$$\|x(T_k) - \widehat{x}\|, \ \|y(T_k) - \widehat{x}\|, \ \|x(-S_k) - \widehat{x}\|, \ \|y(-S_k) - \widehat{x}\| \leq \min\{\delta_k, \ 8^{-k}\kappa\}. \tag{4.39}$$

Set

$$\tau_0 = 0, \ \tau_1 = T_1 + S_1, \ \tau_{k+1} = \tau_k + T_{k+1} + S_{k+1} + 1 \text{ for any integer } k \geq 1. \tag{4.40}$$

By induction, we construct a program $\{u(t), v(t)\}_{t=0}^{\infty}$. Set

$$u(t) = x(t - S_1), \ t = 0, \ldots, \tau_1, \ v(t) = y(t - S_1), \ t = 0 \ldots, \tau_1 - 1. \tag{4.41}$$

Now assume that k is a natural number, we defined $u(t), t = 0, \ldots, \tau_k, v(t), t = 0, \ldots, \tau_k - 1$ such that $(\{u(t)\}_{t=0}^{\tau_k}, \{v(t)\}_{t=0}^{\tau_k - 1})$ is a program,

$$\|u(\tau_k) - \widehat{x}\| \leq 2 \min\{\delta_k, 8^{-k}\kappa\}, \tag{4.42}$$

$$\sum_{t=0}^{\tau_k - 1} \delta(u(t), v(t), u(t+1)) \leq (1 + 4\|\widehat{p}\|) \sum_{j=0}^{k-1} 8^{-j}, \tag{4.43}$$

and if $k \geq 2$, then for $j = 1, \ldots, k - 1$,

$$\max\{\|v(\tau_j + S_{j+1} + 1) - \widehat{x}\|, \ \|u(\tau_j + S_{j+1} + 1) - \widehat{x}\|\} \geq 2^{-1}\kappa. \tag{4.44}$$

(It is not difficult to see that for $k = 1$, all these assumptions hold.)

Define $u(t), t = \tau_k + 1, \ldots, \tau_{k+1}, v(t), t = \tau_k, \ldots, \tau_{k+1} - 1$. First define $u(\tau_k + 1), v(\tau_k)$. By (4.39), (4.42), and (P1), there exist

$$u(\tau_k + 1) \geq x(-S_{k+1}), \ v(\tau_k) \in R^n \tag{4.45}$$

such that

$$(u(\tau_k), u(\tau_k + 1)) \in \Omega, \ v(\tau_k) \in \Lambda(u(\tau_k), u(\tau_k + 1)), \tag{4.46}$$

$$\|u(\tau_k + 1) - \widehat{x}\| \leq 8^{-k}\kappa, \ \|v(\tau_k) - \widehat{x}\| \leq 8^{-k}\kappa, \tag{4.47}$$

$$\delta(u(\tau_k), v(\tau_k), u(\tau_k + 1)) \leq 8^{-k}\kappa. \tag{4.48}$$

For $t = \tau_k + 1, \ldots, \tau_{k+1} - 1$, set

$$v(t) = y(t - \tau_k - S_{k+1} - 1), \tag{4.49}$$

and for $t = \tau_k + 2, \ldots, \tau_{k+1}$, set

$$u(t) = x(t - \tau_k - S_{k+1} - 1) + (1 - d)^{t - \tau_k - 1}[u(\tau_k + 1) - x(-S_{k+1})]. \tag{4.50}$$

By Proposition 4.9, (4.45), (4.46), (4.49), and (4.50), $(\{u(t)\}_{t=0}^{\tau_{k+1}}, \{v(t)\}_{t=0}^{\tau_{k+1} - 1})$ is a program.

By (4.37)–(4.40), (4.47), and (4.50),

$$\|u(\tau_{k+1}) - \widehat{x}\| = \|x(T_{k+1}) + (1-d)^{T_{k+1}+S_{k+1}}[u(\tau_k + 1) - x(-S_{k+1})] - \widehat{x}\|$$

$$\leq \|x(T_{k+1}) - \widehat{x}\| + (1-d)^{T_{k+1}+S_{k+1}}(\|u(\tau_k + 1) - \widehat{x}\| + \|\widehat{x} - x(-S_{k+1})\|)$$

$$\leq \min\{\delta_{k+1},\ 8^{-k-1}\kappa\} + (1-d)^{T_{k+1}+S_{k+1}}(8^{-k}\kappa + 8^{-k}\kappa) \leq 2\min\{\delta_{k+1}, 8^{-k-1}\kappa\}$$

and

$$\|u(\tau_{k+1}) - \widehat{x}\| \leq 2\min\{\delta_{k+1},\ 8^{-k-1}\kappa\}. \tag{4.51}$$

By (4.38), (4.39), (4.47), (4.49) and (4.50),

$$v(\tau_k + 1 + S_{k+1}) = y(0),$$

$$\|u(\tau_k + 1 + S_{k+1}) - x(0)\| = (1-d)^{S_{k+1}}\|u(\tau_k + 1) - x(-S_{k+1})\|$$

$$\leq (1-d)^{S_{k+1}}(\|u(\tau_k + 1) - \widehat{x}\| + \|\widehat{x} - x(-S_{k+1})\|)$$

$$\leq (1-d)^{S_{k+1}}(3 \cdot 8^{-k}\kappa) \leq 8^{-1}\kappa. \tag{4.52}$$

By (4.52),

$$\|v(\tau_k + 1 + S_{k+1}) - \widehat{x}\| = \|y(0) - \widehat{x}\|,$$

$$\|u(\tau_k + 1 + S_{k+1}) - \widehat{x}\| \geq \|\widehat{x} - x(0)\| - \|x(0) - u(\tau_k + 1 + S_{k+1})\|$$

$$\geq \|\widehat{x} - x(0)\| - 8^{-1}\kappa.$$

Together with (4.37) this implies that

$$\max\{\|v(\tau_k + 1 + S_{k+1}) - \widehat{x}\|,\ \|u(\tau_k + 1 + S_{k+1}) - \widehat{x}\|\}$$

$$\geq \max\{\|y(0) - \widehat{x}\| - 8^{-1}\kappa,\ \|\widehat{x} - x(0)\| - 8^{-1}\kappa\} \geq \kappa/2. \tag{4.53}$$

By (4.43) and (4.48),

$$\sum_{t=0}^{\tau_{k+1}-1} \delta(u(t), v(t), u(t+1))$$

$$= \sum_{t=0}^{\tau_k-1} \delta(u(t), v(t), u(t+1)) + \delta(u(\tau_k), v(\tau_k), u(\tau_k + 1))$$

$$+ \sum_{t=\tau_k+1}^{\tau_{k+1}-1} \delta(u(t), v(t), u(t+1))$$

$$\leq (1 + 4\|\widehat{p}\|) \sum_{j=0}^{k-1} 8^{-j} + 8^{-k}\kappa + \sum_{t=\tau_k+1}^{\tau_{k+1}-1} \delta(u(t), v(t), u(t+1)). \qquad (4.54)$$

By (4.8), (4.39), (4.40), (4.45), (4.47), (4.49)–(4.51), and the equality

$$\delta(x(t), y(t), x(t+1)) = 0$$

which holds for all integers t,

$$\sum_{t=\tau_k+1}^{\tau_{k+1}-1} \delta(u(t), v(t), u(t+1)) = \sum_{t=\tau_k+1}^{\tau_{k+1}-1} [w(b\widehat{x}) - w(bv(t)) + \widehat{p}(u(t) - u(t+1))]$$

$$= \sum_{t=\tau_k+1}^{\tau_{k+1}-1} [w(b\widehat{x}) - w(by(t - \tau_k - S_{k+1} - 1))] + \widehat{p}(u(\tau_k + 1) - u(\tau_{k+1}))$$

$$= \sum_{t=-S_{k+1}}^{T_{k+1}-1} [w(b\widehat{x}) - w(by(t)) + \widehat{p}(x(t) - x(t+1))]$$

$$- \widehat{p}(x(-S_{k+1}) - x(T_{k+1})) + \widehat{p}(u(\tau_k + 1) - u(\tau_{k+1}))$$

$$= -\widehat{p}(x(-S_{k+1}) - x(T_{k+1})) + \widehat{p}(u(\tau_k + 1) - u(\tau_{k+1}))$$

$$\leq \|\widehat{p}\|2 \cdot 8^{-k-1}\kappa + \|\widehat{p}\|(2 \cdot 8^{-k-1}\kappa + 8^{-k-1}\kappa) \leq \|\widehat{p}\|8^{-k}\kappa.$$

Together with (4.54) this implies that

$$\sum_{t=0}^{\tau_{k+1}-1} \delta(u(t), v(t), u(t+1)) \leq (1 + 4\|\widehat{p}\|) \sum_{j=0}^{k} 8^{-j}. \qquad (4.55)$$

It follows from (4.51), (4.53), and (4.55) that in such a manner, we constructed by induction a program $\{(u(t), v(t))\}_{t=0}^{\infty}$ such that for each integer $k \geq 1$, (4.42) and (4.43) hold and for each integer $j \geq 1$ (4.44) holds.

By Proposition 4.1(iii) and (4.43), $\{u(t), v(t)\}_{t=0}^{\infty}$ is a good program. In view of (ATP),

$$u(t) \to \widehat{x}, \quad v(t) \to \widehat{x} \text{ as } t \to \infty.$$

This contradicts (4.44) which holds for any integer $j \geq 1$. The contradiction we have reached proves Lemma 4.12.

Lemma 4.13 *Let $\epsilon > 0$. Then there exists $\gamma > 0$ such that for each natural number T and each program $(\{x(t)\}_{t=0}^{T}, \{y(t)\}_{t=0}^{T-1})$ which satisfies*

$$\|x(0) - \widehat{x}\|, \ \|x(T) - \widehat{x}\| \leq \gamma,$$

$$\delta(x(t), y(t), x(t+1)) \leq \gamma, \ t = 0, 1, \ldots, T - 1,$$

the following inequality holds:

$$\|x(t) - \widehat{x}\|, \ \|y(t) - \widehat{x}\| \leq \epsilon, \ t = 0, \ldots, T - 1. \tag{4.56}$$

Proof Let $\{\gamma_k\}_{k=1}^{\infty}$ be a strictly decreasing sequence of positive numbers such that

$$\gamma_1 < 8^{-1}\epsilon, \ \lim_{k \to \infty} \gamma_k = 0. \tag{4.57}$$

Assume that the lemma does not hold. Then for each number k, there exist a natural number T_k and a program $(\{x^{(k)}(t)\}_{t=0}^{T_k}, \{y^{(k)}(t)\}_{t=0}^{T_k-1})$ such that

$$\|x^{(k)}(0) - \widehat{x}\| \leq \gamma_k, \ \|x^{(k)}(T_k) - \widehat{x}\| \leq \gamma_k, \tag{4.58}$$

$$\delta(x^{(k)}(t), y^{(k)}(t), x^{(k)}(t+1)) \leq \gamma_k, \ t = 0, \ldots, T_k - 1, \tag{4.59}$$

and there is a natural number

$$S_k \in [0, T_k - 1] \tag{4.60}$$

such that

$$\max\{\|x^{(k)}(S_k) - \widehat{x}\|, \ \|y^{(k)}(S_k) - \widehat{x}\|\} > \epsilon. \tag{4.61}$$

By (4.58) and Proposition 3.6, there exists $M > 0$ such that for all natural numbers k,

$$x^{(k)}(t) \leq Me \text{ for all } t = 0, \ldots, T_k. \tag{4.62}$$

Extracting a subsequence and re-indexing, we may assume that one of the following cases holds:

(1) $\sup\{T_k : k = 1, 2, \ldots\} < \infty$;
(2) $\sup\{S_k : k = 1, 2, \ldots\} < \infty$, $T_k - S_k \to \infty$ as $k \to \infty$;
(3) $S_k \to \infty$ as $k \to \infty$, $\sup\{T_k - S_k : k = 1, 2, \ldots\} < \infty$.
(4) $S_k \to \infty$, $T_k - S_k \to \infty$ as $k \to \infty$.

Assume that the case (1) holds. Extracting a subsequence and re-indexing, we may assume without loss of generality that

$$T_k = T_1, \ S_k = S_1 \text{ for all natural numbers } k;$$

that for each integer $t \in [0, T_1]$, there exists $x(t) = \lim_{k \to \infty} x^{(k)}(t)$; and that for each $t \in [0, T_1 - 1]$, there is $y(t) = \lim_{k \to \infty} y^{(k)}(t)$. It is not difficult to see that $(\{x(t)\}_{t=0}^{T_1}, \{y(t)\}_{t=0}^{T_1-1})$ is a program.

In view of (4.57)–(4.59) and (4.61),

$$\delta(x(t), y(t), x(t+1)) = 0, \ t = 0, 1, \ldots, T_1 - 1, \tag{4.63}$$

$$x(T_1) = x(0) = \widehat{x},$$

$$\max\{\|x(S_1) - \widehat{x}\|, \ \|y(S_1) - \widehat{x}\|\} \geq \epsilon.$$

Set

$$x(t) = \widehat{x} \text{ for all integers } t \in (-\infty, 0) \cup (T_1, \infty),$$

$$y(t) = \widehat{x} \text{ for all integers } t \in (-\infty, 0) \cup (T_1, \infty).$$

Clearly $\{x(t), y(t)\}_{t=-\infty}^{\infty}$ satisfies the assumptions of Lemma 4.12 which in its turn implies that $x(t) = y(t) = \widehat{x}$ for all integers t. This contradicts (4.63). The contradiction we have reached proves that the case (1) does not hold.

Assume that the case (2) holds. Extracting a subsequence and re-indexing, we may assume without loss of generality that

$$S_k = S_1 \text{ for all natural numbers } k \tag{4.64}$$

and that for each integer $t \geq 0$, there exist

$$x(t) = \lim_{k \to \infty} x^{(k)}(t), \ y(t) = \lim_{k \to \infty} y^{(k)}(t). \tag{4.65}$$

It is clear that $\{x(t), y(t)\}_{t=0}^{\infty}$ is a program. By (4.57)–(4.59), (4.61), (4.62), (4.64), and (4.65),

$$x(t) \leq Me \text{ for all integers } t \geq 0, \tag{4.66}$$

$$\delta(x(t), y(t), x(t+1)) = 0 \text{ for all integers } t \geq 0,$$

$$x(0) = \widehat{x},$$

$$\max\{\|x(S_1) - \widehat{x}\|, \ \|y(S_1) - \widehat{x}\|\} \geq \epsilon.$$

Set

$$x(t) = \widehat{x} \text{ for all integers } t < 0, \quad y(t) = \widehat{x} \text{ for all integers } t < 0.$$

Clearly $\{x(t), y(t)\}_{t=-\infty}^{\infty}$ satisfies the assumptions of Lemma 4.12 which in its turn implies that $x(t) = y(t) = \widehat{x}$ for all integers t. This contradicts (4.66). The contradiction we have reached proves that the case (2) does not hold.

Assume that the case (3) holds. Extracting a subsequence and re-indexing, we may assume that

$$T_k - S_k = T_1 - S_1 \text{ for all natural numbers } k. \tag{4.67}$$

For each natural number k, set

$$\tilde{x}^{(k)}(t) = x^{(k)}(t + T_k), \quad t = -T_k \ldots, 0, \tag{4.68}$$

$$\tilde{y}^{(k)}(t) = y^{(k)}(t + T_k), \quad t = -T_k, \ldots, -1.$$

Extracting a subsequence and re-indexing, we may assume without loss of generality that there exist

$$x(t) = \lim_{k \to \infty} \tilde{x}^{(k)}(t) \text{ for each integer } t \le 0 \text{ and}$$

$$y(t) = \lim_{k \to \infty} \tilde{y}^{(k)}(t) \text{ for each integer } t < 0. \tag{4.69}$$

By (4.57)–(4.59), (4.61), (4.62), and (4.67)–(4.69),

$$x(t) \le Me \text{ for al integers } t \le 0, \tag{4.70}$$

$$\delta(x(t), y(t), x(t+1)) = 0 \text{ for all integers } t < 0,$$

$$x(0) = \widehat{x},$$

$$\max\{\|x(S_1 - T_1) - \widehat{x}\|, \ \|y(S_1 - T_1) - \widehat{x}\|\} \ge \epsilon.$$

Set

$$x(t) = \widehat{x} \text{ for all integers } t > 0,$$

$$y(t) = \widehat{x} \text{ for all integers } t \ge 0.$$

Clearly $\{x(t), y(t)\}_{t=-\infty}^{\infty}$ satisfies the assumptions of Lemma 4.12 which in its turn implies that $x(t) = y(t) = \widehat{x}$ for all integers t. This contradicts (4.70). The contradiction we have reached proves that the case (3) does not hold.

Assume that the case (4) holds. For each natural number k, define sequences $\{\tilde{x}^{(k)}(t)\}_{t=-S_k}^{T_k-S_k}$, $\{\tilde{y}^{(k)}(t)\}_{t=-S_k}^{T_k-S_k-1}$ by

$$\tilde{x}^{(k)}(t) = x^{(k)}(t + S_k), \ t = -S_k, \ldots, T_k - S_k, \tag{4.71}$$

$$\tilde{y}^{(k)}(t) = y^{(k)}(t + S_k), \ t = -S_k, \ldots, T_k - S_k - 1.$$

Extracting a subsequence and re-indexing, we may assume without loss of generality that for each integer t, there exists

$$x_*(t) = \lim_{k\to\infty} \tilde{x}^{(k)}(t), \ y_*(t) = \lim_{k\to\infty} \tilde{y}^{(k)}(t). \tag{4.72}$$

By (4.57), (4.59), (4.61), (4.62), (4.71), and (4.72) for each integer t,

$$x_*(t) \leq Me, \tag{4.73}$$

$$\delta(x_*(t), y_*(t), x_*(t + 1)) = 0,$$

$$\max\{\|x_*(0) - \widehat{x}\|, \ \|y_*(0) - \widehat{x}\|\} \geq \epsilon.$$

Clearly $\{x_*(t), y_*(t)\}_{t=-\infty}^{\infty}$ satisfies the assumptions of Lemma 4.12 which in its turn implies that $x_*(t) = y_*(t) = \widehat{x}$ for all integers t. This contradicts (4.73). The contradiction we have reached proves that the case (4) does not hold.

Therefore in the all four cases, we have reached a contradiction which proves that Lemma 4.13 holds.

Lemma 4.14 *Let $\epsilon > 0$. Then there exists $\gamma > 0$ such that for each natural number T and each program $(\{x(t)\}_{t=0}^{T}, \{y(t)\}_{t=0}^{T-1})$ which satisfies*

$$\|x(0) - \widehat{x}\| \leq \gamma, \ \|x(T) - \widehat{x}\| \leq \gamma, \tag{4.74}$$

$$\sum_{t=0}^{T-1} w(by(t)) \geq U(x(0), x(T), 0, T) - \gamma, \tag{4.75}$$

the following inequality holds: $\sum_{t=0}^{T-1} \delta(x(t), y(t), x(t + 1)) \leq \epsilon$.

Proof Choose a positive number δ_0 such that

$$2(\|\widehat{p}\| + 1)\delta_0 < \epsilon/8, \tag{4.76}$$

$$|w(b\widehat{x}) - w(bz)| < \epsilon/8 \text{ for any } z \in R_+^n \text{ satisfying } \|z - \widehat{x}\| \leq \delta_0.$$

By Proposition 3.13, there is $\gamma \in (0, 1)$ such that $\gamma < \delta_0$ and the following property holds:

(P2) If $x, x' \in R^n_+$ satisfy $\|x - \widehat{x}\|$, $\|x' - \widehat{x}\| \le \gamma$, then there exist $\bar{x} \ge x'$, $y \in R^n$ such that

$$(x, \bar{x}) \in \Omega, \ y \in \Lambda(x, \bar{x}), \ \|y - \widehat{x}\| \le \delta_0, \ \|\bar{x} - \widehat{x}\| \le \delta_0.$$

Let T be a natural number, and let a program $(\{x(t)\}_{t=0}^T, \{y(t)\}_{t=0}^{T-1})$ satisfy (4.74) and (4.75). We construct a program $(\{\tilde{x}(t)\}_{t=0}^T, \{\tilde{y}(t)\}_{t=0}^{T-1})$. Set

$$\tilde{x}(0) = x(0). \tag{4.77}$$

There are two cases: $T = 1$ and $T \ge 2$.

Consider the case with $T = 1$. By (4.74) and (P2), there exist $\tilde{x}(1)$ and $\tilde{y}(0)$ such that

$$\tilde{x}(1) \ge x(1), \ \tilde{y}(0) \in R^n, \tag{4.78}$$

$$(\tilde{x}(0), \tilde{x}(1)) \in \Omega, \ \tilde{y}(0) \in \Lambda(\tilde{x}(0), \tilde{x}(1)), \ \|\tilde{y}(0) - \widehat{x}\| \le \delta_0, \ \|\tilde{x}(1) - \widehat{x}\| \le \delta_0.$$

By (4.8) and (4.74)–(4.78),

$$\delta(x(0), y(0), x(1)) = w(b\widehat{x}) - w(by(0)) + \widehat{p}(x(0) - x(1))$$
$$\le w(b\widehat{x}) - w(b\tilde{y}(0)) + \widehat{p}(x(0) - x(1)) + \gamma$$
$$\le w(b\widehat{x}) - w(b\tilde{y}(0)) + \|\widehat{p}\|2\gamma + \gamma < \delta_0(2\|\widehat{p}\| + 1)$$
$$+ \epsilon/8 < \epsilon.$$

Thus in the case $T = 1$, the lemma holds.

Assume that $T > 1$. By (4.74), (4.75), (4.77), and (P2), there exist $\tilde{x}(1), \tilde{y}(0)$ such that

$$\tilde{x}(1) \ge \widehat{x}, \ \tilde{y}(0) \in R^n, \tag{4.79}$$

$$(\tilde{x}(0), \tilde{x}(1)) \in \Omega, \ \tilde{y}(0) \in \Lambda(\tilde{x}(0), \tilde{x}(1)), \ \|\tilde{y}(0) - \widehat{x}\| \le \delta_0,$$

$$\|\tilde{x}(1) - \widehat{x}\| \le \delta_0.$$

If an integer t satisfies $1 < t \le T - 1$, then we put

$$\tilde{y}(t - 1) = \widehat{x}, \ \tilde{x}(t) = \widehat{x} + (1 - d)^{t-1}(\tilde{x}(1) - \widehat{x}). \tag{4.80}$$

It is clear that $(\{\tilde{x}(t)\}_{t=0}^{T-1}, \{\tilde{y}(t)\}_{t=0}^{T-2})$ is a program.

By (4.74) and (P2), there exist $z \ge x(T), \tilde{y}(T - 1) \in R^n_+$ such that

$$(\widehat{x}, z) \in \Omega, \ \tilde{y}(T - 1) \in \Lambda(\widehat{x}, z),$$

$$\|\tilde{y}(T-1) - \widehat{x}\| \le \delta_0, \ \|z - \widehat{x}\| \le \delta_0. \tag{4.81}$$

Set

$$\tilde{x}(T) = z + (1-d)^{T-1}[\tilde{x}(1) - \widehat{x}]. \tag{4.82}$$

By (4.79)–(4.82), $(\{\tilde{x}(t)\}_{t=0}^{T}, \ \{\tilde{y}(t)\}_{t=0}^{T-1})$ is a program. By (4.77), (4.79), (4.81), and (4.82),

$$\tilde{x}(0) = x(0), \ \tilde{x}(T) \ge x(T). \tag{4.83}$$

In view of (4.75) and (4.83),

$$\sum_{t=0}^{T-1} w(by(t)) \ge \sum_{t=0}^{T-1} w(b\tilde{y}(t)) - \gamma. \tag{4.84}$$

By (4.8), (4.74), (4.76), (4.79)–(4.81), and (4.84),

$$\sum_{t=0}^{T-1} \delta(x(t), y(t), x(t+1)) = \sum_{t=0}^{T-1} [-w(by(t)) + w(b\widehat{x})] + \widehat{p}(x(0) - x(T))$$

$$\le \sum_{t=0}^{T-1} [w(b\widehat{x}) - w(b\tilde{y}(t))] + \gamma + \|\widehat{p}\|(\|x(0) - \widehat{x}\| + \|x(T) - \widehat{x}\|)$$

$$\le |w(b\widehat{x}) - w(b\tilde{y}(0))| + |w(b\widehat{x}) - w(b\tilde{y}(T-1))| + 2\|\widehat{p}\|\gamma$$

$$\le \epsilon/4 + 2\|\widehat{p}\|\delta_0 < \epsilon.$$

Lemma 4.14 is proved.

4.4   Proof of Theorem 4.2

By Proposition 3.6, there is $M_1 > 0$ such that for each natural number T and each program $(\{x(t)\}_{t=0}^{T}, \ \{y(t)\}_{t=0}^{T-1})$ satisfying $x(0) \le Me$, the following inequality holds:

$$x(t) \le M_1 e \text{ for all integers } t \in [0, T]. \tag{4.85}$$

By Proposition 4.7, there exists $M_2 > 0$ such that for each $z_0 \in R_+^n$, each $z_1 \in R_+^n$ satisfying $az_1 \le \Gamma d^{-1}$, and each natural number $T > k(\Gamma)$,

$$U(z_0, z_1, 0, T) \ge Tw(b\widehat{x}) - M_2. \tag{4.86}$$

By Lemma 4.13, there is $\epsilon_1 \in (0, \epsilon)$ such that for each natural number T and each program $(\{x(t)\}_{t=0}^{T}, \{y(t)\}_{t=0}^{T-1})$ which satisfies

$$\|x(0) - \widehat{x}\| \le \epsilon_1, \ \|x(T) - \widehat{x}\| \le \epsilon_1, \tag{4.87}$$

$$\delta(x(t), y(t), x(t+1)) \le \epsilon_1, \ t = 0, \ldots, T-1, \tag{4.88}$$

the following inequality holds:

$$\|x(t) - \widehat{x}\|, \ \|y(t) - \widehat{x}\| \le \epsilon, \ t - 0, \ldots, T-1. \tag{4.89}$$

By Lemma 4.11, there exists a natural number L_0 such that for each program

$$\left(\{x(t)\}_{t=0}^{L_0}, \ \{y(t)\}_{t=0}^{L_0-1}\right)$$

satisfying

$$x(0) \le M_1 e, \tag{4.90}$$

$$\sum_{t=0}^{L_0-1} w(by(t)) \ge L_0 w(b\widehat{y}) - M_2 - M - 4\|\widehat{p}\| n M_1, \tag{4.91}$$

there is an integer $t \in [0, L_0 - 1]$ such that

$$\|y(t) - \widehat{x}\|, \ \|x(t) - \widehat{x}\| \le \epsilon_1. \tag{4.92}$$

Choose a natural number

$$L > 8L_0 + 4k(\Gamma) + (2L_0 + 1)[2 + \epsilon_1^{-1}(M + M_2 + 2M_1 n\|\widehat{p}\|)]. \tag{4.93}$$

Assume that an integer $T > L$,

$$z_0, \ z_1 \in R_+^n, \ z_0 \le Me, \ az_1 \le \Gamma d^{-1} \tag{4.94}$$

and that a program $(\{x(t)\}_{t=0}^{T}, \{y(t)\}_{t=0}^{T-1})$ satisfies

$$x(0) = z_0, \ x(T) \ge z_1, \ \sum_{t=0}^{T-1} w(by(t)) \ge U(z_0, z_1, 0, T) - M. \tag{4.95}$$

In view of (4.94) and (4.95), the relation (4.85) holds. By (4.86), (4.94), (4.95), and the inequality $T > L > k(\Gamma)$,

$$\sum_{t=0}^{T-1} w(by(t)) \geq U(z_0, z_1, 0, T) - M \geq Tw(b\widehat{x}) - M_2 - M. \tag{4.96}$$

It follows from (4.8), (4.85) and (4.96) that

$$\sum_{t=0}^{T-1} \delta(x(t), y(t), x(t+1)) = \sum_{t=0}^{T-1} [w(b\widehat{x}) - w(by(t))] + \widehat{p}(x(0) - x(T))$$

$$\leq M + M_2 + 2M_1 n \|\widehat{p}\|. \tag{4.97}$$

It follows from (4.8), (4.85), and (4.97) that for each pair of integers S_1, S_2 satisfying $0 \leq S_1 < S_2 \leq T$,

$$|\sum_{t=S_1}^{S_2-1} [w(by(t)) - w(b\widehat{x})]| \leq \sum_{t=S_1}^{S_2-1} \delta(x(t), y(t), x(t+1)) + 2nM_1 \|\widehat{p}\|$$

$$\leq M + M_2 + 4nM_1 \|\widehat{p}\|. \tag{4.98}$$

It follows from the choice of L_0 (see (4.85), (4.90)–(4.92)), (4.94), (4.95), and (4.98) that the following property holds:

(P3) For each integer S satisfying $0 \leq S \leq T - L_0$, there is an integer $t \in [S, S + L_0 - 1]$ such that

$$\|y(t) - \widehat{x}\|, \ \|x(t) - \widehat{x}\| \leq \epsilon_1.$$

Set $t_0 = 0$. By (P3), (4.93), and the relation $T > L$, there is an integer $t_1 \in [L_0, 2L_0 - 1]$ such that

$$\|y(t_1) - \widehat{x}\|, \ \|x(t_1) - \widehat{x}\| \leq \epsilon_1.$$

Assume that an integer $j \geq 1$ and we defined integers t_i, $i = 0, \ldots, j$ such that $t_j < T$,

$$t_{i+1} \in [t_i + L_0, t_i + 2L_0 - 1]$$

for $i = 0, \ldots, j - 1$ and that

$$\|y(t_i) - \widehat{x}\|, \ \|x(t_i) - \widehat{x}\| \leq \epsilon_1 \tag{4.99}$$

for $i = 1, \ldots, j$. If $t_j + 2L_0 > T$, then put $t_{j+1} = T$, and complete the construction of the sequence.

If $t_j + 2L_0 \leq T$, then by (P3) there is an integer $t_{j+1} \in [t_j + L_0, t_j + 2L_0 - 1]$ such that

$$\|x(t_{j+1}) - \widehat{x}\|, \ \|y(t_{j+1}) - \widehat{x}\| \leq \epsilon_1.$$

In such a way, we construct by induction a finite sequence of nonnegative integers $\{t_i\}_{i=0}^{k}$ such that $t_0 = 0$, $t_k = T$, $1 \leq t_{i+1} - t_i \leq 2L_0$ for $i = 0, \ldots, k-1$ and that for $i = 1, \ldots, k-1$, the inequalities (4.99) hold.

Assume that an integer i satisfies $1 \leq i \leq k-2$ and that

$$\sum_{j=t_i}^{t_{i+1}-1} \delta(x(t), y(t), x(t+1)) \leq \epsilon_1.$$

Together with (4.99) and the choice of ϵ_1 (see (4.87)–(4.89)), this implies that

$$\|x(t) - \widehat{x}\|, \ \|y(t) - \widehat{x}\| \leq \epsilon, \ t = t_i, \ldots, t_{i+1} - 1. \tag{4.100}$$

By (4.100),

$$\{t \in \{0, \ldots, T-1\} : \max\{\|x(t) - \widehat{x}\|, \ \|y(t) - \widehat{x}\|\} > \epsilon\}$$

$$\subset \{0, \ldots, t_1\} \cup \{t_{k-1}, \ldots, T\} \cup \{\{t_i, \ldots, t_{i+1}\} :$$

$$i \in \{1, \ldots, k-2\} \text{ and } \sum_{j=t_i}^{t_{i+1}-1} \delta(x(t), y(t), x(t+1)) > \epsilon_1\}.$$

This implies that

$$\mathrm{Card}\{t \in \{0, \ldots, T-1\} : \max\{\|x(t) - \widehat{x}\|, \ \|y(t) - \widehat{x}\|\} > \epsilon\}$$

$$\leq 2(2L_0 + 1) + (2L_0 + 1)\mathrm{Card}\{i \in \{1, \ldots, k-2\} :$$

$$\sum_{j=t_i}^{t_{i+1}-1} \delta(x(t), y(t), x(t+1)) > \epsilon_1\}. \tag{4.101}$$

By (4.97),

$$\mathrm{Card}\{i \in \{1, \ldots, k-2\} : \sum_{t=t_i}^{t_{i+1}-1} \delta(x(t), y(t), x(t+1)) > \epsilon_1\}$$

$$\leq \epsilon_1^{-1} \sum_{t=0}^{T-1} \delta(x(t), y(t), x(t+1)) \leq \epsilon_1^{-1}(M + M_2 + 2M_1 n \|\widehat{p}\|).$$

Together with (4.93) and (4.101), this implies that

$$\text{Card}\{t \in \{0, \ldots, T-1\} : \max\{\|x(t) - \widehat{x}\|, \ \|y(t) - \widehat{x}\|\} > \epsilon\}$$

$$\leq (2L_0 + 1)[2 + \epsilon_1^{-1}(M + M_2 + 2M_1 n\|\widehat{p}\|)] < L.$$

This completes the proof of Theorem 4.2.

4.5   Proof of Theorem 4.3

By Proposition 3.6, there is $M_1 > 0$ such that for each natural number T and each program $(\{x(t)\}_{t=0}^{T}, \ \{y(t)\}_{t=0}^{T-1})$ satisfying $x(0) \leq Me$, the following inequality holds:

$$x(t) \leq M_1 e \text{ for all integers } t \in [0, T]. \tag{4.102}$$

By Proposition 4.7, there exists $M_2 > 0$ such that for each $z_0 \in R_+^n$, each $z_1 \in R_+^n$ satisfying $az_1 \leq \Gamma d^{-1}$, and each natural number $T > k(\Gamma)$,

$$U(z_0, z_1, 0, T) \geq Tw(b\widehat{x}) - M_2. \tag{4.103}$$

By Proposition 4.8, there exists $M_3 > 0$ such that for each natural number T and each program $(\{x(t)\}_{t=0}^{T}, \ \{y(t)\}_{t=0}^{T-1})$ satisfying $x(0) \leq M_1 e$, the following inequality holds:

$$\sum_{t=0}^{T-1}[w(by(t)) - w(b\widehat{y})] \leq M_3. \tag{4.104}$$

By Lemma 4.13, there is $\epsilon_1 > 0$ such that for each natural number T and each program $(\{x(t)\}_{t=0}^{T}, \ \{y(t)\}_{t=0}^{T-1})$ which satisfies

$$\|x(0) - \widehat{x}\| \leq \epsilon_1, \ \|x(T) - \widehat{x}\| \leq \epsilon_1, \tag{4.105}$$

$$\delta(x(t), y(t), x(t+1)) \leq \epsilon_1, \ t = 0, \ldots, T-1, \tag{4.106}$$

the following inequality holds:

$$\|x(t) - \widehat{x}\|, \ \|y(t) - \widehat{x}\| \leq \epsilon, \ t = 0, \ldots, T-1. \tag{4.107}$$

By Lemma 4.11, there exists

$$\gamma \in (0, \min\{1, \epsilon, \epsilon_1\}) \tag{4.108}$$

such that for each natural number T and each program $(\{x(t)\}_{t=0}^{T}, \{y(t)\}_{t=0}^{T-1})$ which satisfies

$$\|x(0) - \widehat{x}\| \leq \gamma, \quad \|x(T) - \widehat{x}\| \leq \gamma, \tag{4.109}$$

$$\sum_{t=0}^{T-1} w(by(t)) \geq U(x(0), x(T), 0, T) - \gamma,$$

the following inequality holds:

$$\sum_{t=0}^{T-1} \delta(x(t), y(t), x(t+1)) \leq \epsilon_1. \tag{4.110}$$

By Lemma 4.11, there exists a natural number L_0 such that for each program

$$\left(\{x(t)\}_{t=0}^{L_0}, \{y(t)\}_{t=0}^{L_0-1}\right)$$

satisfying

$$x(0) \leq M_1 e, \quad \sum_{t=0}^{L_0-1} w(by(t)) \geq L_0 w(b\widehat{y}) - M_2 - M_3 - 1, \tag{4.111}$$

there is an integer $t \in [0, L_0 - 1]$ such that

$$\|y(t) - \widehat{x}\|, \quad \|x(t) - \widehat{x}\| \leq \gamma. \tag{4.112}$$

Put

$$L = L_0 + k(\Gamma). \tag{4.113}$$

Assume that an integer $T > 2L$,

$$z_0, z_1 \in R_+^n, \quad z_0 \leq Me, \quad az_1 \leq \Gamma d^{-1} \tag{4.114}$$

and that a program $(\{x(t)\}_{t=0}^{T}, \{y(t)\}_{t=0}^{T-1})$ satisfies

$$x(0) = z_0, \quad x(T) \geq z_1, \tag{4.115}$$

$$\sum_{t=0}^{T-1} w(by(t)) \geq U(z_0, z_1, 0, T) - \gamma. \tag{4.116}$$

In view of (4.114) and (4.115), the relation (4.102) holds. By (4.108), (4.113), (4.114), and (4.116),

$$\sum_{t=0}^{T-1} w(by(t)) \geq U(z_0, z_1, 0, T) - \gamma \geq Tw(b\widehat{x}) - M_2 - 1. \qquad (4.117)$$

It follows from the choice of M_3 (see (4.104)) and (4.102) that

$$\sum_{t=L_0}^{T-1} [w(by(t)) - w(b\widehat{x})] \leq M_3, \quad \sum_{t=0}^{T-L_0-1} [w(by(t)) - w(b\widehat{x})] \leq M_3. \qquad (4.118)$$

By (4.117) and (4.118),

$$\sum_{t=0}^{L_0-1} [w(by(t)) - w(b\widehat{x})] = \sum_{t=0}^{T-1}[w(by(t)) - w(b\widehat{x})] - \sum_{t=L_0}^{T-1} [w(by(t)) - w(b\widehat{x})]$$

$$\geq -M_2 - 1 - M_3, \qquad (4.119)$$

$$\sum_{t=T-L_0}^{T-1} [w(by(t)) - w(b\widehat{x})] = \sum_{t=0}^{T-1}[w(by(t)) - w(b\widehat{x})] \qquad (4.120)$$

$$- \sum_{t=0}^{T-L_0-1} [w(by(t)) - w(b\widehat{x})] \geq -M_2 - 1 - M_3.$$

It follows from (4.102), (4.119), (4.120), and the choice of L_0 that there exist integers $\tau_1 \in [0, L_0 - 1]$, $\tau_2 \in [T - L_0, T - 1]$ such that

$$\|x(\tau_i) - \widehat{x}\|, \ \|y(\tau_i) - \widehat{x}\| \leq \gamma, \ i = 1, 2. \qquad (4.121)$$

(If $\|x(0) - \widehat{x}\| \leq \gamma$, then we put $\tau_1 = 0$, and if $\|x(T) - \widehat{x}\| \leq \gamma$, then we put $\tau_2 = T$). By (4.116) and Proposition 4.10,

$$\sum_{t=\tau_1}^{\tau_2-1} w(by(t)) \geq U(x(\tau_1), x(\tau_2), \tau_1, \tau_2) - \gamma. \qquad (4.122)$$

In view of (4.121), (4.122), and the choice of γ (see (4.109), (4.110))

$$\sum_{t=\tau_1}^{\tau_2-1} \delta(x(t), y(t), x(t+1)) \leq \epsilon_1. \qquad (4.123)$$

It follows from (4.108), (4.121), (4.123), and the choice of ϵ_1 (see (4.105)–(4.107)) that

$$\|x(t) - \widehat{x}\|, \ \|y(t) - \widehat{x}\| \leq \epsilon, \ t = \tau_1, \ldots, \tau_2 - 1.$$

Theorem 4.3 is proved.

4.6 Extensions of Theorem 4.3

In the sequel we use the following helpful result.

Lemma 4.15 *Let a number* $M_0 > \max\{(a_i d)^{-1} : i = 1, \ldots, n\}$, $(x, x') \in \Omega$ *and* $x \leq M_0 e$. *Then* $x' \leq M_0 e$.

Proof Clearly, $a(x' - (1 - d)x) \leq 1$, and for each $i = 1, \ldots, n$,

$$x_i' \leq a_i^{-1} + (1 - d)x_i \leq d(a_i d)^{-1} + (1 - d)M_0 \leq dM_0 + (1 - d)M_0 = M_0.$$

Lemma 4.15 is proved.

In this chapter we prove the following three turnpike results obtained in [108].

Theorem 4.16 *Suppose that for each good program* $\{u(t), v(t)\}_{t=0}^{\infty}$,

$$\lim_{t \to \infty} (u(t), v(t)) = (\widehat{x}, \widehat{x}).$$

Let M, ϵ *be positive numbers and* $\Gamma \in (0, 1)$. *Then there exist a natural number* L *and a positive number* γ *such that for each integer* $T > 2L$, *each* $z_0, z_1 \in R_+^n$ *satisfying* $z_0 \leq Me$ *and* $az_1 \leq \Gamma d^{-1}$, *and each program* $(\{x(t)\}_{t=0}^{T}, \{y(t)\}_{t=0}^{T-1})$ *which satisfies*

$$x(0) = z_0, \ x(T) \geq z_1,$$

$$\sum_{t=\tau}^{\tau+L-1} w(by(t)) \geq U(x(\tau), x(\tau + L), 0, L) - \gamma \ \text{for all } \tau \in \{0, \ldots, T - L\}$$

and

$$\sum_{t=T-L}^{T-1} w(by(t)) \geq U(x(T - L), z_1, 0, L) - \gamma,$$

there are integers τ_1, τ_2 *such that*

$$\tau_1 \in [0, L], \quad \tau_2 \in [T - L, T],$$

$$\|x(t) - \widehat{x}\|, \ \|y(t) - \widehat{x}\| \le \epsilon \text{ for all } t = \tau_1, \ldots, \tau_2 - 1.$$

Moreover if $\|x(0) - \widehat{x}\| \le \gamma$, *then* $\tau_1 = 0$, *and if* $\|x(T) - \widehat{x}\| \le \gamma$, *then* $\tau_2 = T$.

Theorem 4.17 *Suppose that for each good program* $\{u(t), v(t)\}_{t=0}^{\infty}$,

$$\lim_{t \to \infty} (u(t), v(t)) = (\widehat{x}, \widehat{x}).$$

Let M, ϵ *be positive numbers. Then there exist a natural number* L *and a positive number* γ *such that for each integer* $T > 2L$, *each* $z_0 \in R_+^n$ *satisfying* $z_0 \le Me$, *and each program* $(\{x(t)\}_{t=0}^{T}, \{y(t)\}_{t=0}^{T-1})$ *which satisfies*

$$x(0) = z_0,$$

$$\sum_{t=\tau}^{\tau+L-1} w(by(t)) \ge U(x(\tau), x(\tau + L), 0, L) - \gamma \text{ for all } \tau \in \{0, \ldots, T - L\}$$

$$(4.124)$$

and

$$\sum_{t=T-L}^{T-1} w(by(t)) \ge U(x(T - L), L) - \gamma, \quad (4.125)$$

there are integers τ_1, τ_2 *such that*

$$\tau_1 \in [0, L], \quad \tau_2 \in [T - L, T],$$

$$\|x(t) - \widehat{x}\|, \ \|y(t) - \widehat{x}\| \le \epsilon \text{ for all } t = \tau_1, \ldots, \tau_2 - 1.$$

Moreover if $\|x(0) - \widehat{x}\| \le \gamma$, *then* $\tau_1 = 0$, *and if* $\|x(T) - \widehat{x}\| \le \gamma$, *then* $\tau_2 = T$.

Theorem 4.17 implies the following result.

Theorem 4.18 *Suppose that for each good program* $\{u(t), v(t)\}_{t=0}^{\infty}$,

$$\lim_{t \to \infty} (u(t), v(t)) = (\widehat{x}, \widehat{x}).$$

Let M, ϵ *be positive numbers. Then there exist a natural number* L *and a positive number* γ *such that for each integer* $T > 2L$, *each* $z_0 \in R_+^n$ *satisfying* $z_0 \le Me$, *and each program* $(\{x(t)\}_{t=0}^{T}, \{y(t)\}_{t=0}^{T-1})$ *which satisfies*

$$x(0) = z_0, \quad (4.126)$$

$$\sum_{t=0}^{T-1} w(by(t)) \geq U(z_0, T) - \gamma, \tag{4.127}$$

there are integers τ_1, τ_2 such that

$$\tau_1 \in [0, L], \quad \tau_2 \in [T - L, T], \tag{4.128}$$

$$\|x(t) - \widehat{x}\|, \ \|y(t) - \widehat{x}\| \leq \epsilon \text{ for all } t = \tau_1, \ldots, \tau_2 - 1. \tag{4.129}$$

Moreover if $\|x(0) - \widehat{x}\| \leq \gamma$, then $\tau_1 = 0$ and if $\|x(T) - \widehat{x}\| \leq \gamma$ then $\tau_1 = T$.

4.7 Proof of Theorem 4.16

We may assume without loss of generality that

$$M > \max\{(a_i d)^{-1} : i = 1, \ldots, n\}. \tag{4.130}$$

Since $a\widehat{x} = a_\sigma (1 + da_\sigma)^{-1}$, we may assume without loss of generality that

$$a\widehat{x} < \Gamma d^{-1}, \tag{4.131}$$

$$\sup\{ay : y \in R^n \text{ and } \|y - \widehat{x}\| \leq \epsilon\} < \Gamma d^{-1}. \tag{4.132}$$

Theorem 4.3 implies that there are an integer $L_0 \geq 1$ and a number $\gamma > 0$ such that the following property holds:

(P4) for every natural number $T > 2L_0$, every pair of points $z_0, z_1 \in R_+^n$ which satisfy $z_0 \leq Me$, $az_1 \leq \Gamma d^{-1}$, and every program $(\{x(t)\}_{t=0}^T, \{y(t)\}_{t=0}^{T-1})$ satisfying

$$x(0) = z_0, \ x(T) \geq z_1, \tag{4.133}$$

$$\sum_{t=0}^{T-1} w(by(t)) \geq U(z_0, z_1, 0, T) - \gamma, \tag{4.134}$$

there are integers τ_1, τ_2 such that

$$\tau_1 \in [0, L_0], \quad \tau_2 \in [T - L_0, T], \tag{4.135}$$

$$\|x(t) - \widehat{x}\|, \ \|y(t) - \widehat{x}\| \leq \epsilon \text{ for all } t = \tau_1, \ldots, \tau_2 - 1, \tag{4.136}$$

if $\|x(0) - \widehat{x}\| \leq \gamma$, then $\tau_1 = 0$ and if $\|x(T) - \widehat{x}\| \leq \gamma$, then $\tau_2 = T$. \tag{4.137}

Fix

$$L = 3L_0 + 1.$$
(4.138)

Assume that a pair of points $z_0, z_1 \in R_+^n$ satisfy

$$z_0 \le Me, \ az_1 \le \Gamma d^{-1},$$
(4.139)

and let $T > 2L$ be a natural number. Assume that a program

$$\left(\{x(t)\}_{t=0}^T, \{y(t)\}_{t=0}^{T-1} \right)$$

satisfies (4.133); that for every integer $\tau \in \{0, \dots, T - L\}$, we have

$$\sum_{t=\tau}^{\tau+L-1} w(by(t)) \ge U(x(\tau), x(\tau + L), 0, L) - \gamma;$$
(4.140)

and that

$$\sum_{t=T-L}^{T-1} w(by(t)) \ge U(x(T - L), z_1, 0, L) - \gamma.$$
(4.141)

We show that there exist integers $\tau_1 \in [0, L]$ and $\tau_2 \in [T - l, T]$ for which (4.136) and (4.137) are valid.

Lemma 4.15, (4.130), and (4.139) imply that

$$x(t) \le Me, \ t = 0, \dots, T.$$
(4.142)

Consider the program $(\{x(t)\}_{t=T-L}^T, \{y(t)\}_{t=T-L}^{T-1})$. Property (P4), applied to this program, (4.133), (4.138), (4.139), (4.141), and (4.142) imply that

$$\|x(t) - \widehat{x}\|, \ \|y(t) - \widehat{x}\| \le \epsilon, \ t = T - L + L_0, \dots, T - L_0 - 1,$$
(4.143)

if $\|x(T) - \widehat{x}\| \le \gamma$, then

$$\|x(t) - \widehat{x}\|, \ \|y(t) - \widehat{x}\| \le \epsilon, \ t = T - L + L_0, \dots, T - 1.$$
(4.144)

If

$$\|x(t) - \widehat{x}\|, \ \|y(t) - \widehat{x}\| \le \epsilon, \ t = 0, \dots, T - L + L_0,$$

then the assertion of the theorem holds. Hence we may assume that there exists an integer $t_0 \in \{0, \dots, T - L + L_0 - 1\}$ for which

$$\max\{\|x(t_0) - \widehat{x}\|, \ \|y(t_0) - \widehat{x}\|\} > \epsilon. \tag{4.145}$$

We may assume without loss of generality that

$$\|x(t) - \widehat{x}\|, \ \|y(t) - \widehat{x}\| \leq \epsilon$$

for all integers t satisfying $t_0 < t < T - L + L_0$. $\tag{4.146}$

We show that

$$t_0 \leq 3L_0. \tag{4.147}$$

Assume the contrary and consider the program

$$\left(\{x(t)\}_{t=t_0-2L_0}^{t_0+L-2L_0}, \{y(t)\}_{t=t_0-2L_0}^{t_0+L-2L_0-1}\right).$$

It follows from (4.138), (4.143), and (4.146) that

$$\|x(t_0 + L - 2L_0) - \widehat{x}\| \leq \epsilon. \tag{4.148}$$

By (4.132) and (4.148), we have

$$ax(t_0 + L - 2L_0) \leq \Gamma d^{-1}. \tag{4.149}$$

By (4.138), (4.140), (4.142), (4.144), (4.149), and property (P4) applied to the program

$$\left(\{x(t)\}_{t=t_0-2L_0}^{t_0+L-2L_0}, \{y(t)\}_{t=t_0-2L_0}^{t_0+L-2L_0-1}\right),$$

we have

$$\|x(t) - \widehat{x}\|, \ \|y(t) - \widehat{x}\| \leq \epsilon \text{ for all } t = t_0 - 2L_0 + L_0, \dots, t_0 + L - 3L_0 - 1.$$

In view of the relation above, (4.138), (4.143), (4.144), and (4.146), we have

$$\|x(t) - \widehat{x}\|, \ \|y(t) - \widehat{x}\| \leq \epsilon \text{ for all integers } t \text{ satisfying } t_0 - L_0 \leq t \leq T - L_0 - 1.$$

This contradicts inequality (4.145). The contradiction we have reached proves that $t_0 \leq 3L_0$. Combined with relations (4.143), (4.144), and (4.146), this implies that

$$\|x(t) - \widehat{x}\|, \ \|y(t) - \widehat{x}\| \leq \epsilon, \ t = 3L_0 + 1, \dots, T - L_0 - 1, \tag{4.150}$$

if $\|x(T) - \widehat{x}\| \leq \gamma$, then

$$\|x(t) - \widehat{x}\|, \|y(t) - \widehat{x}\| \leq \epsilon, \ t = 3L_0 + 1, \ldots, T - 1. \tag{4.151}$$

Assume that

$$\|x(0) - \widehat{x}\| \leq \gamma. \tag{4.152}$$

Relations (4.132), (4.138), 4.149, and (4.151) imply that

$$\|x(L + L_0) - \widehat{x}\| \leq \epsilon, \ ax(L + L_0) \leq \Gamma d^{-1},$$

$$\|x(L) - \widehat{x}\| \leq \epsilon, \ ax(L) \leq \Gamma d^{-1}. \tag{4.153}$$

It follows from (4.138), (4.140), (4.142), (4.152), (4.153), and property (P4) applied to the programs $(\{x(t)\}_{t=0}^{L}, \{y(t)\}_{t=0}^{L-1})$ and $(\{x(t)\}_{t=L_0}^{L+L_0}, \{y(t)\}_{t=L_0}^{L_0+L-1})$ that

$$\|x(t) - \widehat{x}\|, \|y(t) - \widehat{x}\| \leq \epsilon, \ t = 0, \ldots, L - L_0 - 1,$$

$$\|x(t) - \widehat{x}\|, \|y(t) - \widehat{x}\| \leq \epsilon, \ t = 2L_0, \ldots, L - 1.$$

Together with relations (4.150) and (4.151), this completes the proof of Theorem 4.16.

4.8 Stability Results

For every positive number M and every function $\phi : R_+^n \to R^1$, define

$$\|\phi\|_M = \sup\{|\phi(z)| : \ z \in R^n \text{ and } 0 \leq z \leq Me\}.$$

Let integers T_1, T_2 satisfy $0 \leq T_1 < T_2$ and $w_i : R_+^n \to R^1, i = T_1, \ldots, T_2 - 1$ be bounded on bounded subsets of R_+^n functions. For every pair of points $z_0, z_1 \in R_+^n$, define

$$U\left(\{w_t\}_{t=T_1}^{T_2-1}, z_0, z_1\right) = \sup\left\{ \sum_{t=T_1}^{T_2-1} w_t(y(t)) : \right.$$

$$\left(\{x(t)\}_{t=T_1}^{T_2}, \{y(t)\}_{t=T_1}^{T_2-1}\right) \text{ is a program such that } x(T_1) = z_0, \ x(T_2) \geq z_1 \bigg\},$$

$$U\left(\{w_t\}_{t=T_1}^{T_2-1}, z_0\right) = \sup \left\{ \sum_{t=T_1}^{T_2-1} w_t(y(t)) : \right.$$

$$\left. \left(\{x(t)\}_{t=T_1}^{T_2}, \{y(t)\}_{t=T_1}^{T_2-1}\right) \text{ is a program such that } x(T_1) = z_0 \right\}.$$

(Here we assume that supremum over empty set is $-\infty$.) It is not difficult to see that the following result holds.

Lemma 4.19 *Let integers T_1, T_2 satisfy $0 \leq T_1 < T_2$ and $w_i : R_+^n \to R^1$, $i = T_1, \ldots, T_2 - 1$ be bounded on bounded subsets of R_+^n upper semicontinuous functions. Then the following assertions hold.*

1. *For every point $z_0 \in R_+^n$, there exists a program $(\{x(t)\}_{t=T_1}^{T_2}, \{y(t)\}_{t=T_1}^{T_2-1})$ such that*

$$x(T_1) = z_0, \quad \sum_{t=T_1}^{T_2-1} w_t(y(t)) = U(\{w_t\}_{t=T_1}^{T_2-1}, z_0).$$

2. *For every pair of points $z_0, z_1 \in R_+^n$ such that $U(\{w_t\}_{t=T_1}^{T_2-1}, z_0, z_1)$ is finite, there exists a program $(\{x(t)\}_{t=T_1}^{T_2}, \{y(t)\}_{t=T_1}^{T_2-1})$ such that $x(T_1) = z_0$, $x(T_2) \geq z_1$ and*

$$\sum_{t=T_1}^{T_2-1} w_t(y(t)) = U(\{w_t\}_{t=T_1}^{T_2-1}, z_0, z_1).$$

The following stability results were obtained in [108].

Theorem 4.20 *Suppose that for each good program $\{u(t), v(t)\}_{t=0}^{\infty}$,*

$$\lim_{t \to \infty} (u(t), v(t)) = (\widehat{x}, \widehat{x}).$$

Let $M > \max\{(a_i d)^{-1} : i = 1, \ldots, n\}$, $\epsilon > 0$ and $\Gamma \in (0, 1)$. Then there exist a natural number L and a positive number $\tilde{\gamma}$ such that for each integer $T > 2L$, each $z_0, z_1 \in R_+^n$ satisfying $z_0 \leq Me$ and $az_1 \leq \Gamma d^{-1}$, each finite sequence of functions $w_i : R_+^n \to R^1$, $i = 0, \ldots, T - 1$ which are bounded on bounded subsets of R_+^n and such that

$$\|w_i - w(b(\cdot))\|_M \leq \tilde{\gamma}$$

for every integer $i \in \{0, \ldots, T - 1\}$ and every program $(\{x(t)\}_{t=0}^{T}, \{y(t)\}_{t=0}^{T-1})$ such that

$$x(0) = z_0, \quad x(T) \geq z_1,$$

$$\sum_{t=\tau}^{\tau+L-1} w_t(y(t)) \geq U(\{w_t\}_{t=\tau}^{\tau+L-1}, x(\tau), x(\tau+L)) - \tilde{\gamma}$$

for every $\tau \in \{0, \ldots, T - L\}$ and

$$\sum_{t=T-L}^{T-1} w_t(y(t)) \geq U(\{w_t\}_{t=T-L}^{T-1}, x(T-L), z_1) - \tilde{\gamma},$$

there exist integers τ_1, τ_2 such that

$$\tau_1 \in [0, L], \quad \tau_2 \in [T - L, T],$$

$$\|x(t) - \widehat{x}\|, \ \|y(t) - \widehat{x}\| \leq \epsilon \ \text{for all } t = \tau_1, \ldots, \tau_2 - 1.$$

Moreover if $|x(0) - \widehat{x}| \leq \tilde{\gamma}$, then $\tau_1 = 0$, and if $\|x(T) - \widehat{x}\| \leq \tilde{\gamma}$, then $\tau_2 = T$.

Proof Theorem 4.20 follows easily from Theorem 4.16. Namely, let a natural number L and $\gamma > 0$ be as guaranteed by Theorem 4.16. Set

$$\tilde{\gamma} = \gamma(4^{-1}(L+1))^{-1}.$$

Now it easy to see that the assertion of Theorem 4.20 holds.

Theorem 4.21 *Suppose that for each good program $\{u(t), v(t)\}_{t=0}^{\infty}$,*

$$\lim_{t \to \infty} (u(t), v(t)) = (\widehat{x}, \widehat{x}).$$

Let $M > \max\{(a_i d)^{-1} : i = 1, \ldots, n\}$ and $\epsilon > 0$. Then there exist a natural number L and a positive number $\tilde{\gamma}$ such that for each integer $T > 2L$, each $z_0 \in R_+^n$ satisfying $z_0 \leq Me$, each finite sequence of functions $w_i : R_+^n \to R^1$, $i = 0, \ldots, T - 1$ which are bounded on bounded subsets of R_+^n and such that

$$\|w_i - w(b(\cdot))\|_M \leq \tilde{\gamma}$$

for each $i \in \{0, \ldots, T-1\}$ and each program $(\{x(t)\}_{t=0}^{T}, \{y(t)\}_{t=0}^{T-1})$ which satisfies

$$x(0) = z_0,$$

$$\sum_{t=\tau}^{\tau+L-1} w_t(y(t)) \geq U(\{w_t\}_{t=\tau}^{\tau+L-1}, x(\tau), x(\tau+L)) - \tilde{\gamma},$$

for each integer $\tau \in \{0, \ldots, T - L\}$ and

$$\sum_{t=T-L}^{T-1} w_t(y(t)) \geq U(\{w_t\}_{t=T-L}^{T-1}, x(T-L)) - \tilde{\gamma}$$

there are integers τ_1, τ_2 such that

$$\tau_1 \in [0, L], \quad \tau_2 \in [T - L, T],$$

$$\|x(t) - \widehat{x}\|, \ \|y(t) - \widehat{x}\| \leq \epsilon \ for \ all \ t = \tau_1, \ldots, \tau_2 - 1.$$

Moreover if $\|x(0) - \widehat{x}\| \leq \gamma$, then $\tau_1 = 0$, and if $\|x(T) - \widehat{x}\| \leq \gamma$, then $\tau_2 = T$.

Proof Let a natural number L and $\gamma > 0$ be as guaranteed by Theorem 4.17. Set $\tilde{\gamma} = 4^{-1}\gamma(L + 1)^{-1}$. It is easy now to see that Theorem 4.21 holds.

Theorem 4.20 implies the following result.

Theorem 4.22 *Suppose that for each good program $\{u(t), v(t)\}_{t=0}^{\infty}$,*

$$\lim_{t \to \infty} (u(t), v(t)) = (\widehat{x}, \widehat{x}).$$

Let $M > \max\{(a_i d)^{-1} : i = 1, \ldots, n\}$, $\epsilon > 0$ and $\Gamma \in (0, 1)$. Then there exist a natural number L, a positive number γ and $\lambda > 1$ such that for each integer $T > 2L$, each $z_0, z_1 \in R_+^n$ satisfying $z_0 \leq Me$ and $az_1 \leq \Gamma d^{-1}$, each finite sequence of functions $w_i : R_+^n \to R^1$, $i = 0, \ldots, T - 1$ which are bounded on bounded subsets of R_+^n and such that $\|w_i - w(b(\cdot))\|_M \leq \gamma$ for each $i \in \{0, \ldots, T - 1\}$, each sequence $\{\alpha_i\}_{i=0}^{T-1} \subset (0, 1]$ such that for each $i, j \in \{0, \ldots, T - 1\}$ satisfying $|j - i| \leq L$ the inequality $\alpha_i \alpha_j^{-1} \leq \lambda$ holds and each program $(\{x(t)\}_{t=0}^{T}, \{y(t)\}_{t=0}^{T-1})$ such that

$$x(0) = z_0, \ x(T) \geq z_1,$$

$$\sum_{t=\tau}^{\tau+L-1} \alpha_t w_t(y(t)) \geq U(\{\alpha_t w_t\}_{t=\tau}^{\tau+L-1}, x(\tau), x(\tau + L)) - \gamma \alpha_\tau$$

for each integer $\tau \in \{0, \ldots, T - L\}$ and

$$\sum_{t=T-L}^{T-1} \alpha_t w_t(y(t)) \geq U(\{\alpha_t w_t\}_{t=T-L}^{T-1}, z_1) - \gamma \alpha_{T-L}$$

there are integers τ_1, τ_2 such that

$$\tau_1 \in [0, L], \quad \tau_2 \in [T - L, T],$$

$$\|x(t) - \widehat{x}\|, \ \|y(t) - \widehat{x}\| \leq \epsilon \ for \ all \ t = \tau_1, \ldots, \tau_2 - 1.$$

Moreover if $\|x(0) - \widehat{x}\| \leq \gamma$, *then* $\tau_1 = 0$, *and if* $\|x(0) - \widehat{x}\| \leq \gamma$, *then* $\tau_2 = T$.

Theorem 4.22, applied with $z_1 = 0$, implies the following result.

Theorem 4.23 *Suppose that for each good program* $\{u(t), v(t)\}_{t=0}^{\infty}$,

$$\lim_{t \to \infty} (u(t), v(t)) = (\widehat{x}, \widehat{x}).$$

Let $M > \max\{(a_i d)^{-1} : i = 1, \ldots, n\}$ *and* $\epsilon > 0$. *Then there exist a natural number* L, *a positive number* γ *and* $\lambda > 1$ *such that for each integer* $T > 2L$, *each* $z_0 \in R_+^n$ *satisfying* $z_0 \leq Me$, *each finite sequence of functions* $w_i : R_+^n \to R^1$, $i = 0, \ldots, T - 1$ *which are bounded on bounded subsets of* R_+^n *and such that*

$$\|w_i - w(b(\cdot))\|_M \leq \gamma$$

for each $i \in \{0, \ldots, T - 1\}$, *each sequence* $\{\alpha_i\}_{i=0}^{T-1} \subset (0, 1]$ *such that for each* $i, j \in \{0, \ldots, T - 1\}$ *satisfying* $|j - i| \leq L$ *the inequality* $\alpha_i \alpha_j^{-1} \leq \lambda$ *holds and each program* $(\{x(t)\}_{t=0}^{T}, \{y(t)\}_{t=0}^{T-1})$ *such that*

$$x(0) = z_0,$$

$$\sum_{t=\tau}^{\tau+L-1} \alpha_t w_t(y(t)) \geq U(\{\alpha_t w_t\}_{t=\tau}^{\tau+L-1}, x(\tau), x(\tau + L)) - \gamma \alpha_\tau$$

for each integer $\tau \in \{0, \ldots, T - L\}$ *and*

$$\sum_{t=T-L}^{T-1} \alpha_t w_t(y(t)) \geq U(\{\alpha_t w_t\}_{t=T-L}^{T-1}, x(T - L)) - \gamma \alpha_{T-L}$$

there are integers τ_1, τ_2 *such that*

$$\tau_1 \in [0, L], \quad \tau_2 \in [T - L, T],$$

$$\|x(t) - \widehat{x}\|, \ \|y(t) - \widehat{x}\| \leq \epsilon \text{ for all } t = \tau_1, \ldots, \tau_2 - 1.$$

Moreover if $\|x(0) - \widehat{x}\| \leq \gamma$, *then* $\tau_1 = 0$, *and if* $\|x(T) - \widehat{x}\| \leq \gamma$, *then* $\tau_2 = T$.

Theorem 4.23 implies the following result.

Theorem 4.24 *Suppose that for each good program* $\{u(t), v(t)\}_{t=0}^{\infty}$,

$$\lim_{t \to \infty} (u(t), v(t)) = (\widehat{x}, \widehat{x}).$$

Let $M > \max\{(a_i d)^{-1} : i = 1, \ldots, n\}$ and $\epsilon > 0$. Then there exist a natural number L, a positive number γ and $\lambda > 1$ such that for each integer $T > 2L$, each $z_0 \in R_+^n$ satisfying $z_0 \leq Me$, each finite sequence of upper semicontinuous functions $w_i : R_+^n \to R^1$, $i = 0, \ldots, T-1$ which are bounded on bounded subsets of R_+^n and such that

$$\|w_i - w(b(\cdot))\|_M \leq \gamma$$

for each $i \in \{0, \ldots, T-1\}$, each sequence $\{\alpha_i\}_{i=0}^{T-1} \subset (0, 1]$ such that for each $i, j \in \{0, \ldots, T-1\}$ satisfying $|j - i| \leq L$ the inequality $\alpha_i \alpha_j^{-1} \leq \lambda$ holds and each program $(\{x(t)\}_{t=0}^T, \{y(t)\}_{t=0}^{T-1})$ such that

$$x(0) = z_0$$

and

$$\sum_{t=0}^{T-1} \alpha_t w_t(y(t)) = U(\{\alpha_t w_t\}_{t=0}^{T-1}, x(0))$$

there are integers τ_1, τ_2 such that

$$\tau_1 \in [0, L], \quad \tau_2 \in [T - L, T],$$

$$\|x(t) - \widehat{x}\|, \quad \|y(t) - \widehat{x}\| \leq \epsilon \text{ for all } t = \tau_1, \ldots, \tau_2 - 1.$$

Moreover if $\|x(0) - \widehat{x}\| \leq \gamma$, then $\tau_1 = 0$, and if $\|x(T) - \widehat{x}\| \leq \gamma$, then $\tau_2 = T$.

Theorem 4.20 implies the following result.

Theorem 4.25 *Suppose that for each good program* $\{u(t), v(t)\}_{t=0}^\infty$,

$$\lim_{t \to \infty} (u(t), v(t)) = (\widehat{x}, \widehat{x}).$$

Let $M > \max\{(a_i d)^{-1} : i = 1, \ldots, n\}$, $\epsilon > 0$ and $\Gamma \in (0, 1)$. Then there exist a natural number L and a positive number $\tilde{\gamma}$ such that for each integer $T > 2L$, each $z_0, z_1 \in R_+^n$ satisfying $z_0 \leq Me$ and $a z_1 \leq \Gamma d^{-1}$, each finite sequence of functions $w_i : R_+^n \to R^1$, $i = 0, \ldots, T-1$ which are bounded on bounded subsets of R_+^n and such that

$$\|w_i - w(b(\cdot))\|_M \leq \tilde{\gamma}$$

for each $i \in \{0, \ldots, T-1\}$ and each program $(\{x(t)\}_{t=0}^T, \{y(t)\}_{t=0}^{T-1})$ such that

$$x(0) = z_0, \quad x(T) \geq z_1,$$

$$\sum_{t=0}^{T-1} w_t(y(t)) \geq U(\{w_t\}_{t=0}^{T-1}, z_0, z_1) - \tilde{\gamma}$$

there are integers τ_1, τ_2 such that

$$\tau_1 \in [0, L], \quad \tau_2 \in [T - L, T],$$

$$\|x(t) - \widehat{x}\|, \quad \|y(t) - \widehat{x}\| \leq \epsilon \ for \ all \ t = \tau_1, \ldots, \tau_2 - 1.$$

Moreover if $\|x(0) - \widehat{x}\| \leq \tilde{\gamma}$, *then* $\tau_1 = 0$, *and if* $\|x(T) - \widehat{x}\| \leq \tilde{\gamma}$, *then* $\tau_2 = T$.

Theorem 4.25, applied with $z_1 = 0$, implies the following result.

Theorem 4.26 *Suppose that for each good program* $\{u(t), v(t)\}_{t=0}^{\infty}$,

$$\lim_{t\to\infty} (u(t), v(t)) = (\widehat{x}, \widehat{x}).$$

Let $M > \max\{(a_i d)^{-1} : i = 1, \ldots, n\}$ *and* $\epsilon > 0$. *Then there exist a natural number* L *and a positive number* $\tilde{\gamma}$ *such that for each integer* $T > 2L$, *each* $z_0 \in R_+^n$ *satisfying* $z_0 \leq Me$, *each finite sequence of functions* $w_i : R_+^n \to R^1$, $i = 0, \ldots, T - 1$ *which are bounded on bounded subsets of* R_+^n *and such that*

$$\|w_i - w(b(\cdot))\|_M \leq \tilde{\gamma}$$

for each $i \in \{0, \ldots, T-1\}$ *and each program* $(\{x(t)\}_{t=0}^{T}, \{y(t)\}_{t=0}^{T-1})$ *which satisfies*

$$x(0) = z_0,$$

$$\sum_{t=0}^{T-1} w_t(y(t)) \geq U(\{w_t\}_{t=0}^{T-1}, z_0) - \tilde{\gamma}$$

there are integers τ_1, τ_2 *such that*

$$\tau_1 \in [0, L], \quad \tau_2 \in [T - L, T],$$

$$\|x(t) - \widehat{x}\|, \quad \|y(t) - \widehat{x}\| \leq \epsilon \ for \ all \ t = \tau_1, \ldots, \tau_2 - 1.$$

Moreover if $\|x(0) - \widehat{x}\| \leq \tilde{\gamma}$, *then* $\tau_1 = 0$, *and if* $\|x(T) - \widehat{x}\| \leq \tilde{\gamma}$, *then* $\tau_2 = T$.

The following results were obtained in [114].

Theorem 4.27 *Suppose that for each good program* $\{u(t), v(t)\}_{t=0}^{\infty}$,

$$\lim_{t\to\infty} (u(t), v(t)) = (\widehat{x}, \widehat{x}).$$

Let $M > \max\{(a_i d)^{-1} : i = 1, \ldots, n\}$, $M_0 > 0$, $\epsilon > 0$ and $\Gamma \in (0, 1)$. Then there exist a natural number L and a positive number $\tilde{\gamma}$ such that for each integer $T > L$, each $z_0, z_1 \in R_+^n$ satisfying $z_0 \leq Me$ and $az_1 \leq \Gamma d^{-1}$, each finite sequence of functions $w_i : R_+^n \rightarrow R^1$, $i = 0, \ldots, T - 1$ which are bounded on bounded subsets of R_+^n and such that

$$\|w_i - w(b(\cdot))\|_M \leq \tilde{\gamma}$$

for each $i \in \{0, \ldots, T - 1\}$ and each program $(\{x(t)\}_{t=0}^T, \{y(t)\}_{t=0}^{T-1})$ such that

$$x(0) = z_0, \ x(T) \geq z_1,$$

$$\sum_{t=0}^{T-1} w_t(y(t)) \geq U(\{w_t\}_{t=0}^{T-1}, z_0, z_1) - M_0$$

the following inequality holds:

$$Card(\{t \in \{0, \ldots, T - 1\} : \max\{\|x(t) - \widehat{x}\|, \ \|y(t) - \widehat{x}\|\} > \epsilon\}) \leq L.$$

Theorem 4.27, applied with $z_1 = 0$, implies the following result.

Theorem 4.28 Suppose that for each good program $\{u(t), v(t)\}_{t=0}^\infty$,

$$\lim_{t \to \infty} (u(t), v(t)) = (\widehat{x}, \widehat{x}).$$

Let $M > \max\{(a_i d)^{-1} : i = 1, \ldots, n\}$, $M_0 > 0$ and $\epsilon > 0$. Then there exist a natural number L and a positive number $\tilde{\gamma}$ such that for each integer $T > L$, each $z_0 \in R_+^n$ satisfying $z_0 \leq Me$, each finite sequence of functions $w_i : R_+^n \rightarrow R^1$, $i = 0, \ldots, T - 1$ which are bounded on bounded subsets of R_+^n and such that

$$\|w_i - w(b(\cdot))\|_M \leq \tilde{\gamma}$$

for each $i \in \{0, \ldots, T - 1\}$ and each program $(\{x(t)\}_{t=0}^T, \{y(t)\}_{t=0}^{T-1})$ which satisfies

$$x(0) = z_0,$$

$$\sum_{t=0}^{T-1} w_t(y(t)) \geq U(\{w_t\}_{t=0}^{T-1}, z_0) - M_0$$

the following inequality holds:

$$Card(\{t \in \{0, \ldots, T - 1\} : \max\{\|x(t) - \widehat{x}\|, \ \|y(t) - \widehat{x}\|\} > \epsilon\}) \leq L.$$

4.9   Proof of Theorem 4.27

Clearly,

$$a\widehat{x} = a_\sigma (1 + da_\sigma)^{-1} < d^{-1}. \tag{4.154}$$

In view of (4.154), we may assume without any loss of generality that

$$a\widehat{x} + \epsilon \|a\| < \Gamma d^{-1}. \tag{4.155}$$

By Theorem 4.25, there exist a natural number L_0 and a positive number $\tilde{\gamma}_0 < 1$ such that the following property holds:

(Pi) for each integer $T > 2L_0$, each finite sequence of functions $w_i : R_+^n \to R^1$, $i = 0, \dots, T - 1$ which are bounded on bounded subsets of R_+^n and such that

$$\|w_i - w(b(\cdot))\|_M \le \tilde{\gamma}_0$$

for each $i \in \{0, \dots, T - 1\}$, each $z_0, z_1 \in R_+^n$ satisfying $z_0 \le Me$ and $az_1 \le \Gamma d^{-1}$ and each program $(\{x(t)\}_{t=0}^T, \{y(t)\}_{t=0}^{T-1})$ which satisfies

$$x(0) = z_0, \ x(T) \ge z_1,$$

$$\sum_{t=0}^{T-1} w_t(y(t)) \ge U(\{w_t\}_{t=0}^{T-1}, z_0, z_1) - \tilde{\gamma}_0, \tag{4.156}$$

we have

$$\|x(t) - \widehat{x}\|, \ \|y(t) - \widehat{x}\| \le \epsilon \text{ for all } t = L_0, \dots, T - L_0 - 1.$$

By Theorem 4.2, there exists a natural number $L_1 \ge 2$ such that the following property holds:

(Pii) for each integer $T > L_1$, each $z_0, z_1 \in R_+^n$ satisfying $z_0 \le Me$ and $az_1 \le \Gamma d^{-1}$, and each program $(\{x(t)\}_{t=0}^T, \{y(t)\}_{t=0}^{T-1})$ which satisfies

$$x(0) = z_0, \ x(T) \ge z_1, \ \sum_{t=0}^{T-1} w(by(t)) \ge U(z_0, z_1, 0, T) - M_0 - 4, \tag{4.157}$$

we have

$$\text{Card}(\{t \in \{0, \dots, T - 1\} : \ \max\{\|x(t) - \widehat{x}\|, \ \|y(t) - \widehat{x}\|\} > \epsilon\}) \le L_1.$$

Choose a positive number

$$\tilde{\gamma} < \tilde{\gamma}_0 (8L_1)^{-1} \tag{4.158}$$

and a natural number

$$L > 8L_0 + 11L_1 + (22L_1 + 8L_0 + 5)(2 + M_0 \tilde{\gamma}_0^{-1}). \tag{4.159}$$

Assume that an integer

$$T > L, \tag{4.160}$$

$$z_0, z_1 \in R_+^n, \ z_0 \le Me, \ az_1 \le \Gamma d^{-1}, \tag{4.161}$$

functions $w_i : R_+^n \to R^1, i = 0, \dots, T - 1$ are bounded on bounded subsets of R_+^n and satisfy

$$\|w_i - w(b(\cdot))\|_M \le \tilde{\gamma} \text{ for all } i \in \{0, \dots, T - 1\}, \tag{4.162}$$

and a program $(\{x(t)\}_{t=0}^T, \{y(t)\}_{t=0}^{T-1})$ satisfies

$$x(0) = z_0, \tag{4.163}$$

$$x(T) \ge z_1, \ \sum_{t=0}^{T-1} w_t(y(t)) \ge U(\{w_t\}_{t=0}^{T-1}, z_0, z_1) - M_0. \tag{4.164}$$

In order to complete the proof, it is sufficient to show that

$$\text{Card}(\{t \in \{0, \dots, T - 1\} : \max\{\|x(t) - \hat{x}\|, \ \|y(t) - \hat{x}\|\} > \epsilon\}) \le L. \tag{4.165}$$

By (4.161), (4.163), Lemma 4.15, and the inequality $M > \max\{(a_i d)^{-1} : i = 1, \dots, n\}$, we have

$$x(t) \le Me, \ t = 0, \dots, T. \tag{4.166}$$

In view of (4.159) and (4.160), $T - 4L_1 > 0$. By (4.164),

$$\sum_{t=T-4L_1}^{T-1} w_t(y(t)) \ge U(\{w_t\}_{t=T-4L_1}^{T-1}, x(T - 4L_1), z_1) - M_0, \tag{4.167}$$

By Lemma 4.15 and (4.162), for each program $(\{u(t)\}_{t=T-4L_1}^T, \{v(t)\}_{t=T-4L_1}^{T-1})$ satisfying $u(T - 4L_1) \le Me$,

$$| \sum_{t=T-4L_1}^{T-1} w_t(v(t)) - \sum_{t=T-4L_1}^{T-1} w(bv(t))| \leq 4L_1\tilde{\gamma}. \tag{4.168}$$

In view of (4.166) and (4.168),

$$| \sum_{t=T-4L_1}^{T-1} w_t(y(t)) - \sum_{t=T-4L_1}^{T-1} w(by(t))| \leq 4L_1\tilde{\gamma}, \tag{4.169}$$

$$|U(\{w_t\}_{t=T-4L_1}^{T-1}, x(T-4L_1), z_1) - U(x(T-4L_1), z_1, T-4L_1, T)| \leq 4L_1\tilde{\gamma}. \tag{4.170}$$

By (4.167)–(4.170), we have

$$\sum_{t=T-4L_1}^{T-1} w(by(t)) \geq \sum_{t=T-4L_1}^{T-1} w_t(y(t)) - 4L_1\tilde{\gamma}$$

$$\geq U(\{w_t\}_{t=T-4L_1}^{T-1}, x(T-4L_1), z_1) - M_0 - 4L_1\tilde{\gamma}$$

$$\geq U(x(T-4L_1), z_1, T-4L_1, T) - 4L_1\tilde{\gamma} - M_0 - 4L_1\tilde{\gamma}. \tag{4.171}$$

By (4.158) and (4.171), we have

$$\sum_{t=T-4L_1}^{T-1} w(by(t)) \geq U(x(T-4L_1), z_1, T-4L_1, T) - M_0 - 1. \tag{4.172}$$

By (4.161), (4.164), (4.166), (4.172), and property (Pii) (applied to the program $(\{x(t+T-4L_1)\}_{t=0}^{4L_1}, \{y(t+T-4L_1)\}_{t=0}^{4L_1-1}))$, we have

Card($\{t \in \{T-4L_1, \ldots, T-1\} : \max\{\|x(t) - \hat{x}\|, \|y(t) - \hat{x}\|\} > \epsilon\}) \leq L_1$.

By the inequality above, there is an integer S such that

$$T - 3L_1 \leq S \leq T - L_1, \tag{4.173}$$

$$\max\{\|x(S) - \hat{x}\|, \|y(S) - \hat{x}\|\} \leq \epsilon.$$

Relation (4.155) implies that

$$az < \Gamma d^{-1} \text{ for each } z \in R_+^n \text{ satisfying } \|z - \hat{x}\| \leq \epsilon. \tag{4.174}$$

It follows from (4.164), (4.173), and Proposition 4.9 that

$$\sum_{t=0}^{S-1} w_t(y(t)) \geq U(\{w_t\}_{t=0}^{S-1}, z_0, x(S)) - M_0. \tag{4.175}$$

We show that the following property holds:
(Piii) for each integer τ_0 satisfying

$$4L_1 \leq \tau_0 \leq S, \ \|x(\tau_0) - \widehat{x}\| \leq \epsilon, \tag{4.176}$$

there is an integer τ_1 such that

$$\tau_0 - 3L_1 \leq \tau_1 \leq \tau_0 - L_1,$$

$$\max\{\|x(\tau_1) - \widehat{x}\|, \ \|y(\tau_1) - \widehat{x}\|\} \leq \epsilon.$$

Assume that an integer τ_0 satisfies (4.176). By Lemma 4.15, (4.162) and (4.176), for each program $(\{u(t)\}_{t=\tau_0-4L_1}^{\tau_0}, \{v(t)\}_{t=\tau_0-4L_1}^{\tau_0-1})$ satisfying

$$u(\tau_0 - 4L_1) \leq Me,$$

we have

$$|\sum_{t=\tau_0-4L_1}^{\tau_0-1} w_t(v(t)) - \sum_{t=\tau_0-4L_1}^{\tau_0-1} w(bv(t))| \leq 4L_1\tilde{\gamma}. \tag{4.177}$$

In view of (4.166) and (4.177),

$$|\sum_{t=\tau_0-4L_1}^{\tau_0-1} w_t(y(t)) - \sum_{t=\tau_0-4L_1}^{\tau_0-1} w(by(t))| \leq 4L_1\tilde{\gamma}, \tag{4.178}$$

$$|U(\{w_t\}_{t=\tau_0-4L_1}^{\tau_0-1}, x(\tau_0 - 4L_1), x(\tau_0))$$

$$- U(x(\tau_0 - 4L_1), x(\tau_0), \tau_0 - 4L_1, \tau_0)| \leq 4L_1\tilde{\gamma}. \tag{4.179}$$

By (4.175), (4.176), and Proposition 4.9,

$$\sum_{t=\tau_0-4L_1}^{\tau_0-1} w_t(y(t)) \geq U(\{w_t\}_{t=\tau_0-4L_1}^{\tau_0-1}, x(\tau_0 - 4L_1), x(\tau_0)) - M_0. \tag{4.180}$$

It follows from (4.158) and (4.178)–(4.180) that

$$\sum_{t=\tau_0-4L_1}^{\tau_0-1} w(by(t)) \geq \sum_{t=\tau_0-4L_1}^{\tau_0-1} w_t(y(t)) - 4L_1\tilde{\gamma}$$

$$\geq U(\{w_t\}_{t=\tau_0-4L_1}^{\tau_0-1}, x(\tau_0-4L_1), x(\tau_0)) - M_0 - 4L_1\tilde{\gamma}$$

$$\geq U(x(\tau_0-4L_1), x(\tau_0), \tau_0-4L_1, \tau_0) - M_0 - 8L_1\tilde{\gamma}$$

$$\geq U(x(\tau_0-4L_1), x(\tau_0), \tau_0-4L_1, \tau_0) - M_0 - 1. \qquad (4.181)$$

By (4.166), (4.174), (4.176), (4.181), and the property (Pii) (applied to the program $(\{x(t+\tau_0-4L_1)\}_{t=0}^{4L_1}, \{y(t+\tau_0-4L_1)\}_{t=0}^{4L_1-1}))$, we have

$$\text{Card}(\{t \in \{\tau_0-4L_1, \ldots, \tau_0-1\} : \max\{\|x(t)-\widehat{x}\|, \|y(t)-\widehat{x}\|\} > \epsilon\}) \leq L_1.$$

By the inequality above, there is an integer τ_1 such that

$$\tau_0 - 3L_1 \leq \tau_1 \leq \tau_0 - L_1, \ \max\{\|x(\tau_1)-\widehat{x}\|, \ \|y(\tau_1)-\widehat{x}\|\} \leq \epsilon.$$

Thus the property (Piii) holds.

The property (Piii) and (4.173) imply that there are a natural number k and a strictly increasing sequence of nonnegative integers $\{S_i\}_{i=1}^k$ such that

$$S_k = S, \ S_1 < 4L_1, \ \|x(S_i)-\widehat{x}\| \leq \epsilon, \ i = 1, \ldots, k, \qquad (4.182)$$

$$L_1 \leq S_{i+1} - S_i \leq 3L_1 \text{ for all integers } i \text{ satisfying } 1 \leq i < k. \qquad (4.183)$$

It follows from (4.176), (4.182), and (4.183) that the following property holds:

(Piv) for each integer $t \in [4L_1, S]$, there exist integers $\tilde{t}_1, \tilde{t}_2 \in [S_1, S]$ such that

$$L_1 \leq \tilde{t}_2 - \tilde{t}_1 \leq 3L_1, \ t \in [\tilde{t}_1, \tilde{t}_2],$$

$$\|x(\tilde{t}_i)-\widehat{x}\| \leq \epsilon, \ i = 1, 2.$$

Set $t_0 = 0$. By induction we construct a finite strictly increasing sequence of integers $\{t_i\}_{i=0}^q$ such that

$$t_q = S; \qquad (4.184)$$

(Pv) for each integer i satisfying $0 \leq i < q - 1$,

$$\sum_{t=t_i}^{t_{i+1}-1} w_t(y(t)) < U(\{w_t\}_{t=t_i}^{t_{i+1}-1}, x(t_i), x(t_{i+1})) - \tilde{\gamma}_0; \qquad (4.185)$$

(Pvi) if an integer i satisfies $0 \leq i < q, \ t_{i+1} - t_i \geq 2$ and (4.185), then

$$\sum_{t=t_i}^{t_{i+1}-2} w_t(y(t)) \geq U(\{w_t\}_{t=t_i}^{t_{i+1}-2}, x(t_i), x(t_{i+1}-1)) - \tilde{\gamma}_0. \qquad (4.186)$$

Assume that an integer $p \geq 0$ and we have already defined a strictly increasing sequence of integers $\{t_i\}_{i=0}^{p}$ such that $t_p < S$ and that for each integer i satisfying $0 \leq i < p$, (4.185) and (4.186) hold. (Clearly, for $p = 0$ our assumption holds.) We define t_{p+1}.

There are two cases:

$$\sum_{t=t_p}^{S-1} w_t(y(t)) \geq U(\{w_t\}_{t=t_p}^{S-1}, x(t_p), x(S)) - \tilde{\gamma}_0; \qquad (4.187)$$

$$\sum_{t=t_p}^{S-1} w_t(y(t)) < U(\{w_t\}_{t=t_p}^{S-1}, x(t_p), x(S)) - \tilde{\gamma}_0. \qquad (4.188)$$

If (4.187) holds, then we set $q = p + 1$, $t_q = S$, and in this case, the construction is completed, and (Pv) and (Pvi) hold.

Assume that (4.188) holds. Set

$$t_{p+1} = \min\{\tau \in \{t_p + 1, \ldots, S\} :$$

$$\sum_{t=t_p}^{\tau-1} w_t(y(t)) < U(\{w_t\}_{t=t_p}^{\tau-1}, x(t_p), x(\tau)) - \tilde{\gamma}_0\}.$$

Clearly, t_{p+1} is well-defined and $t_{p+1} > t_p$. If $t_{p+1} = S$, then we set $q = p + 1$, the construction is completed, and it is not difficult to see that (Pv) and (Pvi) hold.

If $t_{p+1} < S$, then it is easy to see that the assumption made for p is also true for $p + 1$.

Clearly our construction is completed after a final number of steps, and let t_q be its last element. It follows from the construction that $t_q = S$ and that the properties (Pv) and (Pvi) hold.

By (4.148), Proposition 4.9, and property (Pv),

$$M_0 \geq U(\{w_t\}_{t=0}^{S-1}, z_0, x(S)) - \sum_{t=0}^{S-1} w_t(y(t))$$

$$\geq \sum \{U(\{w_t\}_{t=t_i}^{t_{i+1}-1}, x(t_i), x(t_{i+1})) - \sum_{t=t_i}^{t_{i+1}-1} w_t(y(t)) :$$

$$i \text{ is an integer}, 0 \leq i < q - 1\} > (q - 1)\tilde{\gamma}_0,$$

$$q < M_0\tilde{\gamma}_0^{-1} + 1. \tag{4.189}$$

Set

$$A = \{i \in \{0, \ldots, q-1\} : t_{i+1} - t_i \geq 12L_1 + 6L_0 + 1\}. \tag{4.190}$$

Let

$$i \in A. \tag{4.191}$$

By (4.190) and (4.191),

$$(t_{i+1} - 1) - t_i \geq 12L_1 + 6L_0. \tag{4.192}$$

By the properties (Pv) and (Pvi) and (4.192), relation (4.186) holds. By (4.192) and (Piv), there exist integers $\tilde{t}_i, \tilde{t}_{i+1}$ such that

$$4L_1 + t_i \leq \tilde{t}_i \leq t_i + 7L_1, \ t_{i+1} - 1 \geq \tilde{t}_{i+1} \geq t_{i+1} - 3L_1 - 1, \tag{4.193}$$

$$\|x(\tilde{t}_i) - \hat{x}\| \leq \epsilon, \ \|x(\tilde{t}_{i+1}) - \hat{x}\| \leq \epsilon. \tag{4.194}$$

In view of (4.192) and (4.193),

$$\tilde{t}_{i+1} - \tilde{t}_i \geq t_{i+1} - t_i - 10L_1 - 1 \geq 2L_1 + 6L_0. \tag{4.195}$$

It follows from (4.186), (4.193) and Proposition 4.9 that

$$\sum_{t=\tilde{t}_i}^{\tilde{t}_{i+1}-1} w_t(y(t)) \geq U(\{w_t\}_{t=\tilde{t}_i}^{\tilde{t}_{i+1}-1}, x(\tilde{t}_i), x(\tilde{t}_{i+1})) - \tilde{\gamma}_0. \tag{4.196}$$

By (4.158), (4.162), (4.166), (4.174), (4.194)–(4.196), and (Pi) (applied to the program $(\{x(t + \tilde{t}_i)\}_{t=0}^{\tilde{t}_{i+1}-\tilde{t}_i}, \{y(t + \tilde{t}_i)\}_{t=0}^{\tilde{t}_{i+1}-\tilde{t}_i-1}))$,

$$\|x(t) - \hat{x}\|, \ \|y(t) - \hat{x}\| \leq \epsilon, \ t = \tilde{t}_i + L_0, \ldots, \tilde{t}_{i+1} - L_0 - 1.$$

Together with (4.193) this implies that

$$\|x(t) - \hat{x}\|, \ \|y(t) - \hat{x}\| \leq \epsilon, \ t = t_i + 7L_1 + L_0, \ldots, t_{i+1} - 3L_1 - L_0 - 2 \tag{4.197}$$

for all $i \in A$.
  By (4.184), (4.190), and (4.197),

$$\{t \in \{0, \ldots, T-1\} : \max\{\|x(t) - \hat{x}\|, \ \|y(t) - \hat{x}\|\} > \epsilon\}$$

$$\subset \{S, \ldots, T-1\} \cup (\cup\{\{t_i, \ldots, t_{i+1}\} : i \in \{0, \ldots, q-1\} \setminus A\})$$

$$\cup (\cup\{\{t_i, \ldots, t_i + 7L_1 + L_0\} \cup \{t_{i+1} - 3L_1 - L_0 - 1, t_{i+1}\} : i \in A\}). \qquad (4.198)$$

By (4.159), (4.173), (4.189), (4.190), and (4.198),

$$\text{Card}(\{t \in \{0, \ldots, T-1\} : \max\{\|x(t) - \widehat{x}\|, \|y(t) - \widehat{x}\|\} > \epsilon\})$$

$$\leq 3L_1 + q(12L_1 + 6L_0 + 2) + q(7L_1 + L_0 + 1) + q(3L_1 + L_0 + 2)$$

$$\leq 3L_1 + q(22L_1 + 8L_0 + 5) \leq 3L_1 + (22L_1 + 8L_0 + 5)(1 + M_0 \widetilde{\gamma}_0^{-1} + 1) < L.$$

This completes the proof of Theorem 4.27.

Chapter 5
The Robinson–Solow–Srinivasan Model with a Nonconcave Utility Function

We study infinite horizon optimal control problems related to the Robinson–Solow–Srinivasan model with a nonconcave utility function. In particular, we establish the existence of good programs and optimal programs using different optimality criterions.

5.1 Good Programs

Let R^1 (R^1_+) be the set of real (nonnegative) numbers, and let R^n be the n-dimensional Euclidean space with nonnegative orthant

$$R^n_+ = \{x = (x_1, \ldots, x_n) \in R^n : x_i \geq 0, \ i = 1, \ldots, n\}.$$

For every pair of vectors $x = (x_1, \ldots, x_n)$, $y = (y_1, \ldots, y_n) \in R^n$, define their inner product by

$$xy = \sum_{i=1}^{n} x_i y_i,$$

and let $x \gg y, x > y, x \geq y$ have their usual meaning.

Let $e(i), i = 1, \ldots, n$, be the ith unit vector in R^n, and e be an element of R^n_+ all of whose coordinates are unity. For every $x \in R^n$, denote by $\|x\|$ its Euclidean norm in R^n.

Let $a = (a_1, \ldots, a_n) \gg 0, b = (b_1, \ldots, b_n) \gg 0$, and $d \in (0, 1]$.

A sequence $\{x(t), y(t)\}_{t=0}^{\infty}$ is called a program if for each integer $t \geq 0$,

$$(x(t), y(t)) \in R^n_+ \times R^n_+, \ x(t+1) \geq (1-d)x(t),$$

© The Author(s), under exclusive license to Springer Nature Switzerland AG 2020
A. J. Zaslavski, *Turnpike Theory for the Robinson–Solow–Srinivasan Model*,
Springer Optimization and Its Applications 166,
https://doi.org/10.1007/978-3-030-60307-6_5

$$0 \leq y(t) \leq x(t), \quad a(x(t+1) - (1-d)x(t)) + ey(t) \leq 1. \tag{5.1}$$

Let T_1, T_2 be integers such that $0 \leq T_1 < T_2$. A pair of sequences

$$\left(\{x(t)\}_{t=T_1}^{T_2}, \{y(t)\}_{t=T_1}^{T_2-1} \right)$$

is called a program if $x(T_2) \in R_+^n$ and for each integer t satisfying $T_1 \leq t < T_2$, relations (5.1) hold.

Let $w : [0, \infty) \to [0, \infty)$ be a continuous strictly increasing function which represents the preferences of the planner.

For every point $x_0 \in R_+^n$ and every natural number T, set

$$U(x_0, T) = \sup \left\{ \sum_{t=0}^{T-1} w(by(t)) : \left(\{x(t)\}_{t=0}^{T}, \{y(t)\}_{t=0}^{T-1} \right) \right.$$

$$\text{is a program such that } x(0) = x_0 \}. \tag{5.2}$$

In the sequel we assume that supremum of empty set is $-\infty$.

Let $x_0, \tilde{x}_0 \in R_+^n$, and let T be a natural number. Set

$$U(x_0, \tilde{x}_0, T) = \sup \left\{ \sum_{t=0}^{T-1} w(by(t)) : \left(\{x(t)\}_{t=0}^{T}, \{y(t)\}_{t=0}^{T-1} \right) \right.$$

$$\text{is a program such that } x(0) = x_0, \ x(T) \geq \tilde{x}_0 \}. \tag{5.3}$$

The next proposition follows immediately from the continuity of w.

Proposition 5.1 *For every point $x_0 \in R_+^n$ and every integer $T > 0$, there exists a program $(\{x(t)\}_{t=0}^{T}, \{y(t)\}_{t=0}^{T-1})$ such that $x(0) = x_0$ and*

$$\sum_{t=0}^{T-1} w(by(t)) = U(x_0, T).$$

Set

$$\Omega = \{ (x, x') \in R_+^n \times R_+^n : x' \geq (1-d)x \text{ and } a(x' - (1-d)x) \leq 1 \}. \tag{5.4}$$

We have a correspondence $\Lambda : \Omega \to R_+^n$ given by

$$\Lambda(x, x') = \{ y \in R_+^n : 0 \leq y \leq x \text{ and}$$

$$ey \leq 1 - a(x' - (1 - d)x)\}, \ (x, x') \in \Omega. \tag{5.5}$$

Let M_0 be a positive number, and let $T \geq 1$ be an integer. Set

$$\widehat{U}(M_0, T) = \sup\{\sum_{t=0}^{T-1} w(by(t)) :$$

$$\left(\{x(t)\}_{t=0}^{T}, \{y(t)\}_{t=0}^{T-1}\right) \text{ is a program such that } x(0) \leq M_0 e\}. \tag{5.6}$$

It is clear that $\widehat{U}(M_0, T)$ is finite. The next proposition follows immediately from the continuity of w.

Proposition 5.2 *For every positive number M_0 and every integer $T \geq 1$, there exists a program $(\{x(t)\}_{t=0}^{T}, \{y(t)\}_{t=0}^{T-1})$ such that $x(0) \leq M_0 e$ and $\sum_{t=0}^{T-1} w(by(t)) = \widehat{U}(M_0, T)$.*

In the sequel we use the following simple auxiliary result.

Lemma 5.3 *Let a number $M_0 > \max\{(a_i d)^{-1} : i = 1, \dots, n\}$, $(x, x') \in \Omega$, and let $x \leq M_0 e$. Then $x' \leq M_0 e$.*

For the proof see Lemma 4.15.

In this chapter we show the existence of a positive constant μ such that the following properties hold:

(a) For each program $\{x(t), y(t)\}_{t=0}^{\infty}$ either the sequence $\{\sum_{t=0}^{T-1}[w(by(t)) - \mu]\}_{T=1}^{\infty}$ is bounded or

$$\lim_{T \to \infty} \sum_{t=0}^{T-1}[w(by(t)) - \mu] = -\infty;$$

(b) for each $x_0 \in R_+^n$ there exists a program $\{x(t), y(t)\}_{t=0}^{\infty}$ such that $x(0) = x_0$ and that the sequence $\{\sum_{t=0}^{T-1}[w(by(t)) - \mu]\}_{T=1}^{\infty}$ is bounded.

For any $(x, x') \in \Omega$, define

$$u(x, x') = \max\{w(by) : y \in \Lambda(x, x')\}.$$

In this section we state several results obtained in [94].
Our first result allows us to define the constant μ.

Theorem 5.4 *Let $M_1, M_2 > \max\{(d a_i)^{-1} : i = 1, \dots, n\}$. Then there exist finite limits*

$$\lim_{p \to \infty} \widehat{U}(M_i, p)/p, \ i = 1, 2$$

and

$$\lim_{p \to \infty} \widehat{U}(M_1, p)/p = \lim_{p \to \infty} \widehat{U}(M_2, p)/p.$$

Theorem 5.4 will be proved in Section 5.4. Define

$$\mu = \lim_{p \to \infty} \widehat{U}(M, p)/p \tag{5.7}$$

where $M > \max\{(da_i)^{-1} : i = 1, \ldots, n\}$. By Theorem 5.4, μ is well-defined and does not depend on M.

The next theorem will be also proved in Section 5.4.

Theorem 5.5 *Let $M_0 > \max\{(da_i)^{-1} : i = 1, \ldots, n\}$. Then there exists $M > 0$ such that*

$$|\widehat{U}(M_0, p) - p\mu| \leq M \text{ for all integers } p \geq 1.$$

Corollary 5.6 *Let $M_0 > \max\{(da_i)^{-1} : i = 1, \ldots, n\}$. Then there exists a positive number M such that for each program $\{x(t), y(t)\}_{t=0}^{\infty}$ satisfying $x(0) \leq M_0 e$ and each integer $T \geq 1$,*

$$\sum_{t=0}^{T-1} [w(by(t)) - \mu] \leq M.$$

Note that Corollary 5.6 easily follows from Theorem 5.5.
The next result will be proved in Section 5.5.

Proposition 5.7 *Let $\{x(t), y(t)\}_{t=0}^{\infty}$ be a program. Then either the sequence $\{\sum_{t=0}^{T-1} [w(by(t)) - \mu]\}_{T=1}^{\infty}$ is bounded or*

$$\lim_{T \to \infty} \sum_{t=0}^{T-1} [w(by(t)) - \mu] = -\infty.$$

A program $\{x(t), y(t)\}_{t=0}^{\infty}$ is called good if there exists $M \in R^1$ such that

$$\sum_{t=0}^{T} (w(y(t)) - \mu) \geq M \text{ for all } T \geq 0.$$

A program is called bad if

$$\lim_{T \to \infty} \sum_{t=0}^{T} (w(y(t)) - \mu) = -\infty.$$

By Proposition 5.7 any program that is not good is bad.
Set

$$x(t) = (2nd \max\{a_i : i = 1, \ldots, n\})^{-1} e,$$

$$y(t) = \min\{(2n)^{-1}, (2nd \max\{a_i : i = 1, \ldots, n\})^{-1}\} e \text{ for all integers } t \geq 0.$$

It is easy to see that $\{x(t), y(t)\}_{t=0}^{\infty}$ is a program. By Corollary 5.6,

$$\mu \geq \lim_{T \to \infty} T^{-1} \sum_{t=0}^{T-1} w(by(t)) > w(0).$$

Thus we have shown that

$$\mu > w(0). \tag{5.8}$$

The following theorem will be proved in Section 5.4.

Theorem 5.8 *Let $M_0 > \max\{(da_i)^{-1} : i = 1, \ldots, n\}$. Then there exists a positive number M such that for every $x_0 \in R_+^n$ which satisfies $x_0 \leq M_0 e$, there exists a program $\{x(t), y(t)\}_{t=0}^{\infty}$ such that $x(0) = x_0$; that for every integer $T_1 \geq 0$ and every natural number $T_2 > T_1$,*

$$\left| \sum_{t=T_1}^{T_2-1} w(by(t)) - \mu(T_2 - T_1) \right| \leq M;$$

and that for every natural number T,

$$\sum_{t=0}^{T-1} w(by(t)) = U(x(0), x(T), T). \tag{5.9}$$

Theorem 5.8 establishes that for every initial state $x_0 \geq 0$, there exists a good program $\{x(t), y(t)\}_{t=0}^{\infty}$ such that $x(0) = x_0$. In addition this program satisfies (5.9) for every natural number T. This leads us to the following definition.

A program $\{x(t), y(t)\}_{t=0}^{\infty}$ is called weakly maximal if equality (5.9) holds for every natural number T.

Our final result which will be proved in Section 5.6 establishes a relation between good programs and weakly maximal programs.

Theorem 5.9 *Let $\{x(t), y(t)\}_{t=0}^{\infty}$ be a weakly maximal program such that $\limsup_{t \to \infty} by(t) > 0$. Then the program $\{x(t), y(t)\}_{t=0}^{\infty}$ is good.*

5.2 Auxiliary Results

Lemma 5.10 *Let $\delta > 0$, and let*

$$M_0 > \max\{(a_i d)^{-1} : i = 1, \dots, n\}. \tag{5.10}$$

Then there exists an integer $T_0 \geq 4$ such that for every natural number $\tau \geq T_0$, every program $(\{x(t)\}_{t=0}^{\tau}, \{y(t)\}_{t=0}^{\tau-1})$ which satisfies

$$x(0) \leq M_0 e, \ by(\tau - 1) \geq \delta, \tag{5.11}$$

and every $\tilde{x}_0 \in R_+^n$ which satisfies

$$\tilde{x}_0 \leq M_0 e, \tag{5.12}$$

there exists a program $(\{\tilde{x}(t)\}_{t=0}^{\tau}, \{\tilde{y}(t)\}_{t=0}^{\tau-1})$ such that

$$\tilde{x}(0) = \tilde{x}_0, \ \tilde{x}(\tau) \geq x(\tau). \tag{5.13}$$

Proof Choose a natural number $T_0 \geq 4$ such that

$$2(1 - d)^{T_0} M_0 \leq \delta n^{-1} (\max\{b_i : i = 1, \dots, n\})^{-1} (\max\{a_i : i = 1, \dots, n\})^{-1}. \tag{5.14}$$

Assume that an integer $\tau \geq T_0$, that a program $(\{x(t)\}_{t=0}^{\tau}, \{y(t)\}_{t=0}^{\tau-1})$ satisfies (5.11), and that a point $\tilde{x}_0 \in R_+^n$ satisfies (5.12). Define

$$\tilde{x}(0) = \tilde{x}_0 \tag{5.15}$$

and for $t = 0, \dots, \tau - 1$, set

$$\tilde{y}(t) = 0, \ \tilde{x}(t + 1) = (1 - d)\tilde{x}(t) + [x(t + 1) - (1 - d)x(t)]$$

$$+ n^{-1}[1 - a(x(t + 1) - (1 - d)x(t))] \left(a_1^{-1}, \dots, a_n^{-1}\right). \tag{5.16}$$

It is easy to see that $(\{\tilde{x}(t)\}_{t=0}^{\tau}, \{\tilde{y}(t)\}_{t=0}^{\tau-1})$ is a program. It follows from (5.16) that for all integers $t = 0, \dots, \tau - 1$, we have

$$\tilde{x}(t + 1) - x(t + 1) \geq (1 - d)(\tilde{x}(t) - x(t)).$$

Together with (5.11), (5.12), and (5.15), the inequality above implies that

$$\tilde{x}(\tau - 1) - x(\tau - 1) \geq (1 - d)^{\tau-1}(\tilde{x}(0) - x(0)) \geq (1 - d)^{\tau-1}(-M_0 e).$$

By this relation, (5.1), (5.10), (5.11), (5.14), and (5.16), we have

$$\tilde{x}(\tau) - x(\tau) = (1 - d)(\tilde{x}(\tau - 1) - x(\tau - 1))$$
$$+ n^{-1}[1 - a(x(\tau) - (1 - d)x(\tau - 1))]\left(a_1^{-1}, \ldots, a_n^{-1}\right)$$
$$\geq (1 - d)^{\tau}(-M_0 e) + n^{-1} e y(\tau - 1)\left(a_1^{-1}, \ldots, a_n^{-1}\right)$$
$$\geq -(1 - d)^{T_0} M_0 e + n^{-1}(by(\tau - 1))(\max\{b_i : i = 1, \ldots, n\})^{-1}$$
$$\times (a_1^{-1}, \ldots, a_n^{-1})$$
$$\geq -(1 - d)^{T_0} M_0 e + \delta n^{-1}(\max\{b_i : i = 1, \ldots, n\})^{-1}$$
$$\times (a_1^{-1}, \ldots, a_n^{-1}) \geq 0.$$

This completes the proof of Lemma 5.10.

Choose a positive number γ such that

$$\gamma < 2^{-1}, \quad 4\gamma < (2n)^{-1} \min\left\{1, a_1^{-1}, \ldots, a_i^{-1}, \ldots, a_n^{-1}\right\} \sum_{i=1}^{n} b_i. \tag{5.17}$$

Lemma 5.11 *Let $M_1 > 0$, and let a number M_0 satisfy (5.10). Then there exists a pair of integers $L_1, L_2 \geq 4$ such that for every natural number $T \geq L_1 + L_2$, every program $(\{x(t)\}_{t=0}^{T}, \{y(t)\}_{t=0}^{T-1})$ satisfying*

$$x(0) \leq M_0 e, \quad \sum_{t=0}^{T-1} w(by(t)) \geq U(x(0), T) - M_1, \tag{5.18}$$

and every integer $\tau \in [L_1, T - L_2]$, the following inequality holds:

$$\max\{by(t) : t = \tau, \ldots, \tau + L_2 - 1\} \geq \gamma. \tag{5.19}$$

Proof Lemma 5.10 implies that there exists an integer $L_1 \geq 4$ such that the following property holds:
(P1) If an integer $S \geq L_1$, a program $(\{u(t)\}_{t=0}^{S}, \{v(t)\}_{t=0}^{S-1})$ satisfies

$$u(0) \leq M_0 e, \quad bv(S - 1) \geq \gamma, \tag{5.20}$$

and if a point $\tilde{u}_0 \in R_+^n$ satisfies $\tilde{u}_0 \leq M_0 e$, then there exists a program $(\{\tilde{u}(t)\}_{t=0}^{S}, \{\tilde{v}(t)\}_{t=0}^{S-1})$ such that

$$\tilde{u}(0) = \tilde{u}_0, \quad \tilde{u}(S) \geq u(S). \tag{5.21}$$

Fix an integer $L_2 \geq 1$ for which

$$L_2 \geq 4(M_1 + w(M_0 eb) + 2L_1 w(1) + 1)$$
$$\times \left[\left(w \left((2n)^{-1} \min \left\{ 1, a_1^{-1}, \ldots, a_i^{-1}, \ldots, a_n^{-1} \right\} eb \right) \right. \right.$$
$$\left. \left. -w(\gamma) \right)^{-1} + 1 \right] + 8(L_1 + 1). \tag{5.22}$$

Assume that an integer $T \geq L_1 + L_2$, a program $(\{x(t)\}_{t=0}^T, \{y(t)\}_{t=0}^{T-1})$ satisfies (5.18), and an integer τ satisfies

$$L_1 \leq \tau \leq T - L_2. \tag{5.23}$$

We show that (5.19) is valid. Let us assume the contrary. Then

$$by(t) < \gamma, \; t = \tau, \ldots, \tau + L_2 - 1. \tag{5.24}$$

Clearly, one of the following cases holds:

$$by(t) < \gamma, \; t = \tau, \ldots, T - 1; \tag{5.25}$$

$$\max\{by(t) : \; t = \tau, \ldots, T - 1\} \geq \gamma. \tag{5.26}$$

Now we define a natural number τ_0 as follows. If (5.25) holds, then we put $\tau_0 = T$. If (5.26) is valid, then in view of (5.23), (5.24), and (5.26), there exists an integer $\tau_0 \geq 1$ for which

$$\tau + L_2 \leq \tau_0 \leq T - 1, \tag{5.27}$$

$$by(\tau_0) \geq \gamma, \tag{5.28}$$

$$by(t) < \gamma, \; t = \tau, \ldots, \tau_0 - 1. \tag{5.29}$$

It is easy to see that (5.29) is valid in both cases and that in both cases

$$\tau_0 - \tau \geq L_2. \tag{5.30}$$

Assume that (5.25) holds. Recall that in this case $\tau_0 = T$. Set

$$\tilde{x}(t) = x(t), \; t = 0, \ldots, \tau, \; \tilde{y}(t) = y(t), \; t = 0, \ldots, \tau - 1, \tag{5.31}$$

$$\tilde{y}(\tau) = 0,$$

$$\tilde{y}(t) = (2n)^{-1} \min \left\{1, a_1^{-1}, \ldots, a_i^{-1}, \ldots, a_n^{-1}\right\} e, \ t = \tau+1, \ldots, T-1, \quad (5.32)$$

$$\tilde{x}(t+1) = (1-d)\tilde{x}(t) + (2n)^{-1}\left(a_1^{-1}, \ldots, a_i^{-1}, \ldots, a_n^{-1}\right), \ t = \tau, \ldots, T-1.$$

It is easy to see that $(\{\tilde{x}(t)\}_{t=0}^{T}, \{\tilde{y}(t)\}_{t=0}^{t-1})$ is a program. By (5.17), (5.18), (5.23), (5.25), (5.31), (5.32), and the inequality $L_2 \geq 4$,

$$M_1 \geq U(x(0), T) - \sum_{t=0}^{T-1} w(by(t)) \geq \sum_{t=0}^{T-1} w(b\tilde{y}(t)) - \sum_{t=0}^{T-1} w(by(t))$$

$$= \sum_{t=\tau}^{T-1} w(b\tilde{y}(t)) - \sum_{t=\tau}^{T-1} w(by(t))$$

$$\geq (T-1-\tau)\left(w((2n)^{-1} \min\left\{1, a_1^{-1}, \ldots, a_i^{-1}, \ldots a_n^{-1}\right\} \sum_{i=1}^{n} b_i)\right)$$

$$- (T-\tau)w(\gamma)$$

$$\geq (T-1-\tau)\left[w\left((2n)^{-1} \min\left\{1, a_1^{-1}, \ldots, a_i^{-1}, \ldots a_n^{-1}\right\} \sum_{i=1}^{n} b_i\right) - w(\gamma)\right]$$

$$- w(\gamma)$$

$$\geq (L_2/2)\left[w\left((2n)^{-1} \min\left\{1, a_1^{-1}, \ldots, a_i^{-1}, \ldots a_n^{-1}\right\} \sum_{i=1}^{n} b_i\right) - w(\gamma)\right]$$

$$- w(1).$$

The relation above implies that

$$L_2 \leq 2(M_1 + w(1))\left[w\left((2n)^{-1} \min\left\{1, a_1^{-1}, \ldots, a_i^{-1}, \ldots a_n^{-1}\right\} \sum_{i=1}^{n} b_i\right) - w(\gamma)\right]^{-1}.$$

This inequality contradicts (5.22). The contradiction we have reached proves that (5.25) does not hold.

Therefore (5.26) holds, and the integer τ_0 satisfies (5.27)–(5.29). Set

$$\tilde{x}(t) = x(t), \ t = 0, \ldots, \tau, \ \tilde{y}(t) = y(t), \ t = 0, \ldots, \tau - 1, \ \tilde{y}(\tau) = 0, \quad (5.33)$$

$$\tilde{y}(t) = (2n)^{-1} \min \left\{1, a_1^{-1}, \ldots, a_i^{-1}, \ldots, a_n^{-1}\right\} e, \ t = \tau + 1, \ldots, \tau_0 - L_1 - 1,$$
$$\quad (5.34)$$

$$\tilde{x}(t+1) = (1-d)\tilde{x}(t) + (2n)^{-1}(a_1^{-1}, \ldots, a_i^{-1}, \ldots, a_n^{-1}), \ t = \tau, \ldots, \tau_0 - L_1 - 1.$$

It is not difficult to see that $(\{\tilde{x}(t)\}_{t=0}^{\tau_0-L_1}, \{\tilde{y}(t)\}_{t=0}^{\tau_0-L_1-1})$ is a program. Lemma 5.3, (5.10), (5.18), and (5.33) imply that

$$x(\tau_0 - L_1), \ \tilde{x}(\tau_0 - L_1) \le M_0 e, \tag{5.35}$$

$$0 \le y(\tau_0) \le x(\tau_0) \le M_0 e.$$

Let us consider the program $(\{(x(t)\}_{t=\tau_0-L_1}^{\tau_0+1}, \{y(t)\}_{t=\tau_0-L_1}^{\tau_0})$. Property (P1), (5.28), and (5.35) imply that there exists a program

$$\left(\{\tilde{x}(t)\}_{t=\tau_0-L_1}^{\tau_0+1}, \{\tilde{y}(t)\}_{t=\tau_0-L_1}^{\tau_0}\right)$$

which satisfies

$$\tilde{x}(\tau_0 + 1) \ge x(\tau_0 + 1). \tag{5.36}$$

It is easy to see that $(\{\tilde{x}(t)\}_{t=0}^{\tau_0+1}, \{\tilde{y}(t)\}_{t=0}^{\tau_0})$ is a program. If $T > \tau_0 + 1$, then we define

$$\tilde{y}(t) = y(t), \ t = \tau_0 + 1, \ldots, T - 1,$$

$$\tilde{x}(t+1) = (1-d)\tilde{x}(t) + x(t+1) - (1-d)x(t), \ t = \tau_0 + 1, \ldots, T - 1. \tag{5.37}$$

By (5.37) and (5.38), $(\{\tilde{x}(t)\}_{t=0}^{T}, \{\tilde{y}(t)\}_{t=0}^{T-1})$ is a program. Lemma 5.4, (5.10), (5.17), (5.18), (5.22), (5.27), (5.29), (5.33)–(5.35), (5.37), and (5.39) imply that

$$M_1 \ge U(x(0), T) - \sum_{t=0}^{T-1} w(by(t)) \ge \sum_{t=0}^{T-1} w(b\tilde{y}(t)) - \sum_{t=0}^{T-1} w(by(t))$$

$$= \sum_{t=\tau}^{T-1} w(b\tilde{y}(t)) - \sum_{t=\tau}^{T-1} w(by(t)) = \sum_{t=\tau}^{\tau_0} w(b\tilde{y}(t)) - \sum_{t=\tau}^{\tau_0} w(by(t))$$

$$\ge \sum_{t=\tau}^{\tau_0-L_1-1} [w(b\tilde{y}(t)) - w(by(t))] - \sum_{t=\tau_0-L_1}^{\tau_0} w(by(t))$$

$$\ge (\tau_0 - L_1 - \tau - 1)\left[w\left((2n)^{-1}\min\left\{1, a_1^{-1}, \ldots, a_i^{-1}, \ldots a_n^{-1}\right\} eb\right) - w(\gamma) \right]$$

$$- w(\gamma) - L_1 w(\gamma) - w(by(\tau_0))$$

$$\ge (L_2 - L_1 - 1)\left[w\left((2n)^{-1}\min\left\{1, a_1^{-1}, \ldots, a_i^{-1}, \ldots a_n^{-1}\right\} be\right) - w(\gamma) \right]$$

$$- (L_1 + 1)w(\gamma) - w(M_0 eb)$$

$$\geq (L_2/2) \left[w \left((2n)^{-1} \min \left\{ 1, a_1^{-1}, \ldots, a_i^{-1}, \ldots a_n^{-1} \right\} eb \right) - w(\gamma) \right]$$
$$- 2L_1 w(1) - w(M_0 eb).$$

By the relation above, we have

$$L_2 \leq 2(M_1 + 2L_1 w(1)$$

$$+ w(M_0 eb))[w((2n)^{-1} \min\{1, a_1^{-1}, \ldots, a_i^{-1}, \ldots a_n^{-1}\} eb) - w(\gamma)]^{-1}.$$

This inequality contradicts (5.22). The contradiction we have reached proves (5.19). This completes the proof of Lemma 5.11.

Lemma 5.12 *Let M_1 be a positive number, and let a number M_0 satisfy (5.10). Then there exist a pair of integers $\bar{L}_1, \bar{L}_2 \geq 1$ and a positive number M_2 such that for every integer $T \geq \bar{L}_1 + \bar{L}_2$ and every program $\{x(t)\}_{t=0}^{T}$, $\{y(t)\}_{t=0}^{T-1}$) satisfying*

$$x(0) \leq M_0 e, \quad \sum_{t=0}^{T-1} w(y(t)) \geq U(x(0), T) - M_1, \tag{5.38}$$

the following assertion holds:
If integers $T_1, T_2 \in [0, T - \bar{L}_2]$ satisfy $\bar{L}_1 \leq T_2 - T_1$, then

$$\sum_{t=T_1}^{T_2-1} w(by(t)) \geq U(x(T_1), T_2 - T_1) - M_2. \tag{5.39}$$

Proof Let a pair of integers $L_1, L_2 \geq 4$ be as guaranteed by Lemma 5.11. Lemma 5.10 implies that there is an integers $L_3 \geq 4$ for which that the following property holds:
(P2) If a natural number $S \geq L_3$, if a program $(\{u(t)\}_{t=0}^{S}, \{v(t)\}_{t=0}^{S-1})$ satisfies

$$u(0) \leq M_0 e, \quad bv(S - 1) \geq \gamma,$$

and if a point $\tilde{u}_0 \in R_+^n$ satisfies $\tilde{u}_0 \leq M_0 e$, then there is a program $(\{\tilde{u}(t)\}_{t=0}^{S}, \{\tilde{v}(t)\}_{t=0}^{S-1})$ satisfying

$$\tilde{u}(0) = \tilde{u}_0, \quad \tilde{u}(S) \geq u(S).$$

Fix a pair of integers $\bar{L}_1, \bar{L}_2 \geq 1$ and a number $M_2 > 0$ which satisfy

$$\bar{L}_1 \geq L_1, \quad \bar{L}_2 > 2(L_2 + L_3 + 1), \tag{5.40}$$

$$M_2 > M_1 + (L_2 + L_3)w\left(M_0 \sum_{i=1}^{n} b_i\right).\tag{5.41}$$

Assume that a natural number $T \geq \bar{L}_1 + \bar{L}_2$, that a program $(\{x(t)\}_{t=0}^{T}, \{y(t)\}_{t=0}^{T-1})$ satisfies (5.38), and that a pair of integers T_1, T_2 satisfies

$$T_1, T_2 \in [0, T - \bar{L}_2], \quad \bar{L}_1 \leq T_2 - T_1.\tag{5.42}$$

We claim that inequality (5.39) holds. Lemma 5.3, (5.10), and (5.38) imply that

$$x(t) \leq M_0 e \text{ for all integers } t = 0, \ldots, T.\tag{5.43}$$

Proposition 5.1 implies that there exists a program

$$\left(\left\{x^{(1)}(t)\right\}_{t=T_1}^{T_2}, \left\{y^{(1)}(t)\right\}_{t=T_1}^{T_2-1}\right)$$

which satisfies

$$x^{(1)}(T_1) = x(T_1), \quad \sum_{t=T_1}^{T_2-1} w\left(by^{(1)}(t)\right) = U(x(T_1), T_2 - T_1).\tag{5.44}$$

Lemma 5.3, (5.10), (5.43), and (5.44) imply that

$$x^{(1)}(t) \leq M_0 e, \quad t = T_1, \ldots, T_2.\tag{5.45}$$

It follows from (5.40) and (5.42) that

$$T_1 + L_1 \leq T_1 + L_3 + \bar{L}_1 \leq L_3 + T_2 \leq T - \bar{L}_2 + L_3 \leq T - 2L_2 - L_3.\tag{5.46}$$

In view of the choice of L_1, L_2, Lemma 5.11, (5.38), (5.40), and (5.46),

$$\max\{by(t) : t = T_2 + L_3, \ldots, T_2 + L_3 + L_2 - 1\} \geq \gamma.$$

Hence there is an integer

$$\tau \in [T_2 + L_3, \ldots, T_2 + L_3 + L_2 - 1]\tag{5.47}$$

for which

$$by(\tau) \geq \gamma.\tag{5.48}$$

Define

$$\tilde{x}(t) = x(t), \ t = 0, \ldots, T_1, \ \tilde{y}(t) = y(t) \text{ for each integer } t$$

$$\text{such that } 0 \le t \le T_1 - 1,$$

$$\tilde{x}(t) = x^{(1)}(t), \ t = T_1 + 1, \ldots, T_2,$$

$$\tilde{y}(t) = y^{(1)}(t), \ t = T_1, \ldots, T_2 - 1.$$

It is not difficult to see that $(\{\tilde{x}(t)\}_{t=0}^{T_2}, \{\tilde{y}(t)\}_{t=0}^{T_2-1})$ is a program. Property (P2), (5.43), and (5.45)–(5.48) imply that there is a program

$$\left(\left\{ x^{(2)}(t) \right\}_{t=T_2}^{\tau+1}, \ \left\{ y^{(2)}(t) \right\}_{t=T_2}^{\tau} \right)$$

such that

$$x^{(2)}(T_2) = x^{(1)}(T_2), \ x^{(2)}(\tau+1) \ge x(\tau+1). \tag{5.49}$$

Set

$$\tilde{x}(t) = x^{(2)}(t), \ t = T_2 + 1, \ldots, \tau + 1, \ \tilde{y}(t) = y^{(2)}(t), \ t = T_2, \ldots, \tau. \tag{5.50}$$

Clearly, $(\{\tilde{x}(t)\}_{t=0}^{\tau+1}, \{\tilde{y}(t)\}_{t=0}^{\tau})$ is a program. It follows from (5.49) and (5.50) that

$$\tilde{x}(\tau + 1) \ge x(\tau + 1). \tag{5.51}$$

Set

$$\tilde{y}(t) = y(t), \ t = \tau + 1, \ldots, T - 1,$$

$$\tilde{x}(t+1) = (1-d)\tilde{x}(t) + x(t+1) - (1-d)x(t), \ t = \tau+1, \ldots, T-1. \tag{5.52}$$

By (5.51) and (5.52), $(\{\tilde{x}(t)\}_{t=0}^{T}, \{\tilde{y}(t)\}_{t=0}^{T-1})$ is a program. Relations (5.38), (5.43), (5.44), (5.47), and (5.52) imply that

$$M_1 \ge U(x(0), T) - \sum_{t=0}^{T-1} w(by(t)) \ge \sum_{t=0}^{T-1} w(b\tilde{y}(t)) - \sum_{t=0}^{T-1} w(by(t))$$

$$= \sum_{t=T_1}^{T-1} w(b\tilde{y}(t)) - \sum_{t=T_1}^{T-1} w(by(t)) \ge \sum_{t=T_1}^{T_2-1} w(b\tilde{y}(t)) - \sum_{t=T_1}^{T_2-1} w(by(t))$$

$$- \sum_{t=T_2}^{\tau} w(by(t))$$

$$\geq U(x(T_1), T_2 - T_1) - \sum_{t=T_1}^{T_2-1} w(by(t)) - (\tau - T_2 + 1)w\left(M_0 \sum_{i=1}^{n} b_i\right).$$

Together with (5.41) and (5.47), the inequality above implies that

$$\sum_{t=T_1}^{T_2-1} w(by(t)) \geq U(x(T_1), T_2 - T_1) - M_1 - (L_3 + L_2)w\left(M_0 \sum_{i=1}^{n} b_i\right)$$

$$\geq U(x(T_1), T_2 - T_1) - M_2.$$

Lemma 5.12 is proved.

5.3   Properties of the Function U

It is not difficult to see that the next result is valid.

Proposition 5.13 *Let $T \geq 1$ be an integer, $\Delta \geq 0$, T_1, T_2 be integers satisfying $0 \leq T_1 < T_2 \leq T$ and let $(\{x(t)\}_{t=0}^{T}, \{y(t)\}_{t=0}^{T-1})$ be a program satisfying*

$$\sum_{t=0}^{T-1} w(by(t)) \geq U(x(0), x(T), T) - \Delta.$$

Then

$$\sum_{t=T_1}^{T_2-1} w(by(t)) \geq U(x(T_1), x(T_2), T_2 - T_1) - \Delta.$$

We use the constant $\gamma > 0$ introduced in Section 5.2 which satisfies relation (5.17).

Lemma 5.14 *Let*

$$M_0 > \max\{(da_i)^{-1} : i = 1, \ldots, n\}. \tag{5.53}$$

Then there are an integer $L \geq 1$ and a positive number M_1 such that for every pair of points $x_0, \tilde{x}_0 \in R_+^n$ which satisfies

$$x_0, \tilde{x}_0 \le M_0 e \tag{5.54}$$

and every natural number $T \ge L$, the following inequality is valid:

$$|U(x_0, T) - U(\tilde{x}_0, T)| \le M_1. \tag{5.55}$$

Proof Let integers $L_1, L_2 \ge 4$ be as guaranteed by Lemma 5.11 with $M_1 = 1$. Lemma 5.10 implies that there is a natural number $L_3 \ge 4$ such that the following property holds:

(P3) If an integer $S \ge L_3$, a program $(\{u(t)\}_{t=0}^S, \{v(t)\}_{t=0}^{S-1})$ satisfies

$$u(0) \le M_0 e, \quad bv(S-1) \ge \gamma,$$

and if $\tilde{u}_0 \in R_+^n$ satisfies $\tilde{u}_0 \le M_0 e$, then there exists a program

$$\left(\{\tilde{u}(t)\}_{t=0}^S, \{\tilde{v}(t)\}_{t=0}^{S-1}\right)$$

which satisfies

$$\tilde{u}(0) = \tilde{u}_0, \quad \tilde{u}(S) \ge u(S).$$

Fix an integer

$$L > 2(L_1 + L_2 + L_3 + 1), \tag{5.56}$$

and set

$$M_1 = (L_1 + L_2 + L_3)w\left(M_0 \sum_{i=1}^n b_i\right). \tag{5.57}$$

Assume that $x_0, \tilde{x}_0 \in R_+^n$ satisfy (5.54) and that a natural number $T \ge L$. Proposition 5.1 implies that there exists a program $(\{x(t)\}_{t=0}^T, \{y(t)\}_{t=0}^{T-1})$ satisfying

$$x(0) = x_0, \quad \sum_{t=0}^{T-1} w(by(t)) = U(x_0, T). \tag{5.58}$$

Lemma 5.5, (5.54) and (5.58) imply that

$$x(t) \le M_0 e \text{ for all } t = 0, \ldots, T. \tag{5.59}$$

By (5.56), we have

$$L_1 + L_3 < L - L_2 \le T - L_2. \tag{5.60}$$

Lemma 5.10, (5.54), (5.56), (5.58), and (5.60) and the choice of L_1, L_2 imply that

$$\max\{by(t) : \ t = L_3 + L_1, \ldots, L_3 + L_1 + L_2 - 1\} \geq \gamma.$$

Thus there exists an integer

$$\tau \in \{L_3 + L_1, \ldots, L_3 + L_1 + L_2 - 1\} \tag{5.61}$$

for which

$$by(\tau) \geq \gamma. \tag{5.62}$$

Let us consider the program $(\{x(t)\}_{t=0}^{\tau+1}, \{y(t)\}_{t=0}^{\tau})$. Property (P3), (5.54), (5.58), (5.59), (5.61), and (5.62) imply that there is a program

$$\left(\{\tilde{x}(t)\}_{t=0}^{\tau+1}, \{\tilde{y}(t)\}_{t=0}^{\tau}\right)$$

which satisfies

$$\tilde{x}(0) = \tilde{x}_0, \ \tilde{x}(\tau + 1) \geq x(\tau + 1). \tag{5.63}$$

Set

$$\tilde{y}(t) = y(t), \ t = \tau + 1, \ldots, T - 1,$$

$$\tilde{x}(t + 1) = (1 - d)\tilde{x}(t) + x(t + 1) - (1 - d)x(t), \ t = \tau + 1, \ldots, T - 1. \tag{5.64}$$

By (5.63) and (5.64), $(\{\tilde{x}(t)\}_{t=0}^{T}, \{\tilde{y}(t)\}_{t=0}^{T-1})$ is a program. Relations (5.57)–(5.59), (5.61), (5.63), and (5.64) imply that

$$U(\tilde{x}_0, T) \geq \sum_{t=0}^{T-1} w(b\tilde{y}(t)) = \sum_{t=0}^{T-1} w(by(t)) - [\sum_{t=0}^{T-1} w(by(t)) - \sum_{t=0}^{T-1} w(b\tilde{y}(t))]$$

$$= U(x_0, T) - \left[\sum_{t=0}^{\tau} w(by(t)) - \sum_{t=0}^{\tau} w(b\tilde{y}(t))\right]$$

$$\geq U(x_0, T) - \sum_{t=0}^{\tau} w(by(t)) \geq U(x_0, T) - (\tau + 1)w\left(M_0 \sum_{i=1}^{n} b_i\right)$$

$$\geq U(x_0, T) - (L_3 + L_1 + L_2)w\left(M_0 \sum_{i=1}^{n} b_i\right) = U(x_0, T) - M_1.$$

Hence we have shown that for every integer $T \geq L$ and every pair of points $x_0, \tilde{x}_0 \in R_+^n$ which satisfy (5.54), we have

$$U(\tilde{x}_0, T) \geq U(x_0, T) - M_1.$$

Lemma 5.14 is proved.

Corollary 5.15 *Let a number M_0 satisfy (5.53). Then there exist a positive number M_1 and an integer $L \geq 1$ such that for every natural number $T \geq L$ and every $x_0 \in R_+^n$ which satisfies $x_0 \leq M_0 e$,*

$$|U(x_0, T) - \widehat{U}(M_0, T)| \leq M_1.$$

The next result follows from Lemmas 5.3 and 5.12, Corollary 5.15, and (5.53).

Lemma 5.16 *Let a number M_0 satisfy (5.53), and let M_1 be a positive number. Then there exist integers $\bar{L}_1, \bar{L}_2 \geq 1$ and a number $M_2 > 0$ such that for every natural number $T \geq \bar{L}_1 + \bar{L}_2$ and every program $(\{x(t)\}_{t=0}^T, \{y(t)\}_{t=0}^{T-1})$ satisfying*

$$x(0) \leq M_0 e, \quad \sum_{t=0}^{T-1} w(by(t)) \geq U(x(0), T) - M_1,$$

the following assertion holds:
If integers $T_1, T_2 \in [0, T - \bar{L}_2]$ satisfy $T_2 - T_1 \geq \bar{L}_1$, then

$$\sum_{t=T_1}^{T_2-1} w(by(t)) \geq \widehat{U}(M_0, T_2 - T_1) - M_2.$$

5.4   Proofs of Theorems 5.4, 5.5 and 5.8

In the sequel we assume that the sum over empty set is zero.
    Choose

$$M_0 > \max\{(da_i)^{-1} : i = 1, \ldots, n\}, \quad M_1 = 1. \tag{5.65}$$

Let natural numbers $\bar{L}_1, \bar{L}_2$ and a positive number M_2 be as guaranteed by Lemma 5.16.

Let $x_0 \in R_+^n$ satisfy

$$x_0 \leq M_0 e. \tag{5.66}$$

Proposition 5.1 implies that for every integer $k \geq 1$, there exists a program $(\{x^{(k)}(t)\}_{t=0}^k, \{y^{(k)}(t)\}_{t=0}^{k-1})$ which satisfies

$$x^{(k)}(0) = x_0, \quad \sum_{t=0}^{k-1} w\left(by^{(k)}(t)\right) = U(x_0, k). \tag{5.67}$$

By Lemma 5.3 and (5.65)–(5.67), for every integer $k \geq 1$,

$$x^{(k)}(t) \leq M_0 e, \ t = 0, \ldots, k. \tag{5.68}$$

Lemma 5.16, the choice of $\bar{L}_1, \bar{L}_2$ and M_2, (5.65), and (5.67) imply that the following property holds:

(i) For every natural number $k \geq \bar{L}_1 + \bar{L}_2$ and every pair of integers $T_1, T_2 \in [0, k - \bar{L}_2]$ satisfying $T_2 - T_1 \geq \bar{L}_1$,

$$\sum_{t=T_1}^{T_2-1} w\left(by^{(k)}(t)\right) \geq \widehat{U}(M_0, T_2 - T_1) - M_2. \tag{5.69}$$

By (5.68), there exists a strictly increasing sequence of natural numbers $\{k_j\}_{j=1}^\infty$ such that for every nonnegative integer t, there exist

$$\widehat{x}(t) = \lim_{j \to \infty} x^{(k_j)}(t), \ \widehat{y}(t) = \lim_{j \to \infty} y^{(k_j)}(t). \tag{5.70}$$

Evidently, $\{\widehat{x}(t), \widehat{y}(t)\}_{t=0}^\infty$ is a program. In view of (5.68) and (5.70),

$$\widehat{x}(t) \leq M_0 e \text{ for all integers } t \geq 0. \tag{5.71}$$

In view of (5.67) and (5.70),

$$\widehat{x}(0) = x_0. \tag{5.72}$$

Property (i), (5.70), and (5.71) imply that for every pair of integers $T_1, T_2 \in [0, \infty)$ which satisfies $T_2 - T_1 \geq \bar{L}_1$, we have

$$\widehat{U}(M_0, T_2 - T_1) \geq \sum_{t=T_1}^{T_2-1} w(b\widehat{y}(t)) \geq \widehat{U}(M_0, T_2 - T_1) - M_2. \tag{5.73}$$

Let p be an integer for which $p \geq \bar{L}_1$. We claim that for all sufficiently large natural numbers T, we have

$$\left| p^{-1}\widehat{U}(M_0, p) - T^{-1}\sum_{t=0}^{T-1} w(b\widehat{y}(t)) \right| \leq 2p^{-1}M_2. \tag{5.74}$$

Assume that $T \geq p$ is a natural number. Then there exist integers q, s for which

$$q \geq 1, \ 0 \leq s < p, \ T = pq + s. \tag{5.75}$$

By (5.75),

$$T^{-1}\sum_{t=0}^{T-1} w(b\widehat{y}(t)) - p^{-1}\widehat{U}(M_0, p)$$

$$= T^{-1}\left(\sum_{t=0}^{pq-1} w(b\widehat{y}(t)) + \sum\{w(b\widehat{y}(t)) : \right.$$

$$\left. t \text{ is an integer such that } pq \leq t \leq T - 1\} \right) - p^{-1}\widehat{U}(M_0, p)$$

$$= T^{-1}\sum\{w(b\widehat{y}(t)) : t \text{ is an integer such that } pq \leq t \leq T - 1\}$$

$$+ \left(T^{-1}pq \right)(pq)^{-1}\sum_{i=0}^{q-1}\sum_{t=ip}^{(i+1)p-1} w(b\widehat{y}(t)) - p^{-1}\widehat{U}(M_0, p)$$

$$= T^{-1}\sum\{w(b\widehat{y}(t)) : t \text{ is an integer such that } pq \leq t \leq T - 1\}$$

$$+ \left(T^{-1}pq \right)(pq)^{-1}\left[\sum_{i=0}^{q-1}\left(\sum_{t=ip}^{(i+1)p-1} w(b\widehat{y}(t)) \right. \right.$$

$$\left. \left. -\widehat{U}(M_0, p) \right) + q\widehat{U}(M_0, p) \right] - p^{-1}\widehat{U}(M_0, p).$$

Together with (5.71), (5.73), (5.75), and the inequality $p \geq \bar{L}$, the equation above implies that

$$\left| T^{-1}\sum_{t=0}^{T-1} w(b\widehat{y}(t)) - p^{-1}\widehat{U}(M_0, p) \right| \leq T^{-1}pw(M_0be) + (pq)^{-1}qM_2$$

$$+ \widehat{U}(M_0, p)|q/T - 1/p| \leq T^{-1} pw(M_0 be)$$
$$+ M_2/p + \widehat{U}(M_0, p)s(pT)^{-1} \to M_2/p \text{ as } T \to \infty.$$

Thus (5.74) is true for all sufficiently large natural numbers T.

Since p is an arbitrary integer satisfying $p \geq \bar{L}_1$, we conclude that

$$\left\{ T^{-1} \sum_{t=0}^{T-1} w(b\widehat{y}(t)) \right\}_{T=1}^{\infty}$$

is a Cauchy sequence. Evidently, there exists

$$\lim_{T \to \infty} T^{-1} \sum_{t=0}^{T-1} w(b\widehat{y}(t)).$$

It follows from (5.74) that for every integer $p \geq \bar{L}_1$, we have

$$|p^{-1}\widehat{U}(M_0, p) - \lim_{T \to \infty} T^{-1} \sum_{t=0}^{T-1} w(b\widehat{y}(t))| \leq p^{-1}(2M_2). \tag{5.76}$$

Since (5.76) is true for every integer $p \geq \bar{L}_1$, we conclude that

$$\lim_{T \to \infty} T^{-1} \sum_{t=0}^{T-1} w(b\widehat{y}(t)) = \lim_{p \to \infty} \widehat{U}(M_0, p)/p. \tag{5.77}$$

Now it is easy to see that Theorem 5.4 is true.

Let μ be defined by (5.7). We have

$$\mu = \lim_{p \to \infty} \widehat{U}(M_0, p)/p. \tag{5.78}$$

It follows from (5.76)–(5.78) that for every natural number $p \geq \bar{L}_1$, we have

$$\left| p^{-1}\widehat{U}(M_0, p) - \mu \right| \leq p^{-1}(2M_2). \tag{5.79}$$

Inequality (5.79) implies the validity of Theorem 5.5.

Now we are ready to complete the proof of Theorem 5.8. In view of (5.73) and (5.79), for every pair of integers $T_1, T_2 \in [0, \infty)$ which satisfies $T_2 - T_1 \geq \bar{L}_1$, we have

$$\left| \sum_{t=T_1}^{T_2-1} w(b\widehat{y}(t)) - \mu(T_2 - T_1) \right| \leq \left| \sum_{t=T_1}^{T_2-1} w(b\widehat{y}(t)) - \widehat{U}(M_0, T_2 - T_1) \right|$$

$$+ \left| \widehat{U}(M_0, T_2 - T_1) - \mu(T_2 - T_1) \right| \leq 3M_2. \tag{5.80}$$

By (5.71), for every nonnegative integer T_1 and every integer $T_2 \in [T_1+1, T_1+\bar{L}_1]$, we have

$$\left| \sum_{t=T_1}^{T_2-1} w(b\widehat{y}(t)) - \mu(T_2 - T_1) \right| \leq \sum_{t=T_1}^{T_2-1} w(b\widehat{y}(t)) + \mu\bar{L}_1 \leq \bar{L}_1 w(M_0 be) + \mu\bar{L}_1.$$

Combined with (5.80) the inequality above implies that for every pair of integers $T_1, T_2 \geq 0$ which satisfies $T_2 > T_1$, we have

$$\left| \sum_{t=T_1}^{T_2-1} w(b\widehat{y}(t)) - \mu(T_2 - T_1) \right| \leq 3M_2 + \bar{L}_1(w(M_0 be) + \mu). \tag{5.81}$$

We claim that for every natural number T,

$$\sum_{t=0}^{T-1} w(b\widehat{y}(t)) = U(x(0), x(T), T). \tag{5.82}$$

Since $\mu > w(0)$ (see (5.8)), it follows from (5.81) that there exists a strictly increasing sequence $\{T_i\}_{i=1}^{\infty}$ such that $T_1 \geq 4$ and that

$$w(b\widehat{y}(T_i - 1)) > 2^{-1}(\mu + w(0)) \text{ for all natural numbers } i. \tag{5.83}$$

Since the function w is continuous, there exists a positive number r_0 such that

$$b\widehat{y}(T_i - 1) > r_0 \text{ for all natural numbers } i.$$

Extracting a subsequence and re-indexing if necessary, we may assume that there exist a positive number r_1 and a natural number $q \in \{1, \ldots, n\}$ such that

$$\widehat{y}_q(T_i - 1) > r_1 \text{ for all natural numbers } i. \tag{5.84}$$

It is clear that it is sufficient to show that (5.82) holds for all $T = T_i - 1$, $i = 1, 2, \ldots$.

Let j be a natural number. We claim that (5.82) is true with $T = T_j - 1$. Assume the contrary. Then there exist a program $(\{\bar{x}(t)\}_{t=0}^{T_j-1}, \{\bar{y}(t)\}_{t=0}^{T_j-2})$ and a number $\Delta > 0$ which satisfy

$$\bar{x}(0) = \widehat{x}(0), \ \bar{x}(T_j - 1) \geq \widehat{x}(T_j - 1), \tag{5.85}$$

$$\sum_{t=0}^{T_j-2} w(b\bar{y}(t)) > \sum_{t=0}^{T_j-2} w(b\widehat{y}(t)) + 2\Delta. \tag{5.86}$$

Since the function w is continuous, there is a number

$$\delta_0 \in (0, r_1/16) \tag{5.87}$$

such that for every pair of points $z_1, z_2 \in R_+^n$ which satisfies $z_1, z_2 \leq M_0 e$ and $\|z_1 - z_2\| \leq \delta_0 n$, we have

$$|w(bz_1) - w(bz_2)| \leq \Delta/(2T_j + 2)^{-1}. \tag{5.88}$$

Fix a positive number

$$\delta < \min\left\{1, \Delta/16, \delta_0, \delta_0(\max\{a_i \ : \ i = 1, \ldots, n\})^{-1}(4n)^{-1}\right\}. \tag{5.89}$$

In view of the construction of the program $\{\widehat{x}(t), \widehat{y}(t)\}_{t=0}^{\infty}$ (see (5.70)), there exists an integer $k > T_j + 1$ such that

$$\|x^{(k)}(t) - \widehat{x}(t)\| \leq \delta, \ t = 0, \ldots, T_j + 1, \tag{5.90}$$

$$\|y^{(k)}(t) - \widehat{y}(t)\| \leq \delta, \ t = 0, \ldots, T_j + 1.$$

Define

$$\tilde{x}(t) = \bar{x}(t), \ t = 0, \ldots, T_j - 1, \tag{5.91}$$

$$\tilde{y}(t) = \bar{y}(t), \ t = 0, \ldots, T_j - 2,$$

and set

$$\tilde{y}(T_j - 1) = \widehat{y}(T_j - 1) - \delta_0 e(q), \tag{5.92}$$

$$\tilde{x}(T_j) = \widehat{x}(T_j) - (1 - d)\widehat{x}(T_j - 1) + (1 - d)\bar{x}(T_j - 1)$$
$$+ \delta_0(\max\{a_i \ : \ i = 1, \ldots, n\})^{-1}(4n)^{-1}e. \tag{5.93}$$

It follows from (5.84), (5.87), and (5.92) that $\tilde{y}(T_j - 1) \geq 0$. By (5.85), (5.91), and (5.92),

$$\tilde{y}(T_j - 1) \leq \widehat{y}(T_j - 1) \leq \widehat{x}(T_j - 1) \leq \bar{x}(T_j - 1) = \tilde{x}(T_j - 1). \tag{5.94}$$

In view of (5.91) and (5.93), we have

$$\tilde{x}(T_j) - (1 - d)\tilde{x}(T_j - 1) = \widehat{x}(T_j) - (1 - d)\widehat{x}(T_j - 1)$$

$$+ \delta_0 \min\{a_i^{-1} : i = 1, \ldots, n\}(4n)^{-1}e. \tag{5.95}$$

By (5.95),

$$\tilde{x}(T_j) \geq (1 - d)\tilde{x}(T_j - 1). \tag{5.96}$$

Relations (5.92) and (5.95) imply that

$$a(\tilde{x}(T_j) - (1 - d)\tilde{x}(T_j - 1)) + e\tilde{y}(T_j - 1)$$

$$\leq a(\widehat{x}(T_j) - (1 - d)\widehat{x}(T_j - 1)) + \delta_0/4 + e\widehat{y}(T_j - 1) - \delta_0 \leq 1.$$

Together with (5.91), (5.94), and (5.96), the relation above implies that $(\{\tilde{x}(t)\}_{t=0}^{T_j}, \{\tilde{y}(t)\}_{t=0}^{T_j-1})$ is a program. It follows from (5.67), (5.72), (5.85), and (5.91) that

$$\tilde{x}(0) = \bar{x}(0) = \widehat{x}(0) = x_0 = x^{(k)}(0). \tag{5.97}$$

By (5.85), (5.89), (5.90), and (5.93),

$$\tilde{x}(T_j) \geq \widehat{x}(T_j) + \delta_0(\max\{a_i : i = 1, \ldots, n\})^{-1}(4n)^{-1}e \geq \widehat{x}(T_j) + \delta e \geq x_{T_j}^{(k)}. \tag{5.98}$$

In view of (5.68), (5.71), (5.89), (5.90), (5.92), and the choice of δ_0 (see (5.88)),

$$|w(by^{(k)}(t)) - w(b\widehat{y}(t))| \leq \Delta(2T_j + 2)^{-1}, \ t = 0, \ldots, T_j + 1,$$

$$|w(b\tilde{y}(T_j - 1)) - w(b\widehat{y}(T_j - 1))| \leq \Delta(2T_j + 2)^{-1}. \tag{5.99}$$

It follows from (5.86), (5.91), and (5.99) that

$$\sum_{t=0}^{T_j-1} \left[w(b\tilde{y}(t)) - w(by^{(k)}(t)) \right]$$

$$= \sum_{t=0}^{T_j-2} \left[w(b\tilde{y}(t)) - w(by^{(k)}(t)) \right] + w(b\tilde{y}(T_j - 1)) - w\left(by^{(k)}(T_j - 1) \right)$$

$$= \sum_{t=0}^{T_j-2} [w(b\bar{y}(t)) - w(b\widehat{y}(t))] + \sum_{t=0}^{T_j-2} \left[w(b\widehat{y}(t)) - w(by^{(k)}(t)) \right]$$

$$+ w(b\tilde{y}(T_j - 1)) - w(b\widehat{y}(T_j - 1)) + w(b\widehat{y}(T_j - 1)) - w(by^{(k)}(T_j - 1))$$

$$\geq 2\Delta - (T_j - 1)\Delta(2T_j + 2)^{-1} - \Delta(2T_j + 2)^{-1} - \Delta(2T_i + 2)^{-1} \geq \Delta$$

and

$$\sum_{t=0}^{T_j-1} w(b\tilde{y}(t)) \geq \sum_{t=0}^{T_j-1} w(by^{(k)}(t)) + \Delta.$$

Combined with (5.97) and (5.98), the inequality above implies that

$$U(x_0, x^{(k)}(T_j), T_j) \geq \sum_{t=0}^{T_j-1} w(by^{(k)}(t)) + \Delta.$$

Combined with Proposition 5.13, the relation above implies that

$$U(x_0, k) \geq \sum_{t=0}^{k-1} w(by^{(k)}(t)) + \Delta/2.$$

This inequality contradicts (5.4). The contradiction we have reached proves that (5.82) holds for $T = T_j - 1$ and for every natural number j. This implies that (5.82) is valid for every natural number T. Now the validity of Theorem 5.8 follows from (5.81) and (5.82).

5.5   Proof of Proposition 5.7

Fix a number $M_0 > 0$ for which

$$M_0 > \|x(0)\| + d^{-1}\max\{a_i^{-1} : i = 1, \ldots, n\}. \tag{5.100}$$

Lemma 5.3 and (5.100) imply that

$$x(t) \leq M_0 e \text{ for all integers } t \geq 0. \tag{5.101}$$

Corollary 5.6 and (5.101) imply that there exists $M > 0$ such that for every nonnegative integer T_1 and every integer $T_2 > T_1$, we have

$$\sum_{t=T_1}^{T_2-1} [w(by(t)) - \mu] \leq M. \tag{5.102}$$

Assume that the sequence $\{\sum_{t=0}^{T-1} [w(by(t)) - \mu]\}_{T=1}^{\infty}$ is not bounded. In view of (5.102),

$$\liminf_{T \to \infty} \sum_{t=0}^{T-1} [w(by(t)) - \mu] = -\infty. \tag{5.103}$$

Let Q be a positive number. It follows from (5.103) that there exists an integer $T_0 \geq 1$ such that

$$\sum_{t=0}^{T_0-1} [w(by(t)) - \mu] < -Q - M.$$

By the inequality above and the choice of M (see (5.102)), for every natural number $T > T_0$,

$$\sum_{t=0}^{T-1} [w(by(t)) - \mu] = \sum_{t=0}^{T_0-1} [w(by(t)) - \mu] + \sum_{t=T_0}^{T-1} [w(by(t)) - \mu]$$

$$< -Q - M + M = -Q.$$

Since Q is an arbitrary positive number, we obtain that

$$\lim_{T \to \infty} \sum_{t=0}^{T-1} [w(by(t)) - \mu] = -\infty.$$

This completes the proof of Proposition 5.7.

5.6   Proof of Theorem 5.9

Assume that a program $\{x(t), y(t)\}_{t=0}^{\infty}$ satisfies

$$\sum_{t=0}^{T-1} w(by(t)) = U(x(0), x(T), T) \text{ for all integers } T \geq 1, \tag{5.104}$$

$$\limsup_{t \to \infty} by(t) > 0. \tag{5.105}$$

In view of (5.105), there exist a number $\Delta > 0$ and a strictly increasing sequence of natural numbers $\{T_i\}_{i=1}^{\infty}$ for which

$$by(T_i - 1) \geq \Delta \text{ for all integers } i \geq 1. \tag{5.106}$$

Theorem 5.8 implies that there exist a program $\{\widehat{x}(t), \widehat{y}(t)\}_{t=0}^{\infty}$ and a positive number $M_1 > 0$ such that

$$\widehat{x}(0) = x(0), \tag{5.107}$$

$$\left| \sum_{t=T_1}^{T_2-1} [w(b\widehat{y}(t)) - (T_2 - T_1)\mu] \right| \leq M_1$$

$$\text{for every integer } T_1 \geq 0 \text{ and every integer } T_2 > T_1. \tag{5.108}$$

Fix a number

$$M_0 > \|x(0)\| + \max\{(a_i d)^{-1} : i = 1, \ldots, n\}. \tag{5.109}$$

Lemma 5.3, (5.107), and (5.109) imply that

$$x(t), \ \widehat{x}(t) \leq M_0 e \text{ for all integers } t \geq 0. \tag{5.110}$$

By (5.109) and Lemma 5.10, there exists an integer $\tau_0 \geq 4$ such that the following property holds:
(P4) If an integer $S \geq \tau_0$, if a program $(\{u(t)\}_{t=0}^{S}, \{v(t)\}_{t=0}^{S-1})$ satisfies

$$u(0) \leq M_0 e, \ bv(S - 1) \geq \Delta,$$

and if $\tilde{u}_0 \in R_+^n$ satisfies $\tilde{u}_0 \leq M_0 e$, then there exists a program

$$\left(\{\tilde{u}(t)\}_{t=0}^{S}, \{\tilde{v}(t)\}_{t=0}^{S-1} \right)$$

such that

$$\tilde{u}(0) = \tilde{u}_0, \ \tilde{u}(S) \geq u(S).$$

Let $i \geq 1$ be an integer satisfying

$$T_i > \tau_0.$$

Consider the program $\left(\{x(t)\}_{t=T_i-\tau_0}^{T_i}, \{y(t)\}_{t=T_i-\tau_0}^{T_i-1} \right)$. Property (P4), (5.106), and (5.110) imply that there exists a program

$$(\{x^{(1)}(t)\}_{t=T_i-\tau_0}^{T_i}, \{y^{(1)}(t)\}_{t=T_i-\tau_0}^{T_i-1})$$

which satisfies

$$x^{(1)}(T_i - \tau_0) = \widehat{x}(T_i - \tau_0), \ x^{(1)}(T_i) \geq x(T_i). \tag{5.111}$$

Set

$$\widetilde{x}(t) \quad = \widehat{x}(t), \ t = 0, \ldots, T_i - \tau_0, \ \widetilde{y}(t) = \widehat{y}(t), \ t = 0, \ldots, T_i - \tau_0 - 1,$$

$$\widetilde{x}(t) = x^{(1)}(t), \ t = T_i - \tau_0 + 1, \ldots, T_i, \ \widetilde{y}(t) = y^{(1)}(t), \ t = T_i - \tau_0, \ldots, T_i - 1. \tag{5.112}$$

It follows from (5.111) and (5.112) that $(\{\widetilde{x}(t)\}_{t=0}^{T_i}, \{\widetilde{y}(t)\}_{t=0}^{T_i-1})$ is a program. By (5.107), (5.111), and (5.112), we have

$$\widetilde{x}(0) = x(0), \ \widetilde{x}(T_i) \geq x(T_i). \tag{5.113}$$

In view of (5.104), (5.108), (5.112), and (5.113), we have

$$0 \leq \sum_{t=0}^{T_i-1} w(by(t)) - \sum_{t=0}^{T_i-1} w(b\widetilde{y}(t))$$

$$= \sum_{t=0}^{T_i-1} w(by(t)) - \sum_{t=0}^{T_i-\tau_0-1} w(b\widehat{y}(t)) - \sum_{t=T_i-\tau_0}^{T_i-1} w(by^{(1)}(t))$$

$$\leq \sum_{t=0}^{T_i-1} w(by(t)) - \mu(T_i - \tau_0) + M_1 = \sum_{t=0}^{T_i-1} w(by(t)) - \mu T_i + M_1 + \mu\tau_0$$

and

$$\sum_{t=0}^{T_i-1} w(by(t)) - \mu T_i \geq -M_1 - \mu\tau_0.$$

Since the inequality above is true for an arbitrary integer $i \geq 1$ for which $T_i > \tau_0$, we obtain that

$$\limsup_{T \to \infty} \sum_{t=0}^{T-1} [w(by(t)) - \mu] \geq M_1 - \mu\tau_0.$$

Together with Proposition 5.7, this implies that the sequence $\{\sum_{t=0}^{T-1}[w(by(t)) - \mu]\}_{t=1}^{\infty}$ is bounded. This completes the proof of Theorem 5.9.

5.7   The RSS Model with Discounting

For every nonnegative integer t, let $w_t : [0, \infty) \rightarrow [0, \infty)$ be a continuous increasing function which represents the preferences of the planner at moment of time t. We suppose that the following assumption holds.

Assumption A For every nonnegative integer $t \geq 0$, $w_t(0) = 0$, and for every positive number M,

$$\lim_{t \to \infty} w_t(M) = 0.$$

For every point $x_0 \in R_+^n$ and every natural number T define

$$U(x_0, T) = \sup \left\{ \sum_{t=0}^{T-1} w_t(by(t)) : \ (\{x(t)\}_{t=0}^{T}, \{y(t)\}_{t=0}^{T-1}) \right\}$$

is a program such that $x(0) = x_0$. \hfill (5.114)

The next proposition follows immediately from the continuity of w_t, $t = 0, 1, \ldots$.

Proposition 5.17 For every $x_0 \in R_+^n$ and every integer $T \geq 1$, there exists a program $(\{x(t)\}_{t=0}^{T}, \{y(t)\}_{t=0}^{T-1})$ which satisfies $x(0) = x_0$ and

$$\sum_{t=0}^{T-1} w_t(by(t)) = U(x_0, T).$$

We prove the following theorem which was obtained in [52].

Theorem 5.18 For every $z \in R_+^n$, there exists a program $\{x_z(t), y_z(t)\}_{t=0}^{\infty}$ such that $x_z(0) = z$ and the following property holds:

For every pair of positive numbers M_0, δ, there exists an integer $L^{(\delta)} \geq 1$ such that for every natural number $S \geq L^{(\delta)}$ and every $z \in R_+^n$ which satisfies $z \leq M_0 e$,

$$\sum_{t=0}^{S-1} w_t(by_z(t)) \geq U(z, S) - \delta.$$

Corollary 5.19 *Let $z \in R_+^n$, and let a program $\{x_z(t), y_z(t)\}_{t=0}^{\infty}$ be as guaranteed by Theorem 5.18. Then for every program $\{x(t), y(t)\}_{t=0}^{\infty}$ satisfying $x(0) = z$, the inequality*

$$\liminf_{T \to \infty} \left[\sum_{t=0}^{T-1} w_t(by_z(t)) - \sum_{t=0}^{T-1} w_t(by(t)) \right] \geq 0$$

holds.

Proof Let a positive number M_0 satisfy $z \leq M_0 e$, $\delta > 0$, and let an integer $L^{(\delta)} \geq 1$ be as guaranteed by Theorem 5.18. Then for every integer $S \geq L^{(\delta)}$, we have

$$\sum_{t=0}^{S-1} w_t(by(t)) - \sum_{t=0}^{S-1} w_t(by_z(t)) \leq U(z, S) - (U(S, z) - \delta) \leq \delta.$$

This completes the proof of Corollary 5.19.

Example 5.20 Let $w : [0, \infty) \to [0, \infty)$ be a continuous increasing function, $w(0) = 0$, $\{\rho_t\}_{t=0}^{\infty} \subset (0, 1)$, $\lim_{t \to \infty} \rho_t = 0$, $w_t = \rho_t w$, $t = 0, 1, \ldots$. Then Assumption A holds.

Assume that $\sum_{t=0}^{\infty} \rho_t = \infty$ and that $w(s) > 0$ for every positive number s. Let $z \in R_+^n$ be given. Set $x(0) = z$, $y(0) = 0$, for every natural number t,

$$y(t) = (2n)^{-1} \min\{1, a_1^{-1}, \ldots, a_i^{-1}, \ldots, a_n^{-1}\}e,$$

for every nonnegative integer t,

$$x(t + 1) = (1 - d)x(t) + (2n)^{-1}(a_1^{-1}, \ldots, a_n^{-1}).$$

Evidently, $\{x(t), y(t)\}_{t=0}^{\infty}$ is a program, and for every natural number T,

$$\sum_{t=0}^{T} w_t(by(t)) = \sum_{t=1}^{T} \rho_t w \left(be(2n)^{-1} \min\{1, a_1^{-1}, \ldots, a_n^{-1}\} \right)$$

$$= \left(\sum_{t=1}^{T} \rho_t \right) w \left(be(2n)^{-1} \min\{1, a_1^{-1}, \ldots, a_n^{-1}\} \right) \to \infty \text{ as } T \to \infty.$$

This implies that $U(z, T) \to \infty$ as $T \to \infty$.

5.8 An Auxiliary Result for Theorem 5.18

Fix $\gamma \in (0, 1)$ which satisfies

$$\gamma < (2n)^{-1} \sum_{i=1}^{n} b_i \min \left\{ 1, a_1^{-1}, \dots, a_i^{-1}, \dots, a_n^{-1} \right\}. \tag{5.115}$$

In the sequel we assume that the sum over empty set is zero.

Lemma 5.21 *Let a number M_0 satisfy*

$$M_0 > \max\{(a_i d)^{-1} : i = 1, \dots, n\}, \tag{5.116}$$

and let δ be a positive number. Then there exists an integer $\bar{L} \geq 1$ such that for every integer $L \geq \bar{L}$, there exists an integer $\tau \geq L$ for which the following assertion holds:

For every natural number $T \geq \tau$ and every program $(\{x(t)\}_{t=0}^{T}, \{y(t)\}_{t=0}^{T-1})$ which satisfies

$$x(0) \leq M_0 e, \quad \sum_{t=0}^{T-1} w_t(by(t)) = U(x(0), T), \tag{5.117}$$

the inequality

$$\sum_{t=0}^{L-1} w_t(by(t)) \geq U(x(0), L) - \delta \tag{5.118}$$

is valid.

Proof Lemma 5.10 implies that there exists an integer $L_0 \geq 4$ such that the following property holds:

(P5) If an integer $S \geq L_0$, if a program $(\{u(t)\}_{t=0}^{S}, \{v(t)\}_{t=0}^{S-1})$ satisfies

$$u(0) \leq M_0 e, \quad bv(S - 1) \geq \gamma,$$

and if $\tilde{u}_0 \in R_+^n$ satisfies $\tilde{u}_0 \leq M_0 e$, then there exists a program

$$\left(\{\tilde{u}(t)\}_{t=0}^{S}, \{\tilde{v}(t)\}_{t=0}^{S-1} \right)$$

which satisfies

$$\tilde{u}(0) = \tilde{u}_0, \quad \tilde{u}(S) \geq u(S).$$

Assumption (A) implies that there exists an integer $\bar{L} \geq 1$ such that for every natural number $L \geq \bar{L}$, we have

$$w_L(M_0be) < (\delta/8)(4L_0)^{-1}. \tag{5.119}$$

Assume that an integer $L \geq \bar{L}$, and fix a natural number

$$\tau \geq L + L_0 + 2. \tag{5.120}$$

Assume that an integer $T \geq \tau$ and that a program $(\{x(t)\}_{t=0}^T, \{y(t)\}_{t=0}^{T-1})$ satisfies (5.117). We claim that (5.118) holds.

Proposition 5.17 implies that there exists a program $(\{\tilde{x}(t)\}_{t=0}^L, \{\tilde{y}(t)\}_{t=0}^{L-1})$ satisfying

$$\tilde{x}(0) = x(0), \quad \sum_{t=0}^{L-1} w_t(b\tilde{y}(t)) = U(x(0), L). \tag{5.121}$$

Lemma 5.3, (5.116), (5.117), and (5.121) imply that

$$\tilde{x}(t) \leq M_0e, \ t = 0, \ldots, L, \ x(t) \leq M_0e, \ t = 0, \ldots, T. \tag{5.122}$$

There are two cases:

$$by(t) < \gamma, \ t = L + L_0, \ldots, T - 1; \tag{5.123}$$

$$\max\{by(t) : \ t = L + L_0, \ldots, T - 1\} \geq \gamma. \tag{5.124}$$

Assume that (5.123) is valid. Set $(\{x^{(1)}(t)\}_{t=0}^T, \{y^{(1)}(t)\}_{t=0}^{T-1})$ as follows:

$$x^{(1)}(t) = \tilde{x}(t), \ t = 0, \ldots, L, \ y^{(1)}(t) = \tilde{y}(t), \ t = 0, \ldots, L - 1, \ \tilde{y}(L) = 0,$$

$$y^{(1)}(t) = (2n)^{-1} \min\{1, a_1^{-1}, \ldots, a_i^{-1}, \ldots, a_n^{-1}\}e, \ t = L + 1, \ldots, T - 1,$$

$$x^{(1)}(t + 1) = (1 - d)x^{(1)}(t) + (2n)^{-1}(a_1^{-1}, \ldots, a_i^{-1}, \ldots, a_n^{-1}), \ t=L, \ldots, T - 1. \tag{5.125}$$

It is clear that $(\{x^{(1)}(t)\}_{t=0}^T, \{y^{(1)}(t)\}_{t=0}^{T-1})$ is a program. It follows from (5.114), (5.115), (5.117), (5.119), (5.121), (5.123), (5.125), assumption (A), and the monotonicity of the functions $w_t, t = 0, 1, \ldots$ that

$$0 \geq \sum_{t=0}^{T-1} w_t\left(by^{(1)}(t)\right) - U(x(0), T) = \sum_{t=0}^{T-1} w_t(by^{(1)}(t)) - \sum_{t=0}^{T-1} w_t(by(t))$$

$$= \sum_{t=0}^{L-1} w_t(b\tilde{y}(t)) + w_L(0)$$

$$+ \sum_{t=L+1}^{T-1} w_t \left((2n)^{-1} \left(\sum_{i=1}^{n} b_i \right) \min\{1, a_1^{-1}, \ldots, a_i^{-1}, \ldots, a_n^{-1}\} \right)$$

$$- \sum_{t=0}^{T-1} w_t(by(t)) \geq U(x(0), L) - \sum_{t=0}^{L-1} w_t(by(t)) - w_L(M_0 be)$$

$$- \sum_{t=L+1}^{L+L_0} w_t(M_0 be)$$

$$+ \sum_{t=L+1}^{T-1} \left[w_t \left((2n)^{-1} (\sum_{i=1}^{n} b_i) \min\{1, a_1^{-1}, \ldots, a_i^{-1}, \ldots, a_n^{-1}\} \right) - w_t(\gamma) \right]$$

$$\geq U(x(0), L) - \sum_{t=0}^{L-1} w_t(by(t)) - \sum_{t=L}^{L+L_0} w_t(M_0 be)$$

$$\geq U(x(0), L) - \sum_{t=0}^{L-1} w_t(by(t)) - \delta$$

and

$$\sum_{t=0}^{L-1} w_t(by(t)) \geq U(x(0), L) - \delta.$$

Therefore if (5.123) holds, then (5.118) is true.

Assume that (5.124) is valid. Then in view of (5.124), there exists an integer S_0 which satisfies

$$L + L_0 \leq S_0 - 1 \leq T - 1,$$

$$by(S_0 - 1) \geq \gamma,$$

$$by(t) < \gamma \text{ for each integer } t \text{ satisfying } L_0 + L \leq t < S_0 - 1. \tag{5.126}$$

Let us define a sequence $(\{x^{(2)}(t)\}_{t=0}^{T}, \{y^{(2)}(t)\}_{t=0}^{T-1})$. Set

$$x^{(2)}(t) = \tilde{x}(t), \ t = 0, \ldots, L, \ y^{(2)}(t) = \tilde{y}(t), \ t = 0, \ldots, L - 1, \ y^{(2)}(L) = 0,$$

$$y^{(2)}(t) = (2n)^{-1} \min\{1, a_1^{-1}, \ldots, a_i^{-1}, \ldots, a_n^{-1}\} e$$

if an integer t satisfies $L < t \leq S_0 - L_0 - 1$,

$$x^{(2)}(t+1) = (1-d)x^{(2)}(t) + (2n)^{-1}(a_1^{-1}, \ldots, a_i^{-1}, \ldots, a_n^{-1})$$

if an integer t satisfies $L \leq t \leq S_0 - L_0 - 1$.   (5.127)

It is easy to see that $(\{x^{(2)}(t)\}_{t=0}^{S_0-L_0}, \{y^{(2)}(t)\}_{t=0}^{S_0-L_0-1})$ is a program. In view of (5.117), (5.121), (5.127), and Lemma 5.3,

$$x(t) \leq M_0 e, \ t = 0, \ldots, T, \ x^{(2)}(t) \leq M_0 e, \ t = 0, \ldots, S_0 - L_0.$$ (5.128)

By (5.126), (5.128), and property (P5), there exists a program

$$\left(\{x^{(2)}(t)\}_{t=S_0-L_0}^{S_0}, \{y^{(2)}(t)\}_{t=S_0-L_0}^{S_0-1}\right)$$

such that

$$x^{(2)}(S_0) \geq x(S_0).$$ (5.129)

It is clear that $(\{x^{(2)}(t)\}_{t=0}^{S_0}, \{y^{(2)}(t)\}_{t=0}^{S_0-1})$ is a program.
Set

$$y^{(2)}(t) = y(t) \text{ for all integers } t \text{ satisfying } S_0 \leq t \leq T - 1,$$

$$x^{(2)}(t+1) = (1-d)x^{(2)}(t) + x(t+1) - (1-d)x(t)$$

$$\text{for all integers } t \text{ satisfying } S_0 \leq t \leq T - 1.$$ (5.130)

It is clear that $(\{x^{(2)}(t)\}_{t=0}^{T}, \{y^{(2)}(t)\}_{t=0}^{T-1})$ is a program. By (5.121) and (5.127),

$$x^{(2)}(0) = x(0).$$ (5.131)

By (5.114), (5.115), (5.117), (5.121), (5.126)–(5.128), (5.130), (5.131), assumption (A), and the monotonicity of the functions $w_t, t = 0, 1, \ldots$, we have

$$0 \geq \sum_{t=0}^{T-1} w_t\left(by^{(2)}(t)\right) - U(x(0), T) = \sum_{t=0}^{T-1} w_t\left(by^{(2)}(t)\right) - \sum_{t=0}^{T-1} w_t(by(t))$$

$$= \sum_{t=0}^{S_0-1} w_t\left(by^{(2)}(t)\right) - \sum_{t=0}^{S_0-1} w_t(by(t))$$

$$= \sum_{t=0}^{L-1} w_t(b\tilde{y}(t)) + w_L(0) + \sum_{t=L+1}^{S_0-1} w_t\left(by^{(2)}(t)\right)$$

$$-\sum_{t=0}^{L-1} w_t(by(t)) - \sum_{t=L}^{S_0-1} w_t(by(t))$$

$$\geq \left[U(x(0), L) - \sum_{t=0}^{L-1} w_t(by(t)) \right] - w_L(M_0be)$$

$$+ \sum_{t=L}^{S_0-L_0-1} w_t(be(2n)^{-1}\min\{1, a_1^{-1}, \ldots, a_i^{-1}, \ldots, a_n^{-1}\})$$

$$+ \sum_{t=S_0-L_0}^{S_0-1} w_t(by^{(2)}(t)) - \sum_{t=L}^{S_0-L_0-1} w_t(by(t)) - \sum_{t=S_0-L_0}^{S_0-1} w_t(by(t))$$

$$\geq \left[U(x(0), L) - \sum_{t=0}^{L-1} w_t(by(t)) \right] - w_L(M_0be)$$

$$+ \sum_{t=L}^{S_0-L_0-1} \left[w_t(be(2n)^{-1}\min\{1, a_1^{-1}, \ldots, a_i^{-1}, \ldots, a_n^{-1}\}) \right]$$

$$- \sum_{t=L}^{S_0-L_0-1} w_t(by(t)) - \sum_{t=S_0-L_0}^{S_0-1} w_t(beM_0)$$

$$\geq U(x(0), L) - \sum_{t=0}^{L-1} w_t(by(t)) - w_L(M_0be)$$

$$+ \sum \left\{ w_t\left(be(2n)^{-1}\min\{1, a_1^{-1}, \ldots, a_i^{-1}, \ldots, a_n^{-1}\}\right) \right.$$

$$\left. -w_t(\gamma) : t \text{ is an integer such that } L + L_0 \leq t \leq S_0 - 1 - L_0 \right\}$$

$$- \sum_{t=L}^{L+L_0-1} w_t(M_0be) - \sum_{t=S_0-L_0}^{S_0-1} w_t(M_0be)$$

$$\geq U(x(0), L) - \sum_{t=0}^{L-1} w_t(by(t)) - w_L(M_0be)$$

$$- \sum_{t=L}^{L+L_0-1} w_t(M_0be) - \sum_{t=S_0-L_0}^{S_0-1} w_t(M_0be).$$

Combined with the choice of $\bar{L}$, (5.129), the relation $L \geq \bar{L}$, and (5.126), the relations above imply that

$$\sum_{t=0}^{L-1} w_t(by(t)) \geq U(x(0), L) - (\delta/8) - 2L_0(\delta/8)(4L_0)^{-1} \geq U(x(0), L) - \delta/2.$$

Hence if (5.124) holds, then (5.118) is valid. Therefore (5.118) is true in both cases. Lemma 5.21 is proved.

5.9 Proof of Theorem 5.18

Proposition 5.17 implies that for every $z \in R_+^n$ and every integer $T \geq 1$, there exists a program $(\{x_{z,T}(t)\}_{t=0}^{T}, \{y_{z,T}(t)\}_{t=0}^{T-1})$ which satisfies

$$x_{z,T}(0) = z, \quad \sum_{t=0}^{T-1} w_t(by_{z,T}(t)) = U(z, T). \tag{5.132}$$

Let

$$M_0 > \max\{(a_i d)^{-1} : i = 1, \ldots, n\},$$

and let δ be a positive number. Lemma 5.21 implies that there exists an integer $L_\delta \geq 1$ for which the following property holds:

(P6) For every integer $L \geq L_\delta$, there exists an integer $\tau_L \geq L$ such that for every integer $T \geq \tau_L$ and every point $z \in R_+^n$ which satisfies $z \leq M_0 e$, we have

$$\sum_{t=0}^{L-1} w_t(by_{z,T}(t)) \geq U(z, L) - \delta/4.$$

Let $z \in R_+^n$ satisfy $z \leq M_0 e$. Lemma 5.3 implies that there exist a strictly increasing sequence of natural numbers $\{T_k\}_{k=1}^{\infty}$ and a program $\{x_z(t), y_z(t)\}_{t=0}^{\infty}$ such that for every nonnegative integer t, we have

$$x_{z,T_k}(t) \to x_z(t), \quad y_{z,T_k}(t) \to y_z(t) \text{ as } k \to \infty. \tag{5.133}$$

Evidently, $x_z(0) = z$.

Let an integer L satisfy $L \geq L_\delta$, and let an integer $\tau_L \geq L$ be as guaranteed by the property (P6). By (5.133), there exists an integer $k \geq 1$ for which

$$T_k \geq \tau_L,$$

$$\left| \sum_{t=0}^{L-1} w_t(by_z(t)) - \sum_{t=0}^{L-1} w_t(by_{z,T_k}(t)) \right| \leq \delta/4. \tag{5.134}$$

Property (P6), (5.134), and the choice of τ_L imply that

$$\sum_{t=0}^{L-1} w_t(by_{z,T_k}(t)) \geq U(z, L) - \delta/4.$$

Combined with (5.134) the inequality above implies that

$$\sum_{t=0}^{L-1} w_t(by_z(t)) \geq U(z, L) - \delta.$$

This completes the proof of Theorem 5.18.

5.10   Weakly Agreeable Programs

A program $\{x^*(t), y^*(t)\}_{t=0}^{\infty}$ is called weakly agreeable if for every nonnegative integer t,

$$u(x^*(t), x^*(t+1)) = w(by^*(t)) \qquad (5.135)$$

and if for every integer $T_0 \geq 1$ and every positive number ϵ, there exists a natural number $T_\epsilon > T_0$ such that for every program $(\{x(t)\}_{t=0}^{T_\epsilon}, \{y(t)\}_{t=0}^{T_\epsilon-1})$ which satisfies $x(0) = x^*(0)$, there exists a program $(\{x'(t)\}_{t=0}^{T_\epsilon}, \{y'(t)\}_{t=0}^{T_\epsilon-1})$ such that

$$x'(0) = x(0), \ x'(t) = x^*(t), \ t = 0, \ldots, T_0,$$

$$\sum_{t=0}^{T_\epsilon-1} w(by'(t)) \geq \sum_{t=0}^{T_\epsilon-1} w(by(t)) - \epsilon.$$

The notion of weakly agreeable programs is a weakened version of the notion of agreeable programs which is well-known in the literature [32–34].

We will prove the following three results obtained in [111].

Theorem 5.22 *Any weakly agreeable program is good.*

Theorem 5.23 *Any weakly agreeable program is weakly maximal.*

Theorem 5.24 *A program $\{x^*(t), y^*(t)\}_{t=0}^{\infty}$ is weakly agreeable if and only if there exist a strictly increasing sequence of natural numbers $\{S_k\}_{k=1}^{\infty}$ and a sequence of programs $(\{x^{(k)}(t)\}_{t=0}^{S_k}, \{y^{(k)}(t)\}_{t=0}^{S_k-1})$, $k = 1, 2, \ldots$ such that*

$$x^{(k)}(0) = x^*(0), \ k = 1, 2, \ldots \qquad (5.136)$$

$$U(x^*(0), S_k) - \sum_{t=0}^{S_k-1} w\left(by^{(k)}(t)\right) \to 0 \text{ as } k \to \infty \tag{5.137}$$

and that for all integers $t \geq 0$,

$$x^*(t) = \lim_{k \to \infty} x^{(k)}(t), \ y^*(t) = \lim_{k \to \infty} y^{(k)}(t). \tag{5.138}$$

Theorem 5.24 easily implies that for every $x_0 \in R_+^n$, there exists a weakly agreeable program $\{x(t), y(t)\}_{t=0}^{\infty}$ satisfying $x(0) = x_0$.

5.11   Proof of Theorem 5.22

Assume that a program $\{x^*(t), y^*(t)\}_{t=0}^{\infty}$ is weakly agreeable. Fix a positive number

$$M_0 > \max\{(da_i))^{-1} : \ i = 1, \ldots, n\} + ||x^*(0)||. \tag{5.139}$$

Lemma 5.3 and (5.139) imply that

$$x^*(t) \leq M_0 e \text{ for every nonnegative integer } t. \tag{5.140}$$

In follows from Theorem 5.5 and (5.139) that there exists a positive number M_1 such that

$$|\widehat{U}(M_0, p) - p\mu| \leq M_1 \text{ for every natural number } p. \tag{5.141}$$

Theorem 5.8 and (5.139) imply that there exist a positive number M_2 and a program $\{\tilde{x}(t), \tilde{y}(t)\}_{t=0}^{\infty}$ which satisfies

$$\tilde{x}(0) = x^*(0), \tag{5.142}$$

$$|\sum_{t=S_1}^{S_2-1} w(b\tilde{y}(t)) - \mu(S_2 - S_1)| \leq M_2 \text{ for all pairs of integers } S_1 \geq 0, S_2 > S_1.$$
$$\tag{5.143}$$

Assume that $T_0 \geq 1$ is an integer. By the definition of a weakly agreeable program, there exists a natural number $T_1 > T_0$ for which the following property holds:

(P7) for every program $(\{x(t)\}_{t=0}^{T_1}, \{y(t)\}_{t=0}^{T_1-1})$ which satisfies $x(0) = x^*(0)$, there exists a program $(\{x'(t)\}_{t=0}^{T_1}, \{y'(t)\}_{t=0}^{T_1-1})$ satisfying

$$x'(0) = x(0), \ x'(t) = x^*(t), \ t = 0, \ldots, T_0,$$

$$\sum_{t=0}^{T_1-1} w(by'(t)) \geq \sum_{t=0}^{T_1-1} w(by(t)) - 1. \tag{5.144}$$

Proposition 5.1 implies that there exists a program $(\{x(t)\}_{t=0}^{T_1}, \{y(t)\}_{t=0}^{T_1-1})$ which satisfies

$$x(0) = x^*(0), \quad \sum_{t=0}^{T_1-1} w(by(t)) = U(x^*(0), T_1). \tag{5.145}$$

Property (P7) and (5.145) imply that there exists a program

$$(\{x'(t)\}_{t=0}^{T_1}, \{y'(t)\}_{t=0}^{T_1-1})$$

such that (5.144) is valid and

$$\sum_{t=0}^{T_1-1} w(by'(t)) \geq \sum_{t=0}^{T_1-1} w(by(t)) - 1. \tag{5.146}$$

By (5.145) and (5.146),

$$\sum_{t=0}^{T_1-1} w(by'(t)) \geq U(x^*(0), T_1) - 1. \tag{5.147}$$

Lemma 5.3, (5.139), and (5.144) imply that

$$x'(t) \leq M_0 e \text{ for all integers } t = 0, \ldots, T_1. \tag{5.148}$$

It follows from (5.135), (5.139), (5.144), and (5.147) that

$$\sum_{t=0}^{T_0-1} w\left(by^*(t)\right) \geq \sum_{t=0}^{T_0-1} w\left(by'(t)\right) = \sum_{t=0}^{T_1-1} w\left(by'(t)\right) - \sum_{t=T_0}^{T_1-1} w\left(by'(t)\right)$$

$$\geq U\left(x^*(0), T_1\right) - 1 - \sum_{t=T_0}^{T_1-1} w\left(by'(t)\right).$$

Combined with (5.141)–(5.143) and (5.148), the relation above implies that

$$\sum_{t=0}^{T_0-1} w(by^*(t)) \geq U(x^*(0), T_1) - 1 - \widehat{U}(M_0, T_1 - T_0)$$

$$\geq U(x^*(0), T_1) - 1 - M_1 - (T_1 - T_0)\mu$$

$$\geq \sum_{t=0}^{T_1-1} w(b\tilde{y}(t)) - 1 - \mu(T_1 - T_0) - M_1 \geq T_0\mu - M_2 - M_1 - 1$$

and

$$\sum_{t=0}^{T_0-1} w(by^*(t)) \geq T_0\mu - 1 - M_1 - M_2.$$

Since the inequality above is valid for every integer $T_0 \geq 1$, Corollary 5.6 implies that the program $\{x^*(t), y^*(t)\}_{t=0}^{\infty}$ is good. This completes the proof of Theorem 5.22.

5.12 Auxiliary Results

It is easy to see that the following auxiliary result holds.

Proposition 5.25 *A program* $\{x^*(t), y^*(t)\}_{t=0}^{\infty}$ *is weakly agreeable if and only if*

$$u(x^*(t), x^*(t+1)) = w(by^*(t)) \text{ for every nonnegative integers } t$$

and for every positive number ϵ and every integer $T_0 \geq 1$, there exist a natural number $T_\epsilon > T_0$ and a program

$$\left(\{x'(t)\}_{t=0}^{T_\epsilon}, \{y'(t)\}_{t=0}^{T_\epsilon-1}\right)$$

which satisfies

$$x'(t) = x^*(t), \ t = 0, \ldots, T_0, \ \sum_{t=0}^{T_\epsilon-1} w\left(by'(t)\right) \geq U(x^*(0), T_\epsilon) - \epsilon.$$

Proposition 5.26 *Let a program* $\{x^*(t), y^*(t)\}_{t=0}^{\infty}$ *be such that for every positive number ϵ and every integer $T_0 \geq 1$, there exist a natural number $T_\epsilon > T_0$ and a program*

$$\left(\{x'(t)\}_{t=0}^{T_\epsilon}, \{y'(t)\}_{t=0}^{T_\epsilon-1}\right)$$

satisfying

$$x'(t) = x^*(t), \ t = 0, \ldots, T_0, \ y'(t) = y^*(t), \ t = 0, \ldots, T_0 - 1, \tag{5.149}$$

$$\sum_{t=0}^{T_\epsilon-1} w\left(by'(t)\right) \geq U\left(x^*(0), T_\epsilon\right) - \epsilon. \tag{5.150}$$

Then the program $\{x^*(t), y^*(t)\}_{t=0}^\infty$ *is weakly agreeable.*

Proof In order to prove the proposition, it is sufficient to show that

$$u(x^*(t), x^*(t+1)) = w(by^*(t)) \text{ for every nonnegative integer } t. \tag{5.160}$$

Let $T_0 \geq 1$ be an integer and let $\epsilon > 0$. Then there exist an integer $T_\epsilon > T_0$ and a program

$$\left(\{x'(t)\}_{t=0}^{T_\epsilon}, \{y'(t)\}_{t=0}^{T_\epsilon-1}\right)$$

such that (5.149) and (5.150) are valid. It follows from (5.149) and (5.150) that for all integers $t = 0, \ldots, T_0 - 1$, we have

$$w\left(by^*(t)\right) = w\left(by'(t)\right) \geq u\left(x'(t), x'(t+1)\right) - \epsilon = u\left(x^*(t), x^*(t+1)\right) - \epsilon.$$

Since ϵ is an arbitrary positive number, we obtain that

$$w\left(by^*(t)\right) = u\left(x^*(t), x^*(t+1)\right) \text{ for all integers } t \in \{0, \ldots, T_0 - 1\}.$$

Since T_0 is an arbitrary natural number, we conclude that (5.166) is true. This completes the proof of Proposition 5.26.

5.13   Proof of Theorem 5.23

Let $\{x^*(t), y^*(t)\}_{t=0}^\infty$ be a weakly agreeable program. We claim that it is weakly maximal. Assume the contrary. Then there exist a natural number T_0 and a positive number ϵ such that

$$U(x^*(0), x^*(T), T_0) > \sum_{t=0}^{T_0-1} w(by^*(t)) + 4\epsilon.$$

This implies that there exists a program $(\{x(t)\}_{t=0}^{T_0}, \{y(t)\}_{t=0}^{T_0-1})$ for which

$$x(0) = x^*(0), \ x(T_0) \geq x^*(T_0), \ \sum_{t=0}^{T_0-1} w(by(t)) > \sum_{t=0}^{T_0-1} w(by^*(t)) + 4\epsilon. \tag{5.161}$$

Proposition 5.25 and (5.135) imply that there exist a natural number $T_1 > T_0$ and a program

$$\left(\{x'(t)\}_{t=0}^{T_1}, \{y'(t)\}_{t=0}^{T_1-1}\right)$$

which satisfies

$$x'(t) = x^*(t), \ t = 0, \dots, T_0, \ y'(t) = y^*(t), \ t = 0, \dots, T_0 - 1, \qquad (5.162)$$

$$\sum_{t=0}^{T_1-1} w\left(by'(t)\right) \geq U\left(x^*(0), T_1\right) - \epsilon. \qquad (5.163)$$

For all integers t satisfying $T_0 \leq t < T_1$, set

$$y(t) = y'(t), \ x(t+1) = (1-d)x(t) + x'(t+1) - (1-d)x'(t). \qquad (5.164)$$

In view of (5.161) and (5.162), $x(t) \geq x'(t)$ for all $t = T_0, \dots, T_1$. Evidently,

$$\left(\{x(t)\}_{t=0}^{T_1}, \{y(t)\}_{t=0}^{T_1-1}\right)$$

is a program. It follows from (5.161), (5.162), and (5.164) that

$$U(x^*(0), T_1) \geq \sum_{t=0}^{T_1-1} w(by(t)) = \sum_{t=0}^{T_0-1} w(by(t)) + \sum_{t=T_0}^{T_1-1} w\left(by'(t)\right)$$

$$> 4\epsilon + \sum_{t=0}^{T_0-1} w\left(by^*(t)\right) + \sum_{t=T_0}^{T_1-1} w\left(by'(t)\right)$$

$$= 4\epsilon + \sum_{t=0}^{T_1-1} w\left(by'(t)\right) \geq U(x^*(0), T_1) + 3\epsilon,$$

a contradiction. The contradiction we have reached proves Theorem 5.23.

5.14   Proof of Theorem 5.24

Assume that $\{x^*(t), y^*(t)\}_{t=0}^{\infty}$ is a weakly agreeable program. Then for all nonnegative integers t, we have

$$u(x^*(t), x^*(t+1)) = w(by^*(t)). \qquad (5.165)$$

Put $S_0 = 0$. By induction we define a strictly increasing sequence of integers $\{S_k\}_{k=1}^{\infty}$ and a sequence of programs $(\{x^{(k)}(t)\}_{t=0}^{S_k}, \{y^{(k)}(t)\}_{t=0}^{S_k-1})$. Assume that $k \geq 1$ is an integer and we defined integers $S_0 < \cdots < S_{k-1}$. Since the program $\{x^*(t), y^*(t)\}_{t=0}^{\infty}$ is weakly agreeable, by Proposition 5.25 there exist an integer $S_k > k + S_{k-1}$ and a program $(\{x^{(k)}(t)\}_{t=0}^{S_k}, \{y^{(k)}(t)\}_{t=0}^{S_k-1})$ which satisfies

$$x^{(k)}(t) = x^*(t), \ t = 0, \ldots, k, \ \sum_{t=0}^{S_k-1} w\left(by^{(k)}(t)\right) \geq U\left(x^*(0), S_k\right) - 1/k.$$

$$(5.166)$$

Evidently, by (5.165), we may assume without loss of generality that

$$y^{(k)}(t) = y^*(t), \ t = 0, \ldots, k - 1. \tag{5.167}$$

In such a manner, we define a sequence of integers $\{S_k\}_{k=1}^{\infty}$ and a sequence of programs $(\{x^{(k)}(t)\}_{t=0}^{S_k}, \{y^{(k)}(t)\}_{t=0}^{S_k-1})$ such that (5.166) and (5.167) hold for all integers $k \geq 1$. Evidently, for all integers $t \geq 0$,

$$x^*(t) = \lim_{k \to \infty} x^{(k)}(t), \ y^*(t) = \lim_{k \to \infty} y^{(k)}(t).$$

Assume now that $\{S_k\}_{k=1}^{\infty}$ is strictly increasing sequence of natural numbers and a sequence of programs $(\{x^{(k)}(t)\}_{t=0}^{S_k}, \{y^{(k)}(t)\}_{t=0}^{S_k-1})$, $k = 1, 2, \ldots$ satisfy (5.136), (5.137), and (5.138).

We show that $(\{x^*(t), y^*(t)\}_{t=0}^{\infty}$ is weakly agreeable. Let ϵ be a positive number and $T_0 \geq 1$ be an integer.

Fix

$$M_0 > \|x^*(0)\| + \max\{(da_i)^{-1} : i = 1, \ldots, n\}. \tag{5.168}$$

By (5.137), we may assume that

$$U\left(x^*(0), S_k\right) - \sum_{t=0}^{S_k-1} w\left(by^{(k)}(t)\right) \leq 1 \text{ for every natural number } k. \tag{5.169}$$

Lemma 5.3, (5.136), and (5.168) imply that

$$x^{(k)}(t) \leq M_0 e, \ t = 0, \ldots, S_k, \ k = 1, 2, \ldots, \ x^*(t) \leq M_0 e, \ t = 0, 1, \ldots.$$

$$(5.170)$$

Theorem 5.8 and (5.168) imply that there exist a program $\{x(t), y(t)\}_{t=0}^{\infty}$ and a positive number M_1 such that

$$x(0) = x^*(0),$$

$$\left| \sum_{t=Q_1}^{Q_2-1} w(by(t)) - \mu(Q_2 - Q_1) \right| \le M_1 \text{ for all integers } Q_1 \ge 0, \ Q_2 > Q_1.$$

(5.171)

Theorem 5.5 implies that there exists $M_2 > 0$ such that

$$\left| \widehat{U}(M_0, p) - p\mu \right| \le M_2 \text{ for all integers } p \ge 1.$$

(5.172)

By (5.169) and (5.171), for every natural number k, we have

$$\sum_{t=0}^{S_k-1} w\left(by^{(k)}(t)\right) \ge U\left(x^*(0), S_k\right) - 1 \ge \sum_{t=0}^{S_k-1} w(by(t)) - 1 \ge \mu S_k - M_1 - 1.$$

(5.173)

By (5.170), (5.172), and (5.173), for every natural number k and every natural number $S < S_k - 1$, we have

$$\sum_{t=0}^{S-1} w\left(by^{(k)}(t)\right) \quad = \sum_{t=0}^{S_k-1} w\left(by^{(k)}(t)\right) - \sum_{t=S}^{S_k-1} w\left(by^{(k)}(t)\right)$$

$$\ge \mu S_k - M_1 - 1 - \widehat{U}(M_0, S_k - S)$$

$$\ge \mu S_k - M_1 - 1 - M_2 - (S_k - S)\mu = \mu S - M_1 - 1 - M_2.$$

(5.174)

In view of (5.138) and (5.174), for all natural numbers S, we have

$$\sum_{t=0}^{S-1} w\left(by^*(t)\right) \ge \mu S - M_1 - 1 - M_2.$$

(5.175)

It follows from (5.8) and (5.175) that there exists a natural number $\tau > T_0 + 4$ such that $y^*(\tau) > 0$. Thus there exists an integer $j \in \{1, \ldots, n\}$ such that

$$y_j^*(\tau) > 0.$$

(5.176)

Fix a number $\delta > 0$ such that

$$4\delta \sum_{i=1}^{n} (1 + a_i) < y_j^*(\tau),$$

(5.177)

$$|w(bz_1) - w(bz_2)| \le \epsilon/4 \text{ for each } z_1, z_2 \in R_+^n$$

satisfying $z_1, z_2 \le M_0 e$ and $\|z_1 - z_2\| \le \delta \sum_{i=1}^{n} (1 + a_i).$

(5.178)

In view of (5.136) and (5.137), there exists an integer $k \geq 1$ such that

$$S_k > \tau + 4, \; U\left(x^*(0), S_k\right) - \sum_{t=0}^{S_k-1} w\left(by^{(k)}\right)(t)) \leq \epsilon/8, \tag{5.179}$$

$$\left| \sum_{t=0}^{\tau} w(by^*(t)) - \sum_{t=0}^{\tau} w(by^{(k)}(t)) \right| \leq \epsilon/8, \tag{5.180}$$

$$\left\| x^*(t) - x^{(k)}(t) \right\|, \; \left\| y^*(t) - y^{(k)}(t) \right\| \leq \delta, \; t = 0, \ldots, \tau + 1. \tag{5.181}$$

Set

$$x'(t) = x^*(t), \; t = 0, \ldots, \tau, \; y'(t) = y^*(t), \; t = 0, \ldots, \tau - 1, \tag{5.182}$$

$$x'(\tau + 1) = x^*(\tau + 1) + \delta e, \; y'(\tau) = y^*(\tau) - \left(\delta \sum_{i=1}^{n} a_i \right) e_j. \tag{5.183}$$

By (5.182) and (5.183), we have

$$x'(\tau + 1) \geq (1 - d)x^*(\tau) = (1 - d)x'(\tau). \tag{5.184}$$

It follows from (5.177) and (5.183) that

$$y'(\tau) \geq 0. \tag{5.185}$$

It follows from (5.182) and (5.183) that

$$ey'(\tau) + a\left(x'(\tau + 1) - (1 - d)x'(\tau)\right)$$
$$= ey^*(\tau) - \left(\delta \sum_{i=1}^{n} a_i \right) + a\left(x^*(\tau + 1) - (1 - d)x^*(\tau)\right) + \delta ae$$
$$= ey^*(\tau) + a\left(x^*(\tau + 1) - (1 - d)x^*(\tau)\right) \leq 1. \tag{5.186}$$

By (5.176), (5.182), and (5.184)–(5.186), $(\{x'(t)\}_{t=0}^{\tau+1}, \{y'(t)\}_{t=0}^{\tau})$ is a program. In view of (5.181) and (5.183), we have

$$x'(\tau + 1) \geq x^{(k)}(\tau + 1). \tag{5.187}$$

For all integers t satisfying $\tau + 1 \leq t < S_k$, set

$$y'(t) = y^{(k)}(t), \; x'(t+1) = (1-d)x'(t) + x^{(k)}(t+1) - (1-d)x^{(k)}(t). \tag{5.188}$$

It follows from (5.187) and (5.188) that

$$x'(t) \geq x^{(k)}(t), \ t = \tau + 1, \ldots, S_k. \tag{5.189}$$

It is easy to see that

$$\left(\{x'(t)\}_{t=0}^{S_k}, \ \{y'(t)\}_{t=0}^{S_k-1} \right)$$

is a program. By (5.170), (5.178), (5.179), (5.180), (5.182), (5.183), and (5.188), we have

$$\sum_{t=0}^{S_k-1} w\left(by'(t)\right) - U\left(x^*(0), S_k\right) \geq \sum_{t=0}^{S_k-1} w\left(by'(t)\right) - \sum_{t=0}^{S_k-1} w\left(by^{(k)}(t)\right) - \epsilon/8$$

$$= \sum_{t=0}^{\tau} w\left(by'(t)\right) - \sum_{t=0}^{\tau} w\left(by^{(k)}(t)\right) - \epsilon/8$$

$$\geq \sum_{t=0}^{\tau} w\left(by'(t)\right) - \sum_{t=0}^{\tau} w\left(by^*(t)\right) - \epsilon/4$$

$$= w\left(by'(\tau)\right) - w\left(by^*(\tau)\right) - \epsilon/4 \geq -\epsilon/2$$

and

$$\sum_{t=0}^{S_k-1} w\left(by'(t)\right) - U\left(x^*(0), S_k\right) \geq -\epsilon/2. \tag{5.190}$$

Therefore for every $\epsilon > 0$ and every integer $T_0 \geq 1$, we showed the existence of a natural number $S_k > T_0 + 4$ and a program $(\{x'(t)\}_{t=0}^{S_k}, \{y'(t)\}_{t=0}^{S_k-1})$ which satisfies

$$y'(t) = y^*(t), \ x'(t) = x^*(t), \ t = 0, \ldots, T_0,$$

(5.179), and (5.190).

Proposition 5.26 implies that $(\{x^*(t)\}_{t=0}^{\infty}, \{y^*(t)\}_{t=0}^{\infty})$ is a weakly agreeable program. Theorem 5.24 is proved.

5.15 Weakly Maximal Programs

We prove three theorems obtained in [103].

We assume for simplicity that $w(0) = 0$ and begin with the following result which establishes the continuity of the function $U(\cdot, \cdot, T)$. This result will be proved in Section 5.16.

Theorem 5.27 *Let* $T > 0$ *be an integer,* $x_0, \tilde{x}_0 \in R_+^n$, $U(x_0, \tilde{x}_0, T) > 0$, *and* $d < 1$.
Then the function $(y, z) \rightarrow U(y, z, T)$, $y, z \in R_+^n$ *is continuous at* $(x_0, \tilde{x}_0)$.

The following result will be proved in Section 5.17.

Theorem 5.28 *Let* $M_0 > \max\{(da_i)^{-1} : i = 1, \ldots, n\}$. *Then there exists* $M_1 > 0$ *such that for each good weakly maximal program* $\{x(t), y(t)\}_{t=0}^{\infty}$ *satisfying* $x(0) \leq M_0 e$ *and each pair of integers* $S_1 \geq 0$ *and* $S_2 > S_1$,

$$\left| \sum_{t=S_1}^{S_2-1} w(by(t)) - \mu(S_2 - S_1) \right| \leq M_1. \tag{5.191}$$

By definition for any good program $\{x(t), y(t)\}_{t=0}^{\infty}$, there is a constant $M_1 > 0$ such that (5.191) holds. In view of Theorem 5.28, the constant M_1 depends only on the constant M_0, and the inequality (5.191) holds for all programs $\{x(t), y(t)\}_{t=0}^{\infty}$ satisfying $x(0) \leq M_0 e$.

The next theorem will be proved in Section 5.18.

Theorem 5.29 *Let* $\{x^{(k)}(t), y^{(k)}(t)\}_{t=0}^{\infty}$, $k = 1, 2, \ldots$ *be good weakly maximal programs. Assume that for any integer* $t \geq 0$, *there exists* $x(t) = \lim_{k\to\infty} x^{(k)}(t)$, $y(t) = \lim_{k\to\infty} y_t^{(k)}$. *Then* $\{x(t), y(t)\}_{t=0}^{\infty}$ *is a good weakly maximal program.*

5.16   Proof of Theorem 5.27

Lemma 5.30 *Let* T *be a natural number,* $x_0, \tilde{x}_0 \in R_+^n$ *satisfy* $U(x_0, \tilde{x}_0, T) > 0$, *and* $\epsilon > 0$. *Then there exists* $\delta > 0$ *such that for each* $y_0, \tilde{y}_0 \in R_+^n$ *satisfying* $\|y_0 - x_0\|$, $\|\tilde{y}_0 - \tilde{x}_0\| \leq \delta$, *the following inequality holds:*

$$U(y_0, \tilde{y}_0, T) < U(x_0, \tilde{x}_0, T) + \epsilon.$$

Proof Let us assume the contrary. Then for each integer $k \geq 1$, there exist $x_0^{(k)}, \tilde{x}_0^{(k)} \in R_+^n$ such that

$$\left\| x_0 - x_0^{(k)} \right\| \leq k^{-1}, \quad \left\| \tilde{x}_0 - \tilde{x}_0^{(k)} \right\|] \leq k^{-1}, \tag{5.192}$$

$$U\left(x_0^{(k)}, \tilde{x}_0^{(k)}, T \right) \geq U(x_0, \tilde{x}_0, T) + \epsilon. \tag{5.193}$$

For every natural number k, there exists a program $(\{x^{(k)}(t)\}_{t=0}^{T}, \{y^{(k)}(t)\}_{t=0}^{T-1})$ such that

$$x^{(k)}(0) = x_0^{(k)}, \quad x_T^{(k)} \geq \tilde{x}_0^{(k)}, \tag{5.194}$$

$$\sum_{t=0}^{T-1} w(by^{(k)}(t)) \geq U(x_0^{(k)}, \tilde{x}_0^{(k)}, T) - k^{-1}. \tag{5.195}$$

By (5.192), (5.193), and Lemma 5.3, the set

$$\left\{ x^{(k)}(t) : t = 0, 1, \ldots, T, \ k = 1, 2, \ldots \right\}$$

is bounded. Extracting a subsequence and re-indexing, we may assume without loss of generality that for $t = 0, \ldots, T$, there exists

$$x(t) = \lim_{k \to \infty} x^{(k)}(t) \tag{5.196}$$

and for $t = 0, \ldots, T - 1$, there exists

$$y(t) = \lim_{k \to \infty} y^{(k)}(t). \tag{5.197}$$

It is clear that $(\{x(t)\}_{t=0}^{T}, \{y(t)\}_{t=0}^{T-1})$ is a program. By (5.192), (5.194), and (5.196),

$$x(0) = x_0, \quad x(T) \geq \tilde{x}_0^{(0)}. \tag{5.198}$$

In view of (5.193), (5.195), (5.197), and (5.198),

$$U(x_0, \tilde{x}_0, T) \geq \sum_{t=0}^{T-1} w(by(t)) = \lim_{k \to \infty} \sum_{t=0}^{T-1} w\left(by^{(k)}(t)\right)$$

$$= \limsup_{k \to \infty} U\left(x_0^{(k)}, \tilde{x}_0^{(k)}, T\right) \geq U(x_0, \tilde{x}_0, T) + \epsilon.$$

The contradiction we have reached proves the lemma.

Lemma 5.31 *Let T be a natural number, $x_0, \tilde{x}_0 \in R_+^n$ be such that*

$$U(x_0, \tilde{x}_0, T) > 0,$$

$d < 1$, and let $\epsilon > 0$. Then there exists $\delta > 0$ such that for each $y_0, \tilde{y}_0 \in R_+^n$ satisfying

$$\| y_0 - x_0 \|, \ \| \tilde{y}_0 - \tilde{x}_0 \| \leq \delta, \qquad (5.199)$$

the following inequality holds:

$$U(y_0, \tilde{y}_0, T) \geq U(x_0, \tilde{x}_0, T) - \epsilon.$$

Proof Since $U(x_0, \tilde{x}_0, T) > 0$, it follows from Lemma 5.3 and the continuity of w that there exists a program $(\{\bar{x}(t)\}_{t=0}^{T}, \{\bar{y}(t)\}_{t=0}^{T-1})$ such that

$$\bar{x}(0) = x_0, \ \bar{x}(T) \geq \tilde{x}_0,$$

$$\sum_{t=0}^{T-1} w(b\bar{y}(t)) = U(x_0, \tilde{x}_0, T) > 0. \qquad (5.200)$$

By (5.200), there exists an integer $\tau_1 \in [0, T-1]$ such that

$$w(b\bar{y}(\tau_1)) > 0;$$

if an integer t satisfies $0 \leq t < \tau_1$ then $w(b\bar{y}(t)) = 0. \qquad (5.201)$

Then

$$\max\{\bar{y}_i(\tau_1) : \ i = 1, \dots, n\} > 0.$$

Fix $i_1 \in \{1, \dots, n\}$ such that

$$\bar{y}_{i_1}(\tau_1) > 0. \qquad (5.202)$$

Choose a positive number Δ such that

$$\Delta < 4^{-1} \bar{y}_{i_1}(\tau_1), \ \Delta < (be)^{-1} b\bar{y}(\tau_1)/8,$$

$$w(b\bar{y}(\tau_1) - \Delta be) > w(b\bar{y}(\tau_1)) - \epsilon/8. \qquad (5.203)$$

We choose $\delta > 0$ such that for all $i = 1, \dots, n$,

$$\delta < (4n)^{-1}(1 + a_i)^{-1}\Delta(1 - d)^T. \qquad (5.204)$$

Assume that $y_0, \tilde{y}_0 \in R_+^n$ satisfy (5.199). Set

$$x(0) = y_0. \qquad (5.205)$$

For each integer t satisfying $0 \leq t < \tau_1$, set

$$x(t+1) = (1-d)x(t) + \bar{x}(t+1) - (1-d)\bar{x}(t), \quad y(t) = 0. \tag{5.206}$$

It is clear that if $\tau_1 > 0$, then $(\{x(t)\}_{t=0}^{\tau_1}, \{y(t)\}_{t=0}^{\tau_1-1})$ is a program. By (5.199), (5.200), (5.205), and (5.206),

$$\|x(\tau_1) - \bar{x}(\tau_1)\| \le \|x(0) - \bar{x}(0)\| = \|y_0 - x_0\| \le \delta. \tag{5.207}$$

Note that (5.207) holds if $\tau_1 > 0$ and if $\tau_1 = 0$. Set

$$y_i(\tau_1) = \max\{\bar{y}_i(\tau_1) - \delta, 0\}, \ i \in \{1, \dots, n\} \setminus \{i_1\}, \ y_{i_1}(\tau_1) = \bar{y}_{i_1}(\tau_1) - \Delta, \tag{5.208}$$

$$x(\tau_1 + 1) = (1-d)x(\tau_1) + \bar{x}(\tau_1 + 1) - (1-d)\bar{x}(\tau_1) + n^{-1}\left(a_1^{-1}, \dots, a_n^{-1}\right)\Delta. \tag{5.209}$$

In view of (5.203), (5.204), and (5.206)–(5.208),

$$0 \le y(\tau_1) \le x(\tau_1). \tag{5.210}$$

Clearly,

$$x(\tau_1 + 1) \ge (1-d)x(\tau_1). \tag{5.211}$$

By (5.204) and (5.208),

$$a(x(\tau_1 + 1) - (1-d)x(\tau_1)) + ey(\tau_1)$$
$$= a(\bar{x}(\tau_1 + 1) - (1-d)\bar{x}(\tau_1)) + \Delta + ey(\tau_1)$$
$$\le a(\bar{x}(\tau_1 + 1) - (1-d)\bar{x}(\tau_1)) + \Delta + e\bar{y}(\tau_1) - \Delta \le 1.$$

Together with (5.210) and (5.211), this implies that $(\{x(t)\}_{t=0}^{\tau_1+1}, \{y(t)\}_{t=0}^{\tau_1})$ is a program. It follows from (5.204), (5.207), and (5.209) that for all $i \in \{1, \dots, n\}$,

$$x_i(\tau_1 + 1) \ge \bar{x}_i(\tau_1 + 1) - \|\bar{x}(\tau_1) - \bar{x}(\tau_1)\| + n^{-1}a_i^{-1}\Delta$$
$$\ge \bar{x}_i(\tau_1 + 1) - \delta + n^{-1}a_i^{-1}\Delta$$
$$\ge \bar{x}_i(\tau_1 + 1) + 2^{-1}n^{-1}a_i^{-1}\Delta$$

and

$$x(\tau_1 + 1) \ge \bar{x}(\tau_1 + 1) + (2n)^{-1}\Delta\left(a_1^{-1}, \dots, a_n^{-1}\right). \tag{5.212}$$

By (5.201), (5.203), (5.204), and (5.208),

$$\sum_{t=0}^{\tau_1} w(by(t)) - \sum_{t=0}^{\tau_1} w(b\bar{y}(t)) = w(by(\tau_1)) - w(b\bar{y}(\tau_1))$$

$$= w\left(\sum_{i=1}^{n} b_i y_i(\tau_1)\right) - w\left(\sum_{i=1}^{n} b_i \bar{y}_i(\tau_1)\right)$$

$$\geq w\left(\max\left\{\sum_{i=1}^{n} b_i(\bar{y}_i(\tau_1) - \Delta), 0\right\}\right) - w\left(\sum_{i=1}^{n} b_i \bar{y}_i(\tau_1)\right)$$

$$\geq w\left(\sum_{i=1}^{n} b_i \bar{y}_i(\tau_1) - \Delta \sum_{i=1}^{n} b_i\right) - w\left(\sum_{i=1}^{n} b_i \bar{y}_i(\tau_1)\right) > -\epsilon/8. \qquad (5.213)$$

If $\tau_1 + 1 = T$, then by (5.199), (5.200), (5.204), and (5.212),

$$x(T) = x(\tau_1 + 1) \geq \bar{x}(\tau_1 + 1) + \delta e \geq \tilde{x}_0 + \delta e \geq \tilde{y}_0 \qquad (5.214)$$

and by (5.200), (5.205), (5.213), and (5.214),

$$U(y_0, \tilde{y}_0, T) \geq \sum_{t=0}^{\tau_1} w(by(t)) \geq \sum_{t=0}^{\tau_1} w(b\bar{y}(t)) - \epsilon/8 = U(x_0, \tilde{x}_0, T) - \epsilon/8.$$

Thus in the case $\tau_1 + 1 = T$, the assertion of the lemma holds.

Assume that $\tau_1 + 1 < T$. For each integer t satisfying $\tau_1 + 1 \leq t < T$, set

$$x(t + 1) = (1 - d)x(t) + \bar{x}(t + 1) - (1 - d)\bar{x}(t), \quad y(t) = \bar{y}(t). \qquad (5.215)$$

In view of (5.212) and (5.215), $(\{x(t)\}_{t=0}^{T}, \{y(t)\}_{t=0}^{T-1})$ is a program. By (5.212), and (5.215),

$$x(T) - \bar{x}(T) = (1 - d)^{T-\tau_1-1}(x(\tau_1 + 1) - \bar{x}(\tau_1 + 1))$$

$$\geq (1 - d)^{T-\tau_1-1}(2n)^{-1}\Delta(a_1^{-1}, \ldots, a_n^{-1})$$

$$\geq (1 - d)^{T}(2n)^{-1}\Delta(a_1^{-1}, \ldots, a_n^{-1}).$$

Together with (5.199), (5.200), and (5.204), this implies that

$$x(T) \geq \bar{x}(T) + (1 - d)^{T}(2n)^{-1}\Delta\left(a_1^{-1}, \ldots, a_n^{-1}\right) \geq \tilde{x}_0 + \delta e \geq \tilde{y}_0. \qquad (5.216)$$

By (5.200), (5.205), (5.213), (5.215), and (5.216)

$$U(y_0, \tilde{y}_0, T) - U(x_0, \tilde{x}_0, T) \geq \sum_{t=0}^{T-1} w(by(t)) - \sum_{t=0}^{T-1} w(b\bar{y}(t))$$

$$= \sum_{t=0}^{\tau_1} w(by(t)) - \sum_{t=0}^{\tau-1} w(b\bar{y}(t)) > -\epsilon/8.$$

Thus the assertion of the lemma holds. Lemma 5.31 is proved.

Theorem 5.27 now easily follows from Lemmas 5.30 and 5.31.

5.17   Proof of Theorem 5.28

There is $\delta > 0$ such that

$$w(t) < \mu/8 \text{ for each } t \in [0, \delta]. \tag{5.217}$$

Let $M > 0$ be as guaranteed by Theorem 5.8. By Corollary 5.6, there is $\tilde{M} > 0$ such that the following property holds:

(P8) for each program $\{x(t), y(t)\}_{t=0}^{\infty}$ satisfying $x(0) \leq M_0 e$ and each integer $T \geq 1$,

$$\sum_{t=0}^{T-1} [w(by(t)) - \mu] \leq \tilde{M}.$$

By Lemma 5.10, there exists a natural number $p \geq 4$ such that the following property holds:

(P9) for each integer $\tau \geq p$, each program $(\{x(t)\}_{t=0}^{\tau}, \{y(t)\}_{t=0}^{\tau-1})$ which satisfies

$$x(0) \leq M_0 e, \ \ by(\tau - 1) \geq \delta,$$

and each $\tilde{x}_0 \in R_+^n$ which satisfies $\tilde{x}_0 \leq M_0 e$, there exists a program

$$\left(\{\bar{x}(t)\}_{t=0}^{\tau}, \{\bar{y}(t)\}_{t=0}^{\tau-1} \right)$$

such that

$$\bar{x}(0) = \tilde{x}_0, \ \ \bar{x}(\tau) \geq x(\tau).$$

Set

$$M_1 = 2\tilde{M} + M + p\mu. \tag{5.218}$$

Assume that $\{x(t), y(t)\}_{t=0}^{\infty}$ is a good program such that

$$x(0) \leq M_0 e \tag{5.219}$$

and that for all integers $T \geq 1$,

$$\sum_{t=0}^{T-1} w(by(t)) = U(x(0), x(T), T). \tag{5.220}$$

By the choice of M and Theorem 5.8, there exists a program $\{\bar{x}(t), \bar{y}(t)\}_{t=0}^{\infty}$ such that

$$\bar{x}(0) = x(0) \tag{5.221}$$

and that for each integer $S_1 \geq 0$ and each integer $S_2 > S_1$,

$$|\sum_{t=S_1}^{S_2-1} w(b\bar{y}(t)) - \mu(S_2 - S_1)| \leq M. \tag{5.222}$$

By (5.222), there exists an increasing sequence of natural numbers $\{S_k\}_{k=1}^{\infty}$ such that

$$w(by(T_k - 1)) \geq \mu/2, \; k = 1, 2, \ldots \tag{5.223}$$

In view of (5.217) and (5.223),

$$by(T_k - 1) > \delta, \; k = 1, 2, \ldots \tag{5.224}$$

We may assume without loss of generality that $T_k > p$ for all integers $k \geq 1$.

Let $k \geq 1$ be an integer, and set

$$\tilde{x}(t) = \bar{x}(t), \; t = 0, \ldots, T_k - p, \; \tilde{y}(t) = \bar{y}(t), \; t = 0, \ldots, T_k - p - 1. \tag{5.225}$$

By (5.219), (5.221), (5.224), (5.225), Lemma 5.3, and property (P9) applied to the program

$$\left(\{x(t)\}_{t=T_k-p}^{T_k}, \{y(t)\}_{t=T_k-p}^{T_k-1} \right),$$

there exists a program $(\{\tilde{x}(t)\}_{t=T_k-p}^{T_k}, \{\tilde{y}(t)\}_{t=T_k-p}^{T_k-1})$ such that

$$\tilde{x}(T_k) \geq x(T_k). \tag{5.226}$$

By (5.220)–(5.222), (5.225), and (5.226),

$$\sum_{t=0}^{T_k-1} w(by(t)) = U(x(0), x(T_k), T_k) \geq \sum_{t=0}^{T_k-1} w(b\tilde{y}(t))$$

$$= \sum_{t=0}^{T_k-p-1} w(b\tilde{y}(t)) + \sum_{t=T_k-p}^{T_k-1} w(b\tilde{y}(t))$$

$$\geq \mu(T_k - p) - M \geq \mu(T_k) - M - p\mu.$$

Thus for each integer $k \geq 1$,

$$\sum_{t=0}^{T_k-1} w(by(t)) \geq \mu(T_k) - M - p\mu. \tag{5.227}$$

Assume that integers $S_1 \geq 0$ and $S_2 > S_1$. Choose a natural number k such that $T_k > S_2$. By (P8), Lemma 5.3, (5.219), and (5.227),

$$\sum_{t=S_1}^{S_2-1} w(by(t)) = \sum_{t=0}^{T_k-1} w(by(t))$$

$$- \sum \{t \text{ is an integer and } 0 \leq t \leq S_1 - 1 : w(by(t))\}$$

$$- \sum_{t=S_2}^{T_k-1} w(by(t))$$

$$\geq \mu T_k - M - p\mu - [\mu S_1 + \tilde{M}] - [T_k - S_2]\mu - \tilde{M} = \mu(S_2 - S_1)$$

$$- M - p\mu - 2\tilde{M} \tag{5.228}$$

(here we assume that the sum over an empty set is zero). In view of (5.219), (P8), and Lemma 5.14,

$$\sum_{t=S_1}^{S_2-1} w(by(t)) \leq \tilde{M} + (S_2 - S_1)\mu.$$

Together with (5.218) and (5.228), this implies that

$$\left| \sum_{t=S_1}^{S_2-1} w(by(t)) - \mu(S_2 - S_1) \right| \leq 2\tilde{M} + M + \mu p = M_1.$$

Theorem 5.28 is proved.

5.18   Proof of Theorem 5.29

Clearly, $\{x(t), y(t)\}_{t=0}^{\infty}$ is a program. By Theorem 5.28, there is $M_1 > 0$ such that for each integer $k \geq 1$ and each pair of integers $S_1 \geq 0$ and $S_2 > S_1$,

$$\left| \sum_{t=S_1}^{S_2-1} w(by^{(k)}(t)) - \mu(S_2 - S_1) \right| \leq M_1. \tag{5.229}$$

This implies that for each pair of integers $S_1 \geq 0$ and $S_2 > S_1$,

$$\left| \sum_{t=S_1}^{S_2-1} (w(by(t)) - \mu(S_2 - S_1) \right| \leq M_1. \tag{5.230}$$

In view of (5.230), the program $\{x(t), y(t)\}_{t=0}^{\infty}$ is good. By (5.230) there is a strictly increasing sequence of natural numbers $\{T_j\}_{j=1}^{\infty}$ such that

$$w(by(T_j - 1)) > \mu/2 \text{ for all natural numbers } j. \tag{5.231}$$

Let $j \geq 1$ be an integer. Applying Theorem 5.27, we obtain that

$$\sum_{t=0}^{T_j-1} w(by(t)) = \lim_{k\to\infty} \sum_{t=0}^{T_j-1} w(by^{(k)}(t))$$

$$= \lim_{k\to\infty} U(x^{(k)}(0), x^{(k)}(T_j), T_j) = U(x(0), x(T_j), T_j).$$

This implies that $\{x(t), y(t)\}_{t=0}^{\infty}$ is weakly maximal. Theorem 5.29 is proved.

Chapter 6
Infinite Horizon Nonautonomous Optimization Problems

In this chapter we study infinite horizon optimal control problems with nonautonomous optimality criterions. The utility functions, which determine the optimality criterion, are nonconcave. The class of models contains, as a particular case, the Robinson–Solow–Srinivasan model. We establish the existence of good programs and optimal programs.

6.1 The Model Description and Main Results

Let R^1 (R_+^1) be the set of real (nonnegative) numbers, and let R^n be the n-dimensional Euclidean space with nonnegative orthant

$$R_+^n = \{x = (x_1, \ldots, x_n) \in R^n : x_i \geq 0, \ i = 1, \ldots, n\}.$$

For every pair of vectors $x = (x_1, \ldots, x_n)$, $y = (y_1, \ldots, y_n) \in R^n$, define their inner product by

$$xy = \sum_{i=1}^{n} x_i y_i,$$

and let $x >> y, x > y, x \geq y$ have their usual meaning.

Let $e(i)$, $i = 1, \ldots, n$, be the ith unit vector in R^n, and e be an element of R_+^n all of whose coordinates are unity. For every $x \in R^n$, denote by $\|x\|_2$ its Euclidean norm in R^n. We assume that $\|\cdot\|$ is a norm in R^n.

For every mapping $a : X \to 2^Y \setminus \{\emptyset\}$, where X, Y are nonempty sets, set

$$\mathrm{graph}(a) = \{(x, y) \in X \times Y : y \in a(x)\}.$$

© The Author(s), under exclusive license to Springer Nature Switzerland AG 2020   179
A. J. Zaslavski, *Turnpike Theory for the Robinson–Solow–Srinivasan Model*,
Springer Optimization and Its Applications 166,
https://doi.org/10.1007/978-3-030-60307-6_6

Let K be a nonempty compact subset of R^n. Denote by $\mathcal{P}(K)$ the set of all nonempty closed subsets of K.

For every pair of nonempty sets $A, B \subset R^n$, define

$$H(A, B) = \sup \left\{ \sup_{x \in A} \inf_{y \in B} \|x - y\|, \ \sup_{y \in B} \inf_{x \in A} \|x - y\| \right\}. \tag{6.1}$$

For every nonnegative integer t, let $a_t : K \to \mathcal{P}(K)$ be such that $\text{graph}(a_t)$ is a closed subset of $R^n \times R^n$.

Assume that there exists a number $\kappa \in (0, 1)$ such that for every pair of points $x, y \in K$ and every nonnegative integer t,

$$H(a_t(x), a_t(y)) \le \kappa \|x - y\| \tag{6.2}$$

and that for every nonnegative integer t, the upper semicontinuous function

$$u_t : \{(x, x') \in K \times K, \ x' \in a_t(x)\} \to [0, \infty)$$

satisfies

$$\sup\{\sup\{u_t(x, x') : (x, x') \in \text{graph}(a_t)\} : t = 0, 1, \dots\} < \infty. \tag{6.3}$$

A sequence $\{x(t)\}_{t=0}^{\infty} \subset K$ is called a program if $x(t + 1) \in a_t(x(t))$ for every nonnegative integer t.

Let T_1, T_2 be integers such that $T_1 < T_2$. A sequence $\{x(t)\}_{t=T_1}^{T_2} \subset K$ is called a program if $x(t + 1) \in a_t(x(t))$ for every integer t satisfying $T_1 \le t < T_2$.

We suppose that the following assumptions hold:

(A1) for every positive number δ, there exists a positive number λ such that if an integer $t \ge 0$ and if $(x, x') \in \text{graph}(a_t)$ satisfies $u_t(x, x') \ge \delta$, then there exists $z \in a_t(x)$ satisfying $z \ge x' + \lambda e$;

(A2) there exist a program $\{\widehat{x}(t)\}_{t=0}^{\infty}$ and a positive number $\widehat{\Delta}$ such that $u_t(\widehat{x}(t), \widehat{x}(t + 1)) \ge \widehat{\Delta}$ for every nonnegative integer t;

(A3) for every nonnegative integer t, every $(x, y) \in \text{graph}(a_t)$, and every $\tilde{x} \in K$ which satisfies $\tilde{x} \ge x$, there exists $\tilde{y} \in a_t(\tilde{x})$ for which

$$\tilde{y} \ge y, \ u_t(\tilde{x}, \tilde{y}) \ge u_t(x, y).$$

In the sequel we assume that supremum of empty set is $-\infty$.

For every point $x_0 \in K$ and every natural number T, define

$$U(x_0, T) = \sup \left\{ \sum_{t=0}^{T-1} u_t(x(t), x(t + 1)) : \right.$$

$$\left. \{x(t)\}_{t=0}^{T-1} \text{ is a program and } x(0) = x_0 \right\}. \tag{6.4}$$

Let $x_0, \tilde{x}_0 \in K$ and let $T \geq 1$ be an integer. Define

$$U(x_0, \tilde{x}_0, T) = \sup \left\{ \sum_{t=0}^{T-1} u_t(x(t), x(t+1)) : \{x(t)\}_{t=0}^{T-1} \text{ is a program such that}\right.$$

$$\left. x(0) = x_0, \ x(T) \geq \tilde{x}_0 \right\}. \tag{6.5}$$

Let $T \geq 1$ be an integer. Define

$$\widehat{U}(T) = \sup \left\{ \sum_{t=0}^{T-1} u_t(x(t), x(t+1)) : \{x(t)\}_{t=0}^{T-1} \text{ is a program}\right\}. \tag{6.6}$$

The results presented in this section were obtained in [99].

Upper semicontinuity of $u_t, t = 0, 1, \ldots$ implies the following two propositions.

Proposition 6.1 *For every $x_0 \in K$ and every integer $T \geq 1$, there exists a program* $\{x(t)\}_{t=0}^{T}$ *which satisfies $x(0) = x_0$ and*

$$\sum_{t=0}^{T-1} u_t(x(t), x(t+1)) = U(x_0, T).$$

Proposition 6.2 *For every integer $T \geq 1$, there exists a program $\{x(t)\}_{t=0}^{T}$ satisfying $\sum_{t=0}^{T-1} u_t(x(t), x(t+1)) = \widehat{U}(T)$.*

For every $x_0 \in K$ and every pair of integers $T_1 < T_2$, define

$$U(x_0, T_1, T_2) = \sup \left\{ \sum_{t=T_1}^{T_2-1} u_t(x(t), x(t+1)) : \right.$$

$$\left. \{x(t)\}_{t=T_1}^{T_2-1} \text{ is a program and } x(T_1) = x_0 \right\}. \tag{6.7}$$

Upper semicontinuity of $u_t, t = 0, 1, \ldots$ implies the following result.

Proposition 6.3 *For every $x_0 \in K$ and every pair of integers $T_1 < T_2$, there exists a program $\{x(t)\}_{t=T_1}^{T_2}$ such that $x(T_1) = x_0$ and*

$$\sum_{t=T_1}^{T_2-1} u_t(x(t), x(t+1)) = U(x_0, T_1, T_2).$$

Let $x_0, \tilde{x}_0 \in K$ and let $T_1 < T_2$ be integers. Define

$$U(x_0, \tilde{x}_0, T_1, T_2) = \sup \left\{ \sum_{t=T_1}^{T_2-1} u_t(x(t), x(t+1)) : \{x(t)\}_{t=T_1}^{T_2} \text{ is a program and} \right.$$

$$\left. x(T_1) = x_0, \ x(T_2) \geq \tilde{x}_0 \right\}. \tag{6.8}$$

Let T_1, T_2 be integers such that $T_1 < T_2$. Define

$$\hat{U}(T_1, T_2) = \sup \left\{ \sum_{t=T_1}^{T_2-1} u_t(x(t), x(t+1)) : \{x(t)\}_{t=T_1}^{T_2} \text{ is a program} \right\}. \tag{6.9}$$

We will prove the following theorem which is the main result of this section.

Theorem 6.4 *There exists a positive number M such that for every $x_0 \in K$, there exists a program $\{\bar{x}(t)\}_{t=0}^{\infty}$ such that $\bar{x}(0) = x_0$ and that for every pair of nonnegative integers T_1, T_2 satisfying $T_1 < T_2$, the inequality*

$$\left| \sum_{t=T_1}^{T_2-1} u_t(\bar{x}(t), \bar{x}(t+1)) - \hat{U}(T_1, T_2) \right| \leq M$$

holds. Moreover, for every natural number T,

$$\sum_{t=0}^{T-1} u_t(\bar{x}(t), \bar{x}(t+1)) = \tilde{U}(\bar{x}(0), \bar{x}(T), 0, T),$$

if the following properties hold:

- *for every nonnegative integer t and every $(z, z') \in \text{graph}(a_t)$ satisfying $u_t(z, z') > 0$, the function u_t is continuous at the point (z, z');*
- *for every nonnegative integer t and each $z, z_1, z_2, z_3 \in K$ which satisfy $z_1 \leq z_2 \leq z_3$ and $z_i \in a_t(z)$, $i = 1, 3$, the inclusion $z_2 \in a_t(z)$ is valid.*

The program $\{\bar{x}(t)\}_{t=0}^{\infty}$ whose existence is guaranteed by Theorem 6.4 in infinite horizon optimal control is considered as an (approximately) optimal program.

Theorem 6.5 *Assume that $\{x(t)\}_{t=0}^{\infty}$ is a program; that there exists a positive number M such that for every natural number T,*

$$\sum_{t=0}^{T-1} u_t(x(t), x(t+1)) \geq U(0, T, x(0), x(T)) - M_0;$$

and that

$$\limsup_{t \to \infty} u_t(x(t), x(t+1)) > 0.$$

Then there exists positive number M_1 such that for every pair of integers $T_1 \geq 0$ satisfying $T_2 > T_1$, the inequality

$$\left| \sum_{t=T_1}^{T_2-1} u_t(x(t), x(t+1)) - \widehat{U}(T_1, T_2) \right| \leq M_1$$

holds.

Theorem 6.4 is proved in Section 6.6, while Theorem 6.5 is obtained in Section 6.7.

Let $M > 0$ be as guaranteed by Theorem 6.4.

Proposition 6.6 *Let $x_0 \in K$, and let a program $\{\bar{x}(t)\}_{t=0}^{\infty}$ be as guaranteed by Theorem 6.4. Assume that $\{x(t)\}_{t=0}^{\infty}$ is a program. Then either the sequence*

$$\left\{ \sum_{t=0}^{T-1} u_t(x(t), x(t+1)) - \sum_{t=0}^{T-1} u_t(\bar{x}(t), \bar{x}(t+1)) \right\}_{T=1}^{\infty}$$

is bounded or

$$\sum_{t=0}^{T-1} u_t(x(t), x(t+1)) - \sum_{t=0}^{T-1} u_t(\bar{x}(t), \bar{x}(t+1)) \to -\infty \text{ as } T \to \infty. \tag{6.10}$$

Proof Assume that the sequence

$$\left\{ \sum_{t=0}^{T-1} u_t(x(t), x(t+1)) - \sum_{t=0}^{T-1} u_t(\bar{x}(t), \bar{x}(t+1)) \right\}_{T=1}^{\infty}$$

is not bounded. Then by Theorem 6.4,

$$\liminf_{T \to \infty} \left[\sum_{t=0}^{T-1} u_t(x(t), x(t+1)) - \sum_{t=0}^{T-1} u_t(\bar{x}(t), \bar{x}(t+1)) \right] = -\infty.$$

Let $Q > 0$. Then there exists a natural number T_0 which satisfies

$$\sum_{t=0}^{T_0-1} u_t(x(t), x(t+1)) - \sum_{t=0}^{T_0-1} u_t(\bar{x}(t), \bar{x}(t+1)) < -Q - M. \qquad (6.11)$$

In view of (6.11), the choice of $\{\bar{x}(t)\}_{t=0}^{\infty}$, and Theorem 6.4, for every natural number $T > T_0$, we have

$$\sum_{t=0}^{T-1} u_t(x(t), x(t+1)) - \sum_{t=0}^{T-1} u_t(\bar{x}(t), \bar{x}(t+1)) = \sum_{t=0}^{T_0-1} u_t(x(t), x(t+1))$$

$$- \sum_{t=0}^{T_0-1} u_t(\bar{x}(t), \bar{x}(t+1)) + \sum_{t=T_0}^{T-1} u_t(x(t), x(t+1)) - \sum_{t=T_0}^{T-1} u_t(\bar{x}(t), \bar{x}(t+1))$$

$$< -Q - M + \widehat{U}(T_0, T) - \sum_{t=T_0}^{T-1} u_t(\bar{x}(t), \bar{x}(t+1)) < -Q.$$

Since Q is an arbitrary positive number, we conclude that (6.10) holds. This completes the proof of Proposition 6.6.

Now assume that $u_t = u_0$ and $a_t = a_0$, $t = 0, 1, \ldots$. Let a positive number $\dot{M}$ be as guaranteed by Theorem 6.4, and set $u = u_0$, $a = a_0$. The following result will be proved in Section 6.8.

Theorem 6.7 *There exists* $\mu = \lim_{p \to \infty} \widehat{U}(0, p)/p$ *and*

$$|p^{-1}\widehat{U}(0, p) - \mu| \le 2M/p \text{ for all natural numbers } p.$$

6.2   Upper Semicontinuity of Cost Functions

For every nonnegative integer t, let $a_t : K \to \mathcal{P}(K)$ be such that graph(a_t) is a closed set, and assume that for every nonnegative integer t, an upper semicontinuous function $\phi_t : R_+^n \to [0, \infty)$ satisfies

$$\sup\{\sup\{\phi_t(z) : z \in (K - K) \cap R_+^n\} : t = 0, 1, \ldots\} < \infty. \qquad (6.12)$$

For every nonnegative integer t and every point $(x, x') \in \text{graph}(a_t)$, set

$$u_t(x, x') = \sup\{\phi_t(z) : z \in R_+^n, \ x' + z \in a(x)\}. \qquad (6.13)$$

By (6.12) and (6.13), u_t, $t = 0, 1, \ldots$ satisfy (6.3). Note that in many models of economic dynamics, cost functions u_t, $t = 0, 1, \ldots$ are defined by (6.13).

Lemma 6.8 *For every nonnegative integer t, the function* $u_t :$ *graph*$(a_t) \to$ *[0, ∞) is upper semicontinuous.*

Proof Let t be a nonnegative integer, and let $\{(x^{(j)}, y^{(j)})\}_{j=1}^{\infty} \subset$ graph(a_t) satisfy

$$\lim_{j \to \infty} (x^{(j)}, y^{(j)}) = (x, y). \tag{6.14}$$

We claim that

$$u_t(x, y) \geq \limsup_{j \to \infty} u(x^{(j)}, y^{(j)}).$$

Extracting a subsequence and re-indexing if necessary, we may assume without loss of generality that there exists

$$\lim_{j \to \infty} u(x^{(j)}, y^{(j)}).$$

In view of (6.13), for every natural number j, there exists $z^{(j)} \in R_+^n$ satisfying

$$y^{(j)} + z^{(j)} \in a_t(x^{(j)}), \quad \phi_t(z^{(j)}) \geq u_t(x^{(j)}, y^{(j)}) - 1/j. \tag{6.15}$$

Evidently, the sequence $\{z^{(j)}\}_{j=1}^{\infty}$ is bounded. Extracting a subsequence and re-indexing, if necessary, we may assume without loss of generality that there exists

$$z = \lim_{j \to \infty} z^{(j)}. \tag{6.16}$$

In view of (6.14)–(6.16), we have

$$z \geq 0$$

and

$$(x, y + z) = \lim_{j \to \infty} (x^{(j)}, y^{(j)} + z^{(j)}) \in \text{graph}(a_t).$$

Together with (6.14)–(6.16), the relation above implies that

$$u_t(x, y) \geq \phi_t(z) \geq \limsup_{j \to \infty} \phi_t(z^{(j)}) \geq \limsup_{j \to \infty} [u_t(x^{(j)}, y^{(j)}) - 1/j]$$

$$= \lim_{j \to \infty} u_t(x^{(j)}, y^{(j)}).$$

This completes the proof of Lemma 6.8.

6.3   The Nonstationary Robinson–Solow–Srinivasan Model

In this section we consider a subclass of the class of infinite horizon optimal control problems considered in Section 6.1. Infinite horizon problems of this subclass correspond to the nonstationary Robinson–Solow–Srinivasan models.

For every nonnegative integer t, let

$$\alpha^{(t)} = (\alpha_1^{(t)}, \ldots, \alpha_n^{(t)}) >> 0, \; b^{(t)} = (b_1^{(t)}, \ldots, b_n^{(t)}) >> 0, \qquad (6.17)$$

$$d^{(t)} = (d_1^{(t)}, \ldots, d_n^{(t)}) \in ((0, 1])^n$$

and for every nonnegative integer t, let $w_t : [0, \infty) \to [0, \infty)$ be a strictly increasing continuous function such that

$$w_t(0) = 0, \; \inf\{w_t(z) : t = 0, 1, \ldots\} > 0 \text{ for all } z > 0 \qquad (6.18)$$

and such that the following assumption holds:

(A4)   for every positive number ϵ, there exists a positive number δ such that for every nonnegative integer t and every $z \in [0, \delta]$, the inequality $w_t(z) \leq \epsilon$ is valid.

Let t be a nonnegative integer. For every $x \in R_+^n$, define

$$a_t(x) = \Big\{ y \in R_+^n : \; y_i \geq (1 - d_i^{(t)})x_i, \; i = 1, \ldots, n,$$

$$\sum_{i=1}^{n} \alpha_i^{(t)}(y_i - (1 - d_i^{(t)})x_i) \leq 1 \Big\}. \qquad (6.19)$$

It is easy to see that for every $x \in R^n$, $a_t(x)$ is a nonempty closed bounded subset of R_+^n and graph(a_t) is a closed subset of $R_+^n \times R_+^n$. Assume that

$$\inf\{d_i^{(t)} : i = 1, \ldots, n, \; t = 0, 1, \ldots\} > 0, \qquad (6.20)$$

$$\inf\{eb^{(t)} : t = 0, 1, \ldots\} > 0, \qquad (6.21)$$

$$\inf\{\alpha_i^{(t)} : i = 1, \ldots, n, \; t = 0, 1, \ldots\} > 0, \qquad (6.22)$$

$$\sup\{b_i^{(t)} : i = 1, \ldots, n, \; t = 0, 1, \ldots\} < \infty, \qquad (6.23)$$

$$\sup\{\alpha_i^{(t)} : i = 1, \ldots, n, \; t = 0, 1, \ldots\} < \infty \qquad (6.24)$$

and that for every positive number M, we have

$$\sup\{w_t(M) : t = 0, 1, \ldots\} < \infty, \ \inf\{w_t(M) : t = 0, 1, \ldots\} > 0. \qquad (6.25)$$

The constraint mappings a_t, $t = 0, 1, \ldots$ have already been defined. Let us now define the cost functions u_t, $t = 0, 1, \ldots$.

For every nonnegative integer t and every $(x, x') \in \text{graph}(a_t)$, put

$$u_t(x, x') = \sup \left\{ w_t(b^{(t)}y) : \ 0 \le y \le x, \right.$$

$$\left. ey + \sum_{i=1}^{n} \alpha_i^{(t)}(x_i' - (1 - d_i^{(t)})x_i) \le 1 \right\}. \qquad (6.26)$$

Fix numbers $\alpha^*, \alpha_* > 0, d_* > 0$ which satisfy

$$\alpha_* < \alpha_i^{(t)} < \alpha^*, \ d_* < d_i^{(t)}, \ i = 1, \ldots, n, \ t = 0, 1, \ldots. \qquad (6.27)$$

Lemma 6.9 *Let a number* $M_0 > (\alpha_* d_*)^{-1}$, *let an integer* $t \ge 0$, *and let* $(x, x') \in \text{graph}(a_t)$ *satisfy* $x \le M_0 e$. *Then* $x' \le M_0 e$.

Proof In view of (6.19),

$$\sum_{i=1}^{n} \alpha_i^{(t)}(x_i' - (1 - d_i^{(t)})x_i) \le 1$$

and by (6.27), for every $i = 1, \ldots, n$, we have

$$x_i' \le (\alpha_i^{(t)})^{-1} + (1 - d_i^{(t)})x_i \le \alpha_*^{-1} + (1 - d_*)x_i \le \alpha_*^{-1} + (1 - d_*)M_0$$

$$\le d_*(\alpha_* d_*)^{-1} + (1 - d_*)M_0 \le d_* M_0 + (1 - d_*)M_0 = M_0.$$

This completes the proof of Lemma 6.9.

Lemma 6.10 *Let t be a nonnegative integer. Then the function*

$$u_t : graph(a_t) \to [0, \infty)$$

is upper semicontinuous. Moreover, if $(x, y) \in graph(a_t)$ and $u_t(x, y) > 0$, then u_t is continuous at (x, y).

Proof Let

$$(x, y) \in \text{graph}(a_t), \ \{(x^{(j)}, y^{(j)})\}_{j=1}^{\infty} \subset \text{graph}(a_t), \ \lim_{j \to \infty} (x^{(j)}, y^{(j)}) = (x, y). \qquad (6.28)$$

We claim that

$$u_t(x, y) \geq \limsup_{j \to \infty} u_t(x^{(j)}, y^{(j)}).$$

Extracting a subsequence and re-indexing, we may assume that there exists $\lim_{j \to \infty} u_t(x^{(j)}, y^{(j)})$. In view of (6.25) and (6.26), for every natural number j, there exists $z^{(j)} \in R_+^n$ which satisfies

$$z^{(j)} \leq x^{(j)}, \;\; ez^{(j)} + \sum_{i=1}^{n} \alpha_i^{(t)}(y_i^{(j)} - (1 - d_i^{(t)})x_i^{(j)}) \leq 1, \tag{6.29}$$

$$w_t(b^{(t)}z^{(j)}) \geq u_t(x^{(j)}, y^{(j)}) - 1/j. \tag{6.30}$$

Extracting a subsequence and re-indexing, we may assume without loss of generality that there exists

$$z = \lim_{j \to \infty} z^{(j)}. \tag{6.31}$$

By (6.28) and (6.31),

$$0 \leq z \leq x. \tag{6.32}$$

It follows from (6.28), (6.29), and (6.31) that

$$ez + \sum_{i=1}^{n} \alpha_i^{(t)}\left(y_i - (1 - d_i^{(t)})x_i\right)$$

$$= \lim_{j \to \infty}\left[ez^{(j)} + \sum_{i=1}^{n} \alpha_i^{(t)}\left(y_i^{(j)} - (1 - d_i^{(t)})x_i^{(j)}\right)\right] \leq 1.$$

Combined with (6.26) and (6.30)–(6.32), the equation above implies that

$$u_t(x, y) \geq w_t(b^{(t)}z) = \lim_{j \to \infty} w_t(b^{(t)}z^{(j)}) = \lim_{j \to \infty} u_t(x^{(j)}, y^{(j)}).$$

Hence u_t is upper lower semicontinuous.

Assume now that $(x, y) \in \text{graph}(a_t)$ satisfies

$$u_t(x, y) > 0, \tag{6.33}$$

and show that u_t is continuous at (x, y). Evidently, it is sufficient to show that u_t is lower semicontinuous at (x, y). Assume that

$(x^{(j)}, y^{(j)}) \in \text{graph}(a_t)$ for all natural numbers j, $\lim_{j\to\infty} (x^{(j)}, y^{(j)}) = (x, y)$.

$$(6.34)$$

Let ϵ be a positive number. It is sufficient to show that

$$\liminf_{j\to\infty} u_t(x^{(j)}, y^{(j)}) \geq u_t(x, y) - \epsilon.$$

It follows from (6.26) and (6.33) that there exists $z \in R_+^n$ for which

$$z \leq x, \ ez + \sum_{i=1}^n \alpha_i^{(t)}(y_i - (1 - d_i^{(t)})x_i) \leq 1, \tag{6.35}$$

$$w_t(b^{(t)}z) > 0, \ w_t(b^{(t)}z) > u_t(x, y) - \epsilon/4. \tag{6.36}$$

By (6.18) and (6.36), there exists $q \in \{1, \ldots, n\}$ satisfying

$$b_q^{(t)}z_q > 0. \tag{6.37}$$

Relations (6.18) and (6.37) imply that there exists $\gamma \in (0, 1)$ which satisfies

$$w_t(b^{(t)}\gamma z) \geq w_t(b^{(t)}z) - \epsilon/4. \tag{6.38}$$

In view of (6.34), (6.35), and (6.37), there exists an integer $j_0 \geq 1$ such that for every natural number $j \geq j_0$, we have

$$\gamma z \leq x^{(j)}, \ e(\gamma z) + \sum_{i=1}^n \alpha_i^{(t)}(y_i^{(j)} - (1 - d_i^{(t)})x_i^{(j)}) \leq 1. \tag{6.39}$$

It follows from (6.26), (6.36), (6.38), and (6.39) that for all natural numbers $j \geq j_0$, we have

$$u_t(x^{(j)}, y^{(j)}) \geq w_t(b^{(t)}\gamma z) \geq w_t(b^{(t)}z) - \epsilon/4 > u_t(x, y) - \epsilon/2.$$

This implies that u_t is lower semicontinuous at (x, y) and completes the proof of Lemma 6.10.

For every vector $x = (x_1, \ldots, x_n) \in R^n$, put

$$\|x\|_1 = \sum_{i=1}^n |x_i|, \ \|x\|_\infty = \max\{|x_i| : i = 1, \ldots, n\}. \tag{6.40}$$

In view of (6.19) and (6.27), for every nonnegative integer t, for every pair of points $x, y \in K$, and for $\|\cdot\| = \|\cdot\|_p$, where $p = 1, 2, \infty$, we have

$$H(a_t(x), a_t(y)) \leq \|((1 - d_i^{(t)})x_i)_{i=1}^n - ((1 - d_i^{(t)})y_i)_{i=1}^n\| \leq (1 - d_*)\|x - y\|$$

$$(6.41)$$

(see (6.2)).

Proposition 6.11 *Let δ be a positive number. Then there exists a positive number λ such that for every nonnegative integer t and every $(x, y) \in graph(a_t)$ satisfying $u_t(x, y) \geq \delta$, the inclusion $y + \lambda e \in a_t(x)$ is valid.*

Proof Assumption (A4) implies that there exists a positive number δ_0 such that for every nonnegative integer t and every $\xi \in R_+^1$ such that $w_t(\xi) \geq \delta/2$, we have

$$\xi \geq \delta_0. \tag{6.42}$$

Put

$$b_* = \sup\{b_i^{(t)} : t = 0, 1, \ldots, i = 1, \ldots, n\} \tag{6.43}$$

(see (6.23)). Fix number λ for which

$$\lambda n \alpha^* < 2^{-1} b_*^{-1} \delta_0. \tag{6.44}$$

Assume that an integer $t \geq 0$,

$$(x, y) \in graph(a_t), \ u_t(x, y) \geq \delta. \tag{6.45}$$

In view of (6.26) and (6.45), there exists $z \in R_+^n$ which satisfies

$$0 \leq z \leq x, \ ez + \sum_{i=1}^n \alpha_i^{(t)}(y_i - (1 - d_i^{(t)})x_i) \leq 1, \ w_t(b^{(t)}z) \geq \delta/2. \tag{6.46}$$

By (6.46) and the choice of δ_0, we have

$$b^{(t)}z \geq \delta_0. \tag{6.47}$$

Relations (6.43) and (6.47) imply that

$$ez = \sum_{i=1}^n z_i = \sum_{i=1}^n (b_i^{(t)})^{-1} b_i^{(t)} z_i \geq b_*^{-1} b^{(t)}z \geq b_*^{-1}\delta_0. \tag{6.48}$$

We claim that $y + \lambda e \in a_t(x)$. Evidently, (see (6.19) and (6.45)) for any $i = 1, \ldots, n$,

$$y_i + \lambda \geq y_i \geq (1 - d_i^{(t)})x_i. \tag{6.49}$$

In view of (6.27), (6.44), (6.46), and (6.48),

$$\sum_{i=1}^{n} \alpha_i^{(t)}((y + \lambda e)_i - (1 - d_i^{(t)})x_i) = \sum_{i=1}^{n} \alpha_i^{(t)}(y_i - (1 - d_i^{(t)})x_i) + \lambda \sum_{i=1}^{n} \alpha_i^{(t)}$$

$$\leq 1 - ez + \lambda \sum_{i=1}^{n} \alpha_i^{(t)} \leq 1 - b_*^{-1}\delta_0 + \lambda n\alpha^* < 1$$

and combined with (6.49), this implies that $y + \lambda e \in a_t(x)$. This completes the proof of Proposition 6.11.

Proposition 6.12 *There exist a program* $\{\widehat{x}(t)\}_{t=0}^{\infty}$ *and* $\widehat{\Delta} > 0$ *such that*

$$u_t(\widehat{x}(t), \widehat{x}(t+1)) \geq \widehat{\Delta} \text{ for all integers } t \geq 0.$$

Proof Fix numbers $\lambda_0 > 0$, $\lambda_1 > 0$ such that

$$\lambda_0 n\alpha^* < 1/2, \quad \lambda_1 < \lambda_0, \quad \lambda_1 n < 1/4. \tag{6.50}$$

In view of (6.21), there exists a positive number ϵ_0 satisfying

$$eb^{(t)} \geq \epsilon_0, \quad t = 0, 1, \dots. \tag{6.51}$$

Set

$$\widehat{\Delta} = \inf\{w_t(\lambda_1\epsilon_0) : t = 0, 1, \dots\}. \tag{6.52}$$

Relation (6.25) implies that $\widehat{\Delta} > 0$. Define

$$\widehat{x}(t) = \lambda_0 e, \quad t = 0, 1, \dots, \quad \widehat{y}(t) = \lambda_1 e, \quad t = 0, 1, \dots \tag{6.53}$$

It follows from (6.27), (6.50), and (6.53) that for $i = 1, \dots, n$ and $t = 0, 1, \dots,$ we have

$$\widehat{x}_i(t+1) - (1 - d_i^{(t)})\widehat{x}_i(t) = \lambda_0 d_i^{(t)} > 0,$$

$$\sum_{i=1}^{n} \alpha_i^{(t)}[\widehat{x}_i(t+1) - (1 - d_i^{(t)})\widehat{x}_i(t)] \tag{6.54}$$

$$= \left(\sum_{i=1}^{n} \alpha_i^{(t)} d_i^{(t)}\right)\lambda_0 \leq \lambda_0 \sum_{i=1}^{n} \alpha_i^{(t)} \leq \lambda_0 n\alpha^* < 1/2 \tag{6.55}$$

and for all $t = 0, 1, \dots,$ we have

$$e\widehat{y}(t) + \sum_{i=1}^{n} \alpha_i^{(t)}[\widehat{x}_i(t+1) - (1 - d_i^{(t)})\widehat{x}_i(t)] \leq \lambda_1 n + 1/2 < 1. \tag{6.56}$$

Thus $\{\widehat{x}(t)\}_{t=0}^{\infty}$ is a program. In view of (6.26), (6.50)–(6.53), and (6.56), for every nonnegative integer t, we have

$$u_t(\widehat{x}(t), \widehat{x}(t+1)) \geq w_t(b^{(t)}\widehat{y}(t)) \geq w_t(\lambda_1 e b^{(t)}) \geq w_t(\lambda_1 \epsilon_0) \geq \widehat{\Delta}.$$

This completes the proof of Proposition 6.12.

Proposition 6.13 *Let t be a nonnegative integer, let $(x, y) \in graph(a_t)$, and let a point $\tilde{x} \in R_+^n$ satisfy $\tilde{x} \geq x$. Then there exists $\tilde{y} \in a_t(\tilde{x})$ which satisfies $\tilde{y} \geq y$ and $u_t(\tilde{x}, \tilde{y}) \geq u_t(x, y)$.*

Proof In view of (6.26), there exists $z \in R_+^n$ which satisfies

$$0 \leq z \leq x, \; ez + \sum_{i=1}^{n} \alpha_i^{(t)}(y_i - (1 - d_i^{(t)})x_i) \leq 1, \; w_t(b^{(t)}z) = u_t(x, y). \tag{6.57}$$

For all integer $i = 1, \ldots, n$, define

$$\tilde{y}_i = \tilde{x}_i(1 - d_i^{(t)}) + y_i - (1 - d_i^{(t)})x_i. \tag{6.58}$$

It follows from (6.19), (6.57), and (6.58) that for $i = 1, \ldots, n$, we have

$$\tilde{y}_i \geq (1 - d_i^{(t)})\tilde{x}_i,$$

$$\sum_{i=1}^{n} \alpha_i^{(t)}(\tilde{y}_i - (1 - d_i^{(t)})\tilde{x}_i) = \sum_{i=1}^{n} \alpha_i^{(t)}(y_i - (1 - d_i^{(t)})x_i) \leq 1 - ez.$$

Thus $\tilde{y} \in a_t(\tilde{x})$. By the inequality $\tilde{x} \geq x$ and (6.58), $\tilde{y} \geq y$. It is not difficult to see that

$$u_t(\tilde{x}, \tilde{y}) \geq w_t(b^{(t)}z) = u_t(x, y).$$

Proposition 6.13 is proved.

It is not difficult to see that the following result holds.

Proposition 6.14 *Let an integer $t \geq 0$, x, x_1, x_2, $x_3 \in R_+^n$, $x_i \in a_t(x)$, $i = 1, 3$, $x_1 \leq x_2 \leq x_3$. Then $x_2 \in a(x_t)$.*

Thus we have defined the mappings a_t and the cost functions u_t, $t = 0, 1, \ldots$. The control system considered in this section is a special case of the control system studied in Section 6.1. As we have already mentioned before, this control system corresponds to the nonstationary Robinson–Solow–Srinivasan model. Note that this control system satisfies the assumptions posed in Section 6.1 and therefore all the results stated there hold for this system. Indeed, choose $M_0 > (\alpha_* d_*)^{-1}$, and put $K = \{z \in R_+^n : z \leq M_0 e\}$. Lemma 6.9 implies that $a_t(K) \subset K$, $t = 0, 1, \ldots$.

Relation (6.2) follows from (6.41). It is clear that (6.3) holds. By Lemma 6.10, u_t is upper semicontinuous for every nonnegative integer t. Proposition 6.11 implies (A1). Assumption (A2) follows from Proposition 6.12, and assumption (A3) follows from Proposition 6.13.

6.4 Auxiliary Results for Theorems 6.4, 6.5, and 6.7

Lemma 6.15 *Let δ be a positive number. Then there exists an integer $T_0 \geq 4$ such that for every natural number $\tau_1 \geq 0$, every natural number $\tau_2 \geq T_0 + \tau_1$, every program $\{x(t)\}_{t=\tau_1}^{\tau_2}$ satisfying*

$$u_{\tau_2-1}(x(\tau_2 - 1), x(\tau_2)) \geq \delta, \tag{6.59}$$

and every $\tilde{x}_0 \in K$, there exists a program $\{\tilde{x}(t)\}_{t=\tau_1}^{\tau_2}$ which satisfies

$$\tilde{x}(\tau_1) = \tilde{x}_0, \ \tilde{x}(\tau_2) \geq x(\tau_2).$$

Proof In view of assumption (A1), there exists a number $\lambda \in (0, 1)$ such that the following property holds:

(P1) For every nonnegative integer t and every $(x, x') \in \text{graph}(a_t)$ which satisfies $u_t(x, x') \geq \delta$, there exists $z \in a_t(x)$ for which $z \geq x' + \lambda e$.

Fix a positive number D_0 such that

$$\|z\| \leq D_0 \text{ for all } z \in K. \tag{6.60}$$

There exists a positive number c_0 such that

$$\|z\|_2 \leq c_0 \|z\| \text{ for all } z \in K. \tag{6.61}$$

Fix an integer $T_0 \geq 4$ satisfying

$$2D_0 c_0 \kappa^{T_0} < \lambda \tag{6.62}$$

(see (6.2)).

Assume that integers $\tau_1 \geq 0$, $\tau_2 \geq T_0 + \tau_1$, that a program $\{x(t)\}_{t=\tau_1}^{\tau_2}$ satisfies (6.59), and that $\tilde{x}_0 \in K$. It follows from (6.59) and property (P1) that there exists $z \in R_+^n$ such that

$$z \in a_{\tau_2-1}(x(\tau_2 - 1)), \ z \geq x(\tau_2) + \lambda e. \tag{6.63}$$

In view of (6.2), there exists a program $\{\tilde{x}(t)\}_{t=\tau_1}^{\tau_2-1}$ such that

$$\tilde{x}(\tau_1) = \tilde{x}_0, \tag{6.64}$$

$$\|\tilde{x}(t+1) - x(t+1)\| \le \kappa \|\tilde{x}(t) - x(t)\|, \ t = \tau_1, \ldots, \tau_2 - 2.$$

By (6.2) and (6.63), there exists

$$\tilde{x}(\tau_2) \in a_{\tau_2 - 1}(\tilde{x}(\tau_2 - 1))$$

which satisfies

$$\|\tilde{x}(\tau_2) - z\| \le \kappa \|x(\tau_2 - 1) - \tilde{x}(\tau_2 - 1)\|. \tag{6.65}$$

Evidently, $\{\tilde{x}(t)\}_{t=\tau_1}^{\tau_2}$ is a program. It follows from (6.60), (6.64), and (6.65) that

$$\|\tilde{x}(\tau_2) - z\| \le \kappa^{\tau_2 - \tau_1} \|\tilde{x}(\tau_1) - x(\tau_1)\| \le \kappa^{\tau_2 - \tau_1}(2D_0) \le \kappa^{T_0}(2D_0)$$

and by (6.61),

$$\|\tilde{x}(\tau_2) - z\|_2 \le 2D_0 c_0 \kappa^{T_0}.$$

The inequality above implies that for all natural numbers $i = 1, \ldots, n$, we have

$$|\tilde{x}_i(\tau_2) - z_i| \le 2D_0 c_0 \kappa^{T_0}$$

and in view of (6.62) and (6.63), we have

$$\tilde{x}(\tau_2) \ge z - 2D_0 c_0 \kappa^{T_0} e \ge x(\tau_2) + [\lambda - 2D_0 c_0 \kappa^{T_0}]e \ge x(\tau_2).$$

This completes the proof of Lemma 6.15.

Fix a number $\gamma > 0$ satisfying

$$\gamma < 1/2 \text{ and } \gamma < 4^{-1}\widehat{\Delta}. \tag{6.66}$$

Lemma 6.16 *Let M_1 be a positive number. Then there exist integers $L_1, L_2 \ge 4$ such that for every pair of integers $T_1 \ge 0$, $T_2 \ge L_1 + L_2 + T_1$, every program $\{x(t)\}_{t=T_1}^{T_2 - 1}$ satisfying*

$$\sum_{t=T_1}^{T_2-1} u_t(x(t), x(t+1)) \ge U(x(T_1), T_1, T_2) - M_1, \tag{6.67}$$

and every integer $\tau \in [T_1 + L_1, T_2 - L_2]$, the following inequality holds:

$$\max\{u_t(x(t), x(t+1)) : t = \tau, \ldots, \tau + L_2 - 1\} \ge \gamma. \tag{6.68}$$

Proof Lemma 6.15 implies that there exists an integer $L_1 \geq 4$ such that the following property holds:

(P2) If integers $S_1 \geq 0$, $S_2 \geq S_1 + L_1$, if a program $\{v(t)\}_{t=S_1}^{S_2}$ satisfies

$$u_{S_2-1}(v(S_2-1), v(S_2)) \geq \gamma,$$

and if $\tilde{v}_0 \in K$, then there exists a program $\{\tilde{v}(t)\}_{t=S_1}^{S_2}$ such that $\tilde{v}(S_1) = \tilde{v}_0$, $\tilde{v}(S_2) \geq v(S_2)$.

Fix a number M_2 for which

$$M_2 > u_t(z, z') \text{ for each integer } t \geq 0 \text{ and each } (z, z') \in \text{graph}(a_t) \qquad (6.69)$$

and an integer $L_2 \geq 1$ which satisfies

$$L_2 > 4(L_1+1) + 16\widehat{\Delta}^{-1}(M_1+L_1\gamma+1) + 16\widehat{\Delta}^{-1}(M_1+M_2+L_1+2). \qquad (6.70)$$

Assume that integers $T_1 \geq 0$, $T_2 \geq L_1 + L_2 + T_1$, that a program $\{x(t)\}_{t=T_1}^{T_2-1}$ satisfies (6.67), and that an integer τ satisfies

$$T_1 + L_1 \leq \tau \leq T_2 - L_2. \qquad (6.71)$$

We claim that (6.68) is valid. Assume the contrary. Then

$$u_t(x(t), x(t+1)) < \gamma, \ t = \tau, \dots, \tau + L_2 - 1. \qquad (6.72)$$

There are two cases:

$$u_t(x(t), x(t+1)) < \gamma, \ t = \tau, \dots, T_2 - 1; \qquad (6.73)$$

$$\max\{u_t(x(t), x(t+1)) : \ t = \tau, \dots, T_2 - 1\} \geq \gamma. \qquad (6.74)$$

Now we define a natural number τ_0 as follows. If (6.73) holds, then we put $\tau_0 = T_2$. If (6.74) is valid, then in view of (6.72), there exists an integer $\tau_0 \geq 1$ such that

$$\tau + L_2 \leq \tau_0 \leq T_2 - 1, \qquad (6.75)$$

$$u_{\tau_0}(x(\tau_0), x(\tau_0 + 1)) \geq \gamma, \qquad (6.76)$$

$$u_t(x(t), x(t+1)) < \gamma, \ t = \tau, \dots, \tau_0 - 1. \qquad (6.77)$$

It is not difficult to see that in both cases (6.77) is valid and that in both cases

$$\tau_0 - \tau \geq L_2. \qquad (6.78)$$

Assume that (6.73) holds. Assumption (A2), the choice of L_1, property (P2), (6.66), (6.70), and (6.71) imply that there exists a program $\{\tilde{x}(t)\}_{t=\tau}^{\tau+L_1}$ which satisfies

$$\tilde{x}(\tau) = x(\tau), \ \tilde{x}(\tau + L_1) \geq \hat{x}(\tau + L_1). \tag{6.79}$$

Define

$$\tilde{x}(t) = x(t), \ t = T_1, \dots, \tau. \tag{6.80}$$

In view of (6.79), (6.80), (A3), and (A2), there exists $\tilde{x}(t) \in K, t = \tau + L_1 + 1, \dots, T_2$ such that $\{\tilde{x}(t)\}_{t=T_1}^{T_2}$ is a program,

$$\tilde{x}(t) \geq \hat{x}(t) \text{ for all integers } t = \tau + L_1, \dots, T_2, \tag{6.81}$$

$$u_t(\tilde{x}(t), \tilde{x}(t+1)) \geq u_t(\hat{x}(t), \hat{x}(t+1)), \ t = \tau + L_1, \dots, T_2 - 1. \tag{6.82}$$

Assumption (A2), (6.66), (6.67), (6.70), (6.71), (6.73), (6.80), and (6.82) imply that

$$M_1 \geq U(x(T_1), T_1, T_2) - \sum_{t=T_1}^{T_2-1} u_t(x(t), x(t+1))$$

$$\geq \sum_{t=T_1}^{T_2-1} u_t(\tilde{x}(t), \tilde{x}(t+1)) - \sum_{t=T_1}^{T_2-1} u_t(x(t), x(t+1))$$

$$= \sum_{t=\tau}^{T_2-1} u_t(\tilde{x}(t), \tilde{x}(t+1)) - \sum_{t=\tau}^{T_2-1} u_t(x(t), x(t+1))$$

$$\geq \sum_{t=\tau}^{\tau+L_1-1} u_t(\tilde{x}(t), \tilde{x}(t+1)) + \sum_{t=\tau+L_1}^{T_2-1} u_t(\hat{x}(t), \hat{x}(t+1))$$

$$- \sum_{t=\tau}^{T_2-1} u_t(x(t), x(t+1))$$

$$\geq \sum_{t=\tau+L_1}^{T_2-1} u_t(\hat{x}(t), \hat{x}(t+1)) - \sum_{t=\tau}^{T_2-1} u_t(x(t), x(t+1))$$

$$\geq (T_2 - \tau - L_1)\hat{\Delta} - (T_2 - \tau)\gamma = (T_2 - \tau - L_1)(\hat{\Delta} - \gamma)$$

$$- L_1\gamma \geq \hat{\Delta}2^{-1}(T_2 - \tau - L_1) - L_1\gamma$$

$$\geq 2^{-1}\hat{\Delta}(L_2 - L_1) - L_1\gamma$$

$$\geq 4^{-1}\hat{\Delta}L_2 - L_1\gamma$$

and

$$L_2 \le 8\widehat{\Delta}^{-1}(M_1 + L_1\gamma).$$

The inequality above contradicts (6.70). The contradiction we have reached proves that (6.73) is not true. Thus (6.74) holds and there exists an integer $\tau_0 \ge 1$ satisfying (6.75)–(6.77). Assumption (A2), the choice of L_1, property (P2), and (6.66) imply that there exists a program $\{\tilde{x}(t)\}_{t=\tau}^{\tau+L_1}$ such that

$$\tilde{x}(\tau) = x(\tau), \ \tilde{x}(\tau + L_1) \ge \widehat{x}(\tau + L_1). \tag{6.83}$$

Put

$$\tilde{x}(t) = x(t), \ t = T_1, \dots, \tau. \tag{6.84}$$

By (A2), (A3), (6.75), (6.76), and (6.83), there exist $\tilde{x}(t) \in K, \ t = \tau + 1 + L_1, \dots, \tau_0 - L_1$ such that $\{\tilde{x}(t)\}_{t=\tau+L_1}^{\tau_0-L_1}$ is a program,

$$\tilde{x}(t) \ge \widehat{x}(t), \ t = \tau + L_1, \dots, \tau_0 - L_1, \tag{6.85}$$

$$u_t(\tilde{x}(t), \tilde{x}(t+1)) \ge u_t(\widehat{x}(t), \widehat{x}(t+1)), \ t = \tau + L_1, \dots, \tau_0 - L_1 - 1. \tag{6.86}$$

Evidently, $\{\tilde{x}(t)\}_{t=T_1}^{\tau_0-L_1}$ is a program. In view of the choice of L_1, property (P2), and (6.76), there exist $\tilde{x}(t) \in K, t = \tau_0 - L_1 + 1, \dots, \tau_0 + 1$ such that $\{\tilde{x}(t)\}_{t=\tau_0-L_1}^{\tau_0+1}$ is a program,

$$\tilde{x}(\tau_0 + 1) \ge x(\tau_0 + 1). \tag{6.87}$$

It is clear that $\{\tilde{x}(t)\}_{t=T_1}^{\tau_0+1}$ is a program. If $T_2 > \tau_0 + 1$, then relation (6.87) and assumption (A3) imply that there exist $\tilde{x}(t) \in K, t = \tau_0 + 2, \dots, T_2$ such that $\{\tilde{x}(t)\}_{t=\tau_0+1}^{T_2}$ is a program,

$$\tilde{x}(t) \ge x(t), \ t = \tau_0 + 1, \dots, T_2, \tag{6.88}$$

$$u_t(\tilde{x}(t), \tilde{x}(t+1)) \ge u_t(x(t), x(t+1)), \ t = \tau_0 + 1, \dots, T_2 - 1. \tag{6.89}$$

It follows from (6.66), (6.67), (6.69)–(6.71), (6.74), (6.77), (6.78), (6.84), (6.86), (6.89), and assumption (A2) that

$$M_1 \ge U(x(T_1), T_1, T_2) - \sum_{t=T_1}^{T_2-1} u_t(x(t), x(t+1))$$

$$\geq \sum_{t=T_1}^{T_2-1} u_t(\tilde{x}(t), \tilde{x}(t+1)) - \sum_{t=T_1}^{T_2-1} u_t(x(t), x(t+1))$$

$$= \sum_{t=\tau}^{T_2-1} u_t(\tilde{x}(t), \tilde{x}(t+1)) - \sum_{t=\tau}^{T_2-1} u_t(x(t), x(t+1))$$

$$\geq \sum_{t=\tau}^{\tau_0} u_t(\tilde{x}(t), \tilde{x}(t+1)) - \sum_{t=\tau}^{\tau_0} u_t(x(t), x(t+1))$$

$$\geq \sum_{t=\tau+L_1}^{\tau_0-L_1-1} u_t(\tilde{x}(t), \tilde{x}(t+1)) - (\tau_0 - \tau)\gamma - u_{\tau_0}(x(\tau_0), x(\tau_0+1))$$

$$\geq \sum_{t=\tau+L_1}^{\tau_0-L_1-1} u_t(\widehat{x}(t), \widehat{x}(t+1)) - (\tau_0 - \tau)\gamma - u_{\tau_0}(x(\tau_0), x(\tau_0+1))$$

$$\geq \widehat{\Delta}(\tau_0 - \tau - 2L_1) - (\tau_0 - \tau)\gamma - M_2 = (\widehat{\Delta} - \gamma)(\tau_0 - \tau - 2L_1)$$

$$-2L_1\gamma - M_2$$

$$\geq (\widehat{\Delta}/2)(\tau_0 - \tau - 2L_1) - 2L_1 - M_2$$

$$\geq (\widehat{\Delta}/2)(L_2 - 2L_1) - 2L_1 - M_2 \geq 4^{-1}L_2\widehat{\Delta} - 2L_1 - M_2$$

and

$$L_2 \leq 4(\widehat{\Delta})^{-1}(M_1 + M_2 + 2L_1).$$

This inequality contradicts (6.70). The contradiction we have reached proves (6.68) and Lemma 6.16 itself.

Lemma 6.17 *Let M_1 be a positive number. Then there exist integers $\bar{L}_1, \bar{L}_2 \geq 1$ and a positive number M_2 such that for every pair of integers $\tau_1 \geq 0$, $\tau_2 \geq \bar{L}_1 + \bar{L}_2 + \tau_1$ and every program $\{x(t)\}_{t=\tau_1}^{\tau_2}$ satisfying*

$$\sum_{t=\tau_1}^{\tau_2-1} u_t(x(t), x(t+1)) \geq U(x(\tau_1), \tau_1, \tau_2) - M_1, \tag{6.90}$$

the following assertion holds.
   If integers $T_1, T_2 \in [\tau_1, \tau_2 - \bar{L}_2]$ satisfy $\bar{L}_1 \leq T_2 - T_1$, then

$$\sum_{t=T_1}^{T_2-1} u_t(x(t), x(t+1)) \geq U(x(T_1), T_1, T_2) - M_2. \tag{6.91}$$

Proof Let integers $L_1, L_2 \geq 4$ be as guaranteed by Lemma 6.16. Lemma 6.15 implies that there exists an integer $L_3 \geq 4$ such that the following property holds:

(P3) If integers $S_1 \geq 0$, $S_2 \geq L_3 + S_1$, if a program $\{v(t)\}_{t=S_1}^{S_2}$ satisfies

$$u_{S_2-1}(v(S_2 - 1), v(S_2)) \geq \gamma,$$

and if $\tilde{v}_0 \in K$, then there exists a program $\{\tilde{v}(t)\}_{t=S_1}^{S_2}$ such that $\tilde{v}(S_1) = \tilde{v}_0$, $\tilde{v}(S_2) \geq v(S_2)$.

Fix a positive a number M_0 for which

$$M_0 > u_t(z, z') \text{ for every nonnegative integer } t \text{ and every } (z, z') \in \text{graph}(a_t),$$
$$\tag{6.92}$$

integers $\bar{L}_1, \bar{L}_2 \geq 1$, and a number $M_2 > 0$ satisfying

$$\bar{L}_1 \geq L_1, \ \bar{L}_2 > 2(L_1 + L_2 + L_3 + 1), \tag{6.93}$$

$$M_2 > M_1 + M_0(L_3 + L_2). \tag{6.94}$$

Assume that integers $\tau_1 \geq 0$, $\tau_2 \geq \bar{L}_1 + \bar{L}_2 + \tau_1$, a program $\{x(t)\}_{t=\tau_1}^{\tau_2}$ satisfies (6.90), and integers T_1, T_2 satisfy

$$T_1, T_2 \in [\tau_1, \tau_2 - \bar{L}_2], \ \bar{L}_1 \leq T_2 - T_1. \tag{6.95}$$

We claim that (6.91) is true. Proposition 6.3 implies that there exists a program $\{x^{(1)}(t)\}_{t=T_1}^{T_2}$ which satisfies

$$x^{(1)}(T_1) = x(T_1), \ \sum_{t=T_1}^{T_2-1} u_t(x^{(1)}(t), x^{(1)}(t + 1)) = U(x(T_1), T_1, T_2). \tag{6.96}$$

It follows from (6.93) and (6.95) that

$$T_1 + L_1 \leq T_1 + \bar{L}_1 + L_3 \leq T_2 + L_3 \leq \tau_2 - \bar{L}_2 + L_3 \leq \tau_2 - 2L_2 - L_3. \tag{6.97}$$

By the choice of the integers L_1, L_2, Lemma 6.16, (6.90), (6.93), and (6.97), we have

$$\max\{u_t(x(t), x(t + 1)) : t = T_2 + L_3, \ldots, T_2 + L_2 + L_3 - 1\} \geq \gamma.$$

Hence there exists an integer $\tau \in [T_2 + L_3, \ldots, T_2 + L_3 + L_2 - 1]$ for which

$$u_\tau(x(\tau), x(\tau + 1)) \geq \gamma. \tag{6.98}$$

Property (P3) and (6.98) imply that there exists a program $\{x^{(2)}(t)\}_{t=T_2}^{\tau+1}$ such that

$$x^{(2)}(T_2) = x^{(1)}(T_2), \ x^{(2)}(\tau + 1) \geq x(\tau + 1). \tag{6.99}$$

Define

$$\tilde{x}(t) = x(t), \ t = \tau_1, \ldots, T_1, \ \tilde{x}(t) = x^{(1)}(t), \ t = T_1 + 1, \ldots, T_2,$$

$$\tilde{x}(t) = x^{(2)}(t), \ t = T_2 + 1, \ldots, \tau + 1. \tag{6.100}$$

It is easy to see that $\{\tilde{x}(t)\}_{t=\tau_1}^{\tau+1}$ is a program. By (6.99) and (6.100), we have

$$\tilde{x}(\tau + 1) \geq x(\tau + 1). \tag{6.101}$$

Assumption (A3) and (6.101) imply that there exist $\tilde{x}(t) \in K, t = \tau + 2, \ldots, \tau_2$ such that $(\tilde{x}(t))_{t=\tau_1}^{\tau_2}$ is a program,

$$\tilde{x}(t) \geq x(t), \ t = \tau + 1, \ldots, \tau_2, \tag{6.102}$$

$$u_t(\tilde{x}(t), \tilde{x}(t + 1)) \geq u_t(x(t), x(t + 1)), \ t = \tau + 1, \ldots, \tau_2 - 1. \tag{6.103}$$

By (6.90), (6.92), (6.94), (6.96), (6.100), (6.103), and the choice of $\bar{L}$,

$$M_1 \geq U(x(\tau_1), \tau_1, \tau_2) - \sum_{t=\tau_1}^{\tau_2-1} u_t(x(t), x(t + 1))$$

$$\geq \sum_{t=\tau_1}^{\tau_2-1} u_t(\tilde{x}(t), \tilde{x}(t + 1)) - \sum_{t=\tau_1}^{\tau_2-1} u_t(x(t), x(t + 1))$$

$$= \sum_{t=T_1}^{\tau_2-1} u_t(\tilde{x}(t), \tilde{x}(t + 1)) - \sum_{t=T_1}^{\tau_2-1} u_t(x(t), x(t + 1))$$

$$\geq \sum_{t=T_1}^{\tau} u_t(\tilde{x}(t), \tilde{x}(t + 1)) - \sum_{t=T_1}^{\tau} u_t(x(t), x(t + 1))$$

$$\geq \sum_{t=T_1}^{T_2-1} u_t(\tilde{x}(t), \tilde{x}(t + 1)) - \sum_{t=T_1}^{T_2-1} u_t(x(t), x(t + 1)) - \sum_{t=T_2}^{\tau} u_t(x(t), x(t + 1))$$

$$\geq U(x(T_1), T_1, T_2) - \sum_{t=T_1}^{T_2-1} u_t(x(t), x(t + 1)) - (\tau - T_2 + 1)M_0$$

and

$$\sum_{t=T_1}^{T_2-1} u_t(x(t), x(t+1)) \geq U(x(T_1), T_1, T_2) - M_1 - M_0(L_3 + L_2)$$

$$> U(x(T_1), T_1, T_2) - M_2.$$

This completes the proof of Lemma 6.17.

6.5   Properties of the Function U

It is not difficult to see that the following auxiliary result holds.

Proposition 6.18 *Let $\tau_1 \geq 0$, $\tau_1 > \tau_1$ be integers, let $\Delta \geq 0$, T_1, T_2 be integers such that $\tau_1 \leq T_1 < T_2 \leq \tau_2$, and let $\{x(t)\}_{t=\tau_1}^{\tau_2}$ be a program satisfying*

$$\sum_{t=\tau_1}^{\tau_2-1} u_t(x(t), x(t+1)) \geq U(x(\tau_1), x(\tau_2), \tau_1, \tau_2) - \Delta.$$

Then

$$\sum_{t=T_1}^{T_2-1} u_t(x(t), x(t+1)) \geq U(x(T_1), x(T_2), T_1, T_2) - \Delta.$$

Lemma 6.19 *There exist an integer $L \geq 1$ and a positive number M_1 such that for every pair of points $x_0, \tilde{x}_0 \in K$ and every pair of integers $T_1 \geq 0$, $T_2 \geq T_1 + L$, the inequality*

$$|U(x_0, T_1, T_2) - U(\tilde{x}_0, T_1, T_2)| \leq M_1$$

is valid.

Proof Let integers $L_1, L_2 \geq 4$ be as guaranteed by Lemma 6.16 with $M_1 = 1$. Lemma 6.15 implies that there exists a natural number $L_3 \geq 4$ such that the following property holds:

(P4) If integers $S_1 \geq 0$, $S_2 \geq S_1 + L_3$, if a program $\{v(t)\}_{t=S_1}^{S_2}$ satisfies

$$u_{S_2-1}(v(S_2 - 1), v(S_2)) \geq \gamma,$$

and if $\tilde{v}_0 \in K$, then there exists a program $\{\tilde{v}(t)\}_{t=S_1}^{S_2}$ such that $\tilde{v}(S_1) = \tilde{v}_0$, $\tilde{v}(S_2) \geq v(S_2)$.

Fix an integer

$$L > 2(L_1 + L_2 + L_3 + 1), \tag{6.104}$$

a number

$$M_0 > u_t(z, z'), \quad t = 0, 1, \ldots, \quad (z, z') \in \text{graph}(a_t) \tag{6.105}$$

and set

$$M_1 = M_0(L_1 + L_2 + L_3). \tag{6.106}$$

Assume that $x_0, \tilde{x}_0 \in K$ and that integers $T_1 \geq 0$, $T_2 \geq T_1 + L$. Proposition 6.3 implies that there exists a program $\{x(t)\}_{t=T_1}^{T_2}$ which satisfies

$$x(T_1) = x_0, \quad \sum_{t=T_1}^{T_2-1} u_t(x(t), x(t+1)) = U(x_0, T_1, T_2). \tag{6.107}$$

By (6.104), we have

$$T_1 + L_1 + L_3 < T_1 + L - L_2 \leq T_2 - L_2. \tag{6.108}$$

By the choice of L_1, L_2, Lemma 6.16, (6.104), and (6.107),

$$\max\{u_t(x(t), x(t+1)) : \ t = L_3 + L_1 + T_1, \ldots, L_3 + L_1 + L_2 + T_1 - 1\} \geq \gamma.$$

Thus there exists an integer

$$\tau \in \{T_1 + L_1 + L_3, \ldots, T_1 + L_3 + L_1 + L_2 - 1\} \tag{6.109}$$

such that

$$u_\tau(x(\tau), x(\tau+1)) \geq \gamma. \tag{6.110}$$

Property (P4), the choice of L_3, (6.109), and (6.110) imply that there exists a program $\{\tilde{x}(t)\}_{t=T_1}^{\tau+1}$ satisfying

$$\tilde{x}(T_1) = \tilde{x}_0, \quad \tilde{x}(\tau+1) \geq x(\tau+1). \tag{6.111}$$

Assumption (A3) and (6.111) imply that there exist $\tilde{x}(t) \in K$, $t = \tau + 2, \ldots, T_2$ such that $\{\tilde{x}(t)\}_{t=\tau+1}^{T_2}$ is a program,

$$\tilde{x}(t) \geq x(t), \quad t = \tau + 1, \ldots, T_2, \tag{6.112}$$

$$u_t(\tilde{x}(t), \tilde{x}(t+1)) \geq u_t(x(t), x(t+1)), \quad t = \tau + 1, \ldots, T_2 - 1. \tag{6.113}$$

Evidently, $\{\tilde{x}(t)\}_{t=T_1}^{T_2}$ is a program. In view of (6.105)–(6.107), (6.109), (6.111), and (6.113), we have

$$U(\tilde{x}_0, T_1, T_2) \geq \sum_{t=T_1}^{T_2-1} u_t(\tilde{x}(t), \tilde{x}(t+1)) = \sum_{t=T_1}^{T_2-1} u_t(x(t), x(t+1))$$

$$- \left[\sum_{t=T_1}^{T_2-1} u_t(x(t), x(t+1)) - \sum_{t=T_1}^{T_2-1} u_t(\tilde{x}(t), \tilde{x}(t+1)) \right]$$

$$\geq U(x_0, T_1, T_2) - \left[\sum_{t=T_1}^{\tau} u_t(x(t), x(t+1)) - \sum_{t=T_1}^{\tau} u_t(\tilde{x}(t), \tilde{x}(t+1)) \right]$$

$$\geq U(x_0, T_1, T_2) - \sum_{t=T_1}^{\tau} u_t(x(t), x(t+1)) \geq U(x_0, T_1, T_2)$$

$$-(\tau - T_1)M_0$$

$$\geq U(x_0, T_1, T_2) - (L_1 + L_2 + L_3)M_0 = U(x_0, T_1, T_2) - M_1.$$

Therefore we have shown that for each $x_0, \tilde{x}_0 \in K$ and each pair of integers $T_1 \geq 0$, $T_2 \geq T_1 + L$,

$$U(\tilde{x}_0, T_1, T_2) \geq U(x_0, T_1, T_2) - M_1.$$

Lemma 6.19 is proved.

Corollary 6.20 *There exist a positive number M_1 and an integer $L \geq 1$ such that for every pair of integers $T_1 \geq 0$, $T_2 \geq T_1 + L$ and every $x_0 \in K$, the inequality*

$$|U(x_0, T_1, T_2) - \widehat{U}(T_1, T_2)| \leq M_1$$

holds.

Lemmas 6.16 and 6.17 and Corollary 6.20 imply the following auxiliary result.

Lemma 6.21 *Let $M_1 > 0$. Then there exist natural numbers $\bar{L}_1, \bar{L}_2$ and $M_2 > 0$ such that for each pair of integers $\tau_1 \geq 0$, $\tau_2 \geq \tau_1 + \bar{L}_1 + \bar{L}_2$ and each program $\{x(t)\}_{t=\tau_1}^{\tau_2}$ which satisfies $\sum_{t=\tau_1}^{\tau_2-1} u_t(x(t), x(t+1)) \geq U(x(\tau_1), \tau_1, \tau_2) - M_1$, the following assertion holds:*
If integers $T_1, T_2 \in [\tau_1, \tau_2 - \bar{L}_2]$ satisfy $\bar{L}_1 \leq T_2 - T_1$, then

$$\sum_{t=T_1}^{T_2-1} u_t(x(t), x(t+1)) \geq \widehat{U}(T_1, T_2) - M_2.$$

6.6   Proof of Theorem 6.4

Let $M_1 = 1$, and let integers $\bar{L}_1, \bar{L}_2 \geq 1$ and a positive number M_2 be as guaranteed by Lemma 6.21.

Let $x_0 \in K$ be given. Proposition 6.3 implies that for every integer $k \geq 1$, there exists a program $\{x^{(k)}(t)\}_{t=0}^k$ satisfying

$$x^{(k)}(0) = x_0, \quad \sum_{t=0}^{k-1} u_t(x^{(k)}(t), x^{(k)}(t+1)) = U(x_0, 0, k). \tag{6.114}$$

By the choice of $\bar{L}_1, \bar{L}_2, M_2$, and Lemma 6.21, the following property holds:

(i) For every natural number $k \geq \bar{L}_1 + \bar{L}_2$ and every pair of integers $T_1, T_2 \in [0, k - \bar{L}_2]$ satisfying $\bar{L}_1 \leq T_2 - T_1$, $\sum_{t=T_1}^{T_2-1} u_t(x^{(k)}(t), x^{(k)}(t+1)) \geq \widehat{U}(T_1, T_2) - M_2$.

Evidently, there exists a strictly increasing sequence of natural numbers $\{k_j\}_{j=1}^\infty$ such that for every nonnegative integer t, there exists

$$\bar{x}(t) = \lim_{j \to \infty} x^{(k_j)}(t). \tag{6.115}$$

It is clear that $\{\bar{x}(t)\}_{t=0}^\infty$ is a program. By (6.114) and (6.115), we have

$$\bar{x}(0) = x_0. \tag{6.116}$$

In view of (6.115), property (i), and upper semicontinuity of the functions u_t, $t = 0, 1, \ldots$, the following property holds:

(ii) for each pair of nonnegative integers T_1, T_2 which satisfy $T_2 - T_1 \geq \bar{L}_1$,

$$\left| \sum_{t=T_1}^{T_2-1} u_t(\bar{x}(t), \bar{x}(t+1)) - \widehat{U}(T_1, T_2) \right| \leq M_2. \tag{6.117}$$

Fix number $M_0 > 0$ such that

$$M_0 > u_t(z, z') \text{ for every integer } t \geq 0 \text{ and every } (z, z') \in \text{graph}(a_t). \tag{6.118}$$

Put

$$M = M_2 + M_0 \bar{L}_1. \tag{6.119}$$

Assume that nonnegative integers T_1, T_2 satisfy $T_1 < T_2$. If $T_2 - T_1 \geq \bar{L}_1$, then property (ii), (6.117), and (6.118) imply

$$\left| \sum_{t=T_1}^{T_2-1} u_t(\bar{x}(t), \bar{x}(t+1)) - \widehat{U}(T_1, T_2) \right| \leq M_2 \leq M.$$

If $T_2 - T_1 \leq \bar{L}_1$, then in view of (6.118) and (6.119), we have

$$\left| \sum_{t=T_1}^{T_2-1} u_t(\bar{x}(t), \bar{x}(t+1)) - \widehat{U}(T_1, T_2) \right| \leq (T_2 - T_1)M_0 \leq M_0\bar{L}_1 < M.$$

Thus in both cases

$$\left| \sum_{t=T_1}^{T_2-1} u_t(\bar{x}(t), \bar{x}(t+1)) - \widehat{U}(T_1, T_2) \right| \leq M. \tag{6.120}$$

Assume now that the following properties hold:
(iii) for every nonnegative integer t and every $(z, z') \in \mathrm{graph}(a_t)$ satisfying $u_t(z, z') > 0$, the function u_t is continuous at (z, z');
(iv) if an integer $t \geq 0$ and $z, z_1, z_2, z_3 \in K$ satisfy $z_i \in a_t(z)$, $i = 1, 3$ and $z_1 \leq z_2 \leq z_3$, then $z_2 \in a_t(z)$.

In order to complete the proof of the theorem, it is sufficient to show that for every natural number T, we have

$$\sum_{t=0}^{T-1} u_t(\bar{x}(t), \bar{x}(t+1)) = U(x(0), x(T), 0, T). \tag{6.121}$$

Denote by E the set of all natural numbers τ such that

$$u_{\tau-1}(\bar{x}(\tau-1), \bar{x}(\tau)) > 0. \tag{6.122}$$

Assumption (A2) and (6.120) imply that the set E is infinite. By Proposition 6.18, it is sufficient to show that (6.121) is valid for all $T = \tau - 1$, where $\tau \in E$.

Let $\tau \in E$ and $T = \tau - 1$. We claim that (6.121) holds. Assume the contrary. Then there exist a program $\{x(t)\}_{t=0}^{T}$ and a number $\Delta > 0$ such that

$$x(0) = \bar{x}(0), \quad x(T) \geq \bar{x}(T), \tag{6.123}$$

$$\sum_{t=0}^{T-1} u_t(x(t), x(t+1)) \geq \sum_{t=0}^{T-1} u_t(\bar{x}(t), \bar{x}(t+1)) + 2\Delta. \tag{6.124}$$

In view of the inclusion $\tau \in E$ and the definition of E,

$$u_T(\bar{x}(T), \bar{x}(T+1)) = u_{\tau-1}(\bar{x}(\tau-1), \bar{x}(\tau)) > 0. \tag{6.125}$$

It follows from (6.125) and (A1) that there exists a number $\lambda_0 \in (0, 1)$ and

$$z_0 \in a_{\tau-1}(\bar{x}(\tau-1)) = a_T(\bar{x}(T)) \tag{6.126}$$

such that

$$z_0 \geq \bar{x}(\tau) + \lambda_0 e = \bar{x}(T+1) + \lambda_0 e. \tag{6.127}$$

There exists a number $c_0 > 1$ such that

$$\|y\| \leq c_0 \|y\|_2 \leq c_0^2 \|y\| \text{ for all } y \in R^n. \tag{6.128}$$

In view of (6.125), (6.127), and properties (iii) and (iv), we may assume without loss of generality that

$$|u_{\tau-1}(\bar{x}(\tau-1), z_0) - u_{\tau-1}(\bar{x}(\tau-1), \bar{x}(\tau))| \leq \Delta/4. \tag{6.129}$$

Assumption (A3), (6.123), and (6.126) imply that there exists $z_1 \in a_T(x(T))$ such that

$$z_1 \geq z_0, \quad u_T(x(T), z_1) \geq u_T(\bar{x}(T), z_0). \tag{6.130}$$

Fix a positive number

$$\delta < \min\{1, \lambda_0, \Delta\tau^{-1}\}. \tag{6.131}$$

By the construction of the program $\{\bar{x}(t)\}_{t=0}^{\infty}$ (see (6.115)) and upper semicontinuity of u_t, $t = 0, 1, \ldots$, there is a natural number $k > \tau + 4$ such that

$$\|x^{(k)}(t) - \bar{x}(t)\|_2 \leq \delta, \ t = 0, \ldots, \tau+2, \tag{6.132}$$

$$u_t(x^{(k)}(t), x^{(k)}(t+1)) \leq u_t(\bar{x}(t), \bar{x}(t+1)) + \delta, \ t = 0, \ldots, \tau+2. \tag{6.133}$$

Define

$$\tilde{x}(t) = x(t), \ t = 0, \ldots, \tau-1. \tag{6.134}$$

We claim that $z_1 \geq x^{(k)}(\tau)$. In view of (6.132),

$$\|x^{(k)}(\tau) - \bar{x}(\tau)\|_2 \leq \delta. \tag{6.135}$$

By (6.127), (6.130), (6.131), and (6.135), we have

$$x^{(k)}(\tau) \leq \bar{x}(\tau) + \delta e \leq \bar{x}(\tau) + \lambda_0 e \leq z_0 \leq z_1. \tag{6.136}$$

Put

$$\tilde{x}(\tau) = z_1. \tag{6.137}$$

Since $z_1 \in a_T(x_T) = a_{\tau-1}(\tilde{x}_{\tau-1})$, $\{\tilde{x}(t)\}_{t=0}^{\tau}$ is a program. In view of (6.116), (6.123), (6.134), (6.136), and (6.137),

$$\tilde{x}(0) = \bar{x}(0) = x_0, \ \tilde{x}(\tau) \geq x^{(k)}(\tau). \tag{6.138}$$

By (6.124), (6.129), (6.130), (6.134), (6.137), and the equality $T = \tau - 1$,

$$\sum_{t=0}^{\tau-1} u_t(\tilde{x}(t), \tilde{x}(t+1)) - \sum_{t=0}^{\tau-1} u_t(\bar{x}(t), \bar{x}(t+1))$$

$$\geq \sum_{t=0}^{\tau-2} u_t(x(t), x(t+1)) + u_{\tau-1}(\tilde{x}(\tau-1), \tilde{x}(\tau)) - \sum_{t=0}^{\tau-1} u_t(\bar{x}(t), \bar{x}(t+1))$$

$$\geq \sum_{t=0}^{\tau-2} u_t(\bar{x}(t), \bar{x}(t+1)) + 2\Delta + u_{\tau-1}(x(\tau-1), z_1) - \sum_{t=0}^{\tau-1} u_t(\bar{x}(t), \bar{x}(t+1))$$

$$\geq 2\Delta + \sum_{t=0}^{\tau-2} u_t(\bar{x}(t), \bar{x}(t+1)) + u_{\tau-1}(\bar{x}(\tau-1), z_0) - \sum_{t=0}^{\tau-1} u_t(\bar{x}(t), \bar{x}(t+1))$$

$$\geq 2\Delta + u_{\tau-1}(\bar{x}(\tau-1), z_0) - u_{\tau-1}(\bar{x}(\tau-1), \bar{x}(\tau)) \geq (3/2)\Delta. \tag{6.139}$$

It follows from (6.131), (6.133), and (6.139) that

$$\sum_{t=0}^{\tau-1} u_t(\tilde{x}(t), \tilde{x}(t+1)) - \sum_{t=0}^{\tau-1} u_t(x^{(k)}(t), x^{(k)}(t+1))$$

$$= \sum_{t=0}^{\tau-1} u_t(\tilde{x}(t), \tilde{x}(t+1)) - \sum_{t=0}^{\tau-1} u_t(\bar{x}(t), \bar{x}(t+1)) + \sum_{t=0}^{\tau-1} u_t(\bar{x}(t), \bar{x}(t+1))$$

$$- \sum_{t=0}^{\tau-1} u_t(x^{(k)}(t), x^{(k)}(t+1)) \geq (3/2)\Delta - \delta\tau \geq \Delta/2. \tag{6.140}$$

In view of (6.138) and (6.140), we have

$$U(x_0, x^{(k)}(\tau), 0, \tau) \geq \sum_{t=0}^{\tau-1} u_t(x^{(k)}(t), x^{(k)}(t+1)) + \Delta/2.$$

This inequality contradicts (6.114). The contradiction we have reached proves that (6.121) holds for all $T = \tau - 1$ where $\tau \in E$. Theorem 6.4 is proved.

6.7 Proof of Theorem 6.5

In the sequel we assume that the sum over empty set is zero. There exist a positive number Δ and a strictly increasing sequence of natural numbers $\{\tau_i\}_{i=1}^{\infty}$ such that $\tau_1 \geq 4$ and that

$$u_{\tau_i - 1}(x(\tau_i - 1), x(\tau_i)) \geq \Delta \text{ for all natural numbers } i. \tag{6.141}$$

Let a positive number M be as guaranteed by Theorem 6.4. Lemma 6.15 implies that there exists an integer $L_0 \geq 4$ such that the following property holds:

(P5) For every nonnegative integer S_1, every integer $S_2 \geq S_1 + L_0$, every program $\{v(t)\}_{t=S_1}^{S_2}$ satisfying

$$u_{S_2 - 1}(v(S_2 - 1), v(S_2)) \geq \Delta,$$

and every $\tilde{v}_0 \in K$, there exists a program $\{\tilde{v}(t)\}_{t=S_1}^{S_2}$ such that $\tilde{v}(S_1) = \tilde{v}_0$, $\tilde{v}(S_2) \geq v(S_2)$.

Corollary 6.20 and (6.3) imply that there exists a positive number M_* such that

$$|U(v_0, T_1, T_2) - \widehat{U}(T_1, T_2)| \leq M_*$$

$$\text{for every } v_0 \in K \text{ and every pair of integers } T_1 < T_2, \tag{6.142}$$

$$u_t(z, z') \leq M_* \text{ for every nonnegative integer } t, \text{ and every } (z, z') \in \text{graph}(a_t). \tag{6.143}$$

Fix a number

$$M_1 > L_0 M_* + M_0 + 3M. \tag{6.144}$$

Theorem 6.4 implies that there exists a program $\{\bar{x}(t)\}_{t=0}^{\infty}$ such that

$$\bar{x}(0) = x(0) \tag{6.145}$$

and that for each pair of integers S_1, S_2 satisfying $S_1 < S_2$,

$$\left| \sum_{t=S_1}^{S_2 - 1} u_t(\bar{x}(t), \bar{x}(t + 1)) - \widehat{U}(S_1, S_2) \right| \leq M. \tag{6.146}$$

Assume that T_1, T_2 are integers such that $0 \leq T_1 < T_2$. We claim that

$$\left| \sum_{t=T_1}^{T_2-1} u_t(x(t), x(t+1)) - \widehat{U}(T_1, T_2) \right| \leq M_1. \tag{6.147}$$

If $T_2 \leq T_1 + L_0$, then this inequality follows from (6.143) and (6.144).

Assume that $T_2 > T_1 + L_0$. There exists an integer $i \geq 1$ such that

$$\tau_i > T_2 + 2L_0. \tag{6.148}$$

By (6.141), (6.148), and (P5), there exists a program $\{\tilde{x}(t)\}_{t=\tau_i - L_0}^{\tau_i}$ such that

$$\bar{x}(\tau_i - L_0) = \bar{x}(\tau_i - L_0), \; \tilde{x}(\tau_i) \geq x(\tau_i). \tag{6.149}$$

Define

$$\tilde{x}(t) = \bar{x}(t), \; t = 0, \dots, \tau_i - L_0 - 1. \tag{6.150}$$

Evidently, $\{\tilde{x}(t)\}_{t=0}^{\tau_i}$ is a program, and by (6.149), (6.150), and (6.145),

$$\sum_{t=0}^{\tau_i-1} u_t(x(t), x(t+1)) \geq \sum_{t=0}^{\tau_i-1} u_t(\tilde{x}(t), \tilde{x}(t+1)) - M_0. \tag{6.151}$$

By (6.143) and (6.151),

$$\sum_{t=0}^{\tau_i-1} u_t(x(t), x(t+1)) \geq \sum_{t=0}^{\tau_i-L_0-1} u_t(\tilde{x}(t), \tilde{x}(t+1)) - M_0$$

$$\geq \sum_{t=0}^{\tau_i-1} u_t(\bar{x}(t), \bar{x}(t+1)) - M_0 - L_0 M_*.$$

Together with (6.146) this implies that

$$-(M_0 + L_0 M_*) \leq \sum_{t=0}^{\tau_i-1} u_t(x(t), x(t+1)) - \sum_{t=0}^{\tau_i-1} u_t(\bar{x}(t), \bar{x}(t+1))$$

$$\leq \sum \{u_t(x(t), x(t+1)) : 0 \leq t < T_1\}$$

$$- \sum \{u_t(\bar{x}(t), \bar{x}(t+1)) : 0 \leq t < T_1\}$$

$$+ \sum_{t=T_1}^{T_2-1} u_t(x(t), x(t+1)) - \sum_{t=T_1}^{T_2-1} u_t(\bar{x}(t), \bar{x}(t+1))$$

$$+ \sum_{t=T_2}^{\tau_i-1} u_t(x(t), x(t+1)) - \sum_{t=T_2}^{\tau_i-1} u_t(\bar{x}(t), \bar{x}(t+1))$$

$$\leq M + \sum_{t=T_1}^{T_2-1} u_t(x(t), x(t+1)) - (\widehat{U}(T_1, T_2) - M) + \widehat{U}(T_2, \tau_i)$$

$$- \sum_{t=T_2}^{\tau_i} u_t(\bar{x}(t), \bar{x}(t+1))$$

$$\leq \sum_{t=T_1}^{T_2-1} u_t(x(t), x(t+1)) - \widehat{U}(T_1, T_2) + 3M$$

and combined with (6.144) this implies that

$$\sum_{t=T_1}^{T_2-1} u_t(x(t), x(t+1)) - \widehat{U}(T_1, T_2) \geq -3M - (M_0 + L_0 M_*) > -M_1.$$

This completes the proof of Theorem 6.5.

6.8   Proof of Theorem 6.7

Let $x_0 \in K$, $M_1 > 0$ and let $\{\bar{x}(t)\}_{t=0}^{\infty}$ be as guaranteed by Theorem 6.4. Then for every pair of integers $T_1, T_2 \geq 0$ satisfying $T_1 < T_2$,

$$\left| \sum_{t=T_1}^{T_2-1} u_t(\bar{x}(t), \bar{x}(t+1)) - \widehat{U}(T_1, T_2) \right| \leq M. \tag{6.152}$$

Fix a positive number Δ such that

$$\Delta > u(z, z') \text{ for each } (z, z') \in \text{graph}(a). \tag{6.153}$$

Let p be a natural number. We show that for all sufficiently large natural numbers T,

$$\left| p^{-1}\widehat{U}(0, p) - T^{-1} \sum_{t=0}^{T-1} u(\bar{x}(t), \bar{x}(t+1)) \right| \leq 2M/p. \tag{6.154}$$

Assume that $T \geq p$ is a natural number. Then there exist integers q, s such that

$$q \geq 1, \ 0 \leq s < p, \ T = pq + s. \tag{6.155}$$

By (6.155) we have

$$T^{-1} \sum_{t=0}^{T-1} u(\bar{x}(t), \bar{x}(t+1)) - p^{-1}\widehat{U}(0, p) = T^{-1} \left(\sum_{t=0}^{pq-1} u(\bar{x}(t), \bar{x}(t+1)) \right.$$

$$+ \sum \{ u(\bar{x}(t), \bar{x}(t+1)) : t \text{ is an integer such that } pq \leq t \leq T-1 \} \Bigg)$$

$$- p^{-1}\widehat{U}(0, p)$$

$$= T^{-1} \sum \{ u(\bar{x}(t), \bar{x}(t+1)) : t \text{ is an integer such that } pq \leq t \leq T-1 \}$$

$$+ (T^{-1}pq)(pq)^{-1} \sum_{i=0}^{q-1} \sum_{t=ip}^{(i+1)p-1} u(\bar{x}(t), \bar{x}(t+1)) - p^{-1}\widehat{U}(0, p)$$

$$= (T^{-1}pq)(pq)^{-1} \left[\sum_{l=0}^{q-1} \left(\sum_{t=ip}^{(i+1)p-1} u(\bar{x}(t), \bar{x}(t+1)) - \widehat{U}(0, p) \right) \right.$$

$$+ q\widehat{U}(0, p) \Bigg] - p^{-1}\widehat{U}(0, p)$$

$$+ T^{-1} \left\{ \sum u(\bar{x}(t), \bar{x}(t+1)) : t \text{ is an integer such that } pq \leq t \leq T-1 \right\}. \tag{6.156}$$

In view of (6.152), (6.153), (6.155), and (6.156),

$$\left| T^{-1} \sum_{t=0}^{T-1} u(\bar{x}(t), \bar{x}(t+1)) - p^{-1}\widehat{U}(0, p) \right|$$

$$\leq T^{-1}p\Delta + (pq)^{-1}qM + \widehat{U}(0, p)|q/T - 1/p|$$

$$\leq T^{-1}p\Delta + M/p + \widehat{U}(0, p)s(pT)^{-1} \to M/p \text{ as } T \to \infty.$$

Since p is an arbitrary natural number, we conclude that

$$T^{-1} \sum_{t=0}^{T-1} u(\bar{x}(t), \bar{x}(t+1)))\}_{T=1}^{\infty}$$

is a Cauchy sequence. Evidently, there exists

$$\lim_{T \to \infty} T^{-1} \sum_{t=0}^{T-1} u(\bar{x}(t), \bar{x}(t+1))$$

and for every integer $p \geq 1$, we have

$$\left| p^{-1} \widehat{U}(0, p) - \lim_{T \to \infty} T^{-1} \sum_{t=0}^{T-1} u(\bar{x}(t), \bar{x}(t+1)) \right| \leq 2M/p. \qquad (6.157)$$

Since (6.157) is true for every integer $p \geq 1$, we conclude that

$$\lim_{T \to \infty} T^{-1} \sum_{t=0}^{T-1} u(\bar{x}(t), \bar{x}(t+1)) = \lim_{p \to \infty} \widehat{U}(0, p)/p. \qquad (6.158)$$

Define

$$\mu = \lim_{p \to \infty} \widehat{U}(0, p)/p. \qquad (6.159)$$

It follows from (6.157)–(6.159) that, for every integer $p \geq 1$, we have

$$|p^{-1} \widehat{U}(0, p) - \mu| \leq 2M/p.$$

This completes the proof of Theorem 6.7.

6.9 Overtaking Optimal Programs

In this section we study the existence of overtaking optimal solutions for a large class of infinite horizon discrete-time optimal control problems. This class contains optimal control problems arising in economic dynamics which describe a model proposed by Robinson, Solow, and Srinivasan with nonconcave utility functions representing the preferences of the planner. The results of this section were obtained in [104].

We continue to use the notation and definitions introduced in Section 6.1. In particular, we assume that for every nonnegative integer t, $a_t : K \to \mathcal{P}(K)$ is such that graph(a_t) is a closed subset of $R^n \times R^n$.

We also suppose that there exists $\kappa \in (0, 1)$ such that for every pair of points $x, y \in K$ and every nonnegative integer t, we have

$$H(a_t(x), a_t(y)) \leq \kappa \|x - y\| \tag{6.160}$$

and that for every nonnegative integer t, the upper semicontinuous function $u_t :$ graph$(a_t) \rightarrow [0, \infty)$ satisfies

$$\lim_{t \to \infty} \sup\{u_t(x, x') : (x, x') \in \text{graph}(a_t)\} = 0. \tag{6.161}$$

In Section 6.1 we introduced assumptions (A1)–(A3). Here we assume that (A3) holds, but we do not assume (A1) and (A2). Namely, we assume:

(A3) for every nonnegative integer t, every $(x, y) \in$ graph(a_t), and every $\tilde{x} \in K$ which satisfies $\tilde{x} \geq x$, there exists $\tilde{y} \in a_t(\tilde{x})$ for which

$$\tilde{y} \geq y, \quad u_t(\tilde{x}, \tilde{y}) \geq u_t(x, y).$$

We also suppose that the following assumptions hold:
(A5) there exist a positive γ and a sequence of positive numbers $\{\Delta_t\}_{t=0}^{\infty}$ such that:

(i) for every nonnegative integer s and every point $z_0 \in K$, there exists a sequence $\{x(t)\}_{t=s}^{\infty} \subset K$ such that $x(s) = z_0$, $x(t + 1) \in a_t(x(t))$ for all integers $t \geq s$ and that $u_t(x(t), x(t + 1)) \geq \Delta_t$ for every natural number $t \geq s + 1$;

(ii) for every nonnegative integer t if $(x, x') \in$ graph(a_t) satisfies $u_t(x, x') \geq \Delta_t$, then there exists $z \in a_t(x)$ such that $z \geq x' + \gamma e$;

For every point $x_0 \in K$ and every natural number T, define

$$U(x_0, T) = \sup \left\{ \sum_{t=0}^{T-1} u_t(x(t), x(t + 1)) : \right.$$

$$\left. \{x(t)\}_{t=0}^{T-1} \text{ is a program and } x(0) = x_0 \right\}. \tag{6.162}$$

Upper semicontinuity of u_t, $t = 0, 1, \ldots$ implies the following result.

Proposition 6.22 *For every $x_0 \in K$ and every integer $T \geq 1$, there exists a program $\{x(t)\}_{t=0}^{T}$ such that $x(0) = x_0$ and*

$$\sum_{t=0}^{T-1} u_t(x(t), x(t + 1)) = U(x_0, T).$$

We prove the following theorem.

Theorem 6.23 *For every* $z \in K$, *there exists a program* $\{x_z(t)\}_{t=0}^{\infty}$ *such that* $x_z(0) = z$ *and the following assertion holds:*

For every positive number δ, *there exists an integer* $L^{(\delta)} \geq 1$ *such that for every natural number* $S \geq L^{(\delta)}$ *and every* $z \in K$,

$$\sum_{t=0}^{S-1} u_t(x_z(t), x_z(t+1)) \geq U(z, S) - \delta.$$

Theorem 6.23 easily implies the following corollary.

Corollary 6.24 *Let* $z \in K$, *and let a program* $\{x_z(t)\}_{t=0}^{\infty}$ *be as guaranteed by Theorem 6.23. Then for every program* $\{x(t)\}_{t=0}^{\infty}$ *satisfying* $x(0) = z$,

$$\limsup_{T \to \infty} \left[\sum_{t=0}^{T-1} u_t(x(t), x(t+1)) - \sum_{t=0}^{T-1} u_t(x_z(t), x_z(t+1)) \right] \leq 0.$$

6.10   A Subclass of Infinite Horizon Problems

In this section we consider a subclass of the class of infinite horizon optimal control problems considered in Section 6.9. Infinite horizon problems of this subclass correspond to the nonstationary Robinson–Solow–Srinivasan models.

For every nonnegative integer t, let

$$\alpha^{(t)} = (\alpha_1^{(t)}, \ldots, \alpha_n^{(t)}) >> 0, \ b^{(t)} = (b_1^{(t)}, \ldots, b_n^{(t)}) >> 0, \tag{6.163}$$

$$d^{(t)} = (d_1^{(t)}, \ldots, d_n^{(t)}) \in ((0, 1])^n$$

and for every nonnegative integer t, let $w_t : [0, \infty) \to [0, \infty)$ be a strictly increasing continuous function such that

$$w_t(0) = 0, \ \lim_{t \to \infty} w_t(z) = 0 \text{ for all } z > 0. \tag{6.164}$$

Let $t \geq 0$ be an integer. For every $x \in R_+^n$, define

$$a_t(x) = \left\{ y \in R_+^n : \ y_i \geq (1 - d_i^{(t)})x_i, \ i = 1, \ldots, n, \right.$$

$$\left. \sum_{i=1}^{n} \alpha_i^{(t)}(y_i - (1 - d_i^{(t)})x_i) \leq 1 \right\}. \tag{6.165}$$

It is not difficult to see that for every $x \in R^n$, $a_t(x)$ is a nonempty closed bounded subset of R_+^n and graph(a_t) is a closed subset of $R_+^n \times R_+^n$. Suppose that

$$\inf\{d_i^{(t)} : i = 1, \ldots, n, \ t = 0, 1, \ldots\} > 0, \tag{6.166}$$

$$\inf\{eb^{(t)} : t = 0, 1, \ldots\} > 0, \tag{6.167}$$

$$\inf\{\alpha_i^{(t)} : i = 1, \ldots, n, \ t = 0, 1, \ldots\} > 0, \tag{6.168}$$

$$\sup\{b_i^{(t)} : i = 1, \ldots, n, \ t = 0, 1, \ldots\} < \infty, \tag{6.169}$$

$$\sup\{\alpha_i^{(t)} : i = 1, \ldots, n, \ t = 0, 1, \ldots\} < \infty. \tag{6.170}$$

The constraint mappings a_t, $t = 0, 1, \ldots$ have already been defined. Let us now define the cost functions u_t, $t = 0, 1, \ldots$.

For every nonnegative integer t and every $(x, x') \in$ graph(a_t), define

$$u_t(x, x') = \sup \left\{ w_t(b^{(t)}y) : \ 0 \leq y \leq x, \right.$$

$$\left. ey + \sum_{i=1}^n \alpha_i^{(t)}(x_i' - (1 - d_i^{(t)})x_i) \leq 1 \right\}. \tag{6.171}$$

Fix α^*, $\alpha_* > 0$, $d_* > 0$ such that

$$\alpha_* < \alpha_i^{(t)} < \alpha^*, \ d_* < d_i^{(t)}, \ i = 1, \ldots, n, \ t = 0, 1, \ldots. \tag{6.172}$$

Lemma 6.25 *Let a number $M_0 > (\alpha_* d_*)^{-1}$, let an integer $t \geq 0$, and let $(x, x') \in$ graph(a_t) satisfy $x \leq M_0 e$. Then $x' \leq M_0 e$.*

For the proof see Lemma 6.9.

Lemma 6.26 *Let $t \geq 0$ be an integer. Then the function $u_t : $ graph$(a_t) \rightarrow [0, \infty)$ is upper semicontinuous.*

For the proof see Lemma 6.10.

Proposition 6.27 *Let $t \geq 0$ be an integer, let $(x, y) \in$ graph(a_t), and let $\tilde{x} \in R_+^n$ satisfy $\tilde{x} \geq x$. Then there is $\tilde{y} \in a_t(\tilde{x})$ such that $\tilde{y} \geq y$ and $u_t(\tilde{x}, \tilde{y}) \geq u_t(x, y)$.*

For the proof see Proposition 6.13.

Set

$$\beta = \inf\{b^{(t)}e : t = 0, 1, \ldots\}, \ b^* = \sup\{b_i^{(t)} : i = 1, \ldots, n, \ t = 0, 1, \ldots\}, \ . \tag{6.173}$$

In view of (6.167) and (6.169), $\beta > 0$ and $b^* < \infty$.

Proposition 6.28 *There exist $\gamma > 0$ and a sequence of positive numbers $\{\Delta_t\}_{t=0}^{\infty}$ such that:*

(i) *for every nonnegative integer s and every $x_0 \in R_+^n$, there exists a sequence $\{x(t)\}_{t=s}^{\infty} \subset R_+^n$ such that $x(s) = x_0$, $x(t+1) \in a_t(x(t))$ for every integer $t \geq s$ and $u_t(x(t), x(t+1)) \geq \Delta_t$ for every integer $t \geq s + 1$;*

(ii) *for every nonnegative integer t if $(x, x') \in graph(a_t)$ satisfies $u_t(x, x') \geq \Delta_t$, then $x' + \gamma e \in a_t(x)$.*

Proof Choose numbers $\lambda_0, \lambda_1 \in (0, 1)$ for which

$$\lambda_0 n(1 + \alpha^*) < 1 \text{ and } \lambda_1 < \lambda_0 \tag{6.174}$$

and set

$$\gamma = (b^*)^{-1}\beta\lambda_1 n^{-1}\alpha_*^{-1}, \ \Delta_t = w_t(\beta\lambda_1), \ t = 0, 1, \ldots. \tag{6.175}$$

Let an integer $s \geq 0$ and $x_0 \in R_+^n$. Define

$$x(s) = x_0, \ y(s) = 0,$$

$$x_i(s + 1) = (1 - d_i^{(s)})x_i(s) + \lambda_0 \text{ for } i = 1, \ldots, n, \tag{6.176}$$

and for every integer $t \geq s + 1$, define

$$x_i(t + 1) = (1 - d_i^{(t)})x_i(t) + \lambda_0, \ i = 1, \ldots, n, \ y(t) = \lambda_1 e. \tag{6.177}$$

It follows from (6.165), (6.172), (6.174), (6.176), and (6.177) that

$$x(t + 1) \in a_t(x(t)) \text{ for all integers } t \geq s.$$

By (6.172), (6.174), and (6.177), for all integers $t > s$,

$$ey(t) + \sum_{i=1}^{n} \alpha_i^{(t)}[x_i(t + 1) - (1 - d_i^{(t)})x_i(t)]$$

$$= \sum_{i=1}^{n} \alpha_i^{(t)}\lambda_0 + \lambda_1 n \leq \alpha^*\lambda_0 n + \lambda_1 n \leq \lambda_0 n(1 + \alpha^*) < 1.$$

Combined with (6.171), (6.173), (6.175), and (6.177), this implies that for all integers $t \geq s + 1$,

$$u_t(x(t), x(t + 1)) \geq w_t(b^{(t)}\lambda_1 e) \geq w_t(\lambda_1\beta) = \Delta_t.$$

Therefore (i) is true.

Let us now prove that the property (ii) is true. Assume that an integer $t \geq 0$ and that

$$(x, x') \in \text{graph}(a_t), \ u_t(x, x') \geq \Delta_t. \tag{6.178}$$

In follows from (6.165), (6.171), and (6.178) that

$$x'_i \geq (1 - d_i^{(t)})x_i, \ i = 1, \ldots, n \tag{6.179}$$

and there exists $y \in R_+^n$ satisfying

$$y \leq x, \ ey + \sum_{i=1}^{n} \alpha_i^{(t)}(x'_i - (1 - d_i^{(t)})x_i) \leq 1,$$

$$w_t(b^{(t)}y) = u_t(x, x'). \tag{6.180}$$

In view of (6.175), (6.178), and (6.180), we have

$$w_t(b^{(t)}y) \geq \Delta_t \geq w_t(\beta\lambda_1),$$

and since the function w_t is strictly increasing, relation (6.173) implies that

$$\beta\lambda_1 \leq b^{(t)}y \leq b^* ey$$

and

$$ey \geq (b^*)^{-1}\beta\lambda_1. \tag{6.181}$$

We claim that $x' + \gamma e \in a_t(x)$. In view of (6.179), for $i = 1, \ldots, n$,

$$x'_i + \gamma \geq x'_i \geq (1 - d_i^{(t)})x_i. \tag{6.182}$$

It follows from (6.172), (6.175), (6.180), and (6.181) that for integers $i = 1, \ldots, n$, we have

$$\sum_{i=1}^{n} \alpha_i^{(t)}(x'_i + \gamma - (1 - d_i^{(t)})x_i) = \sum_{i=1}^{n} \alpha_i^{(t)}(x'_i - (1 - d_i^{(t)})x_i) + \gamma \sum_{i=1}^{n} \alpha_i^{(t)}$$

$$\leq \sum_{i=1}^{n} \alpha_i^{(t)}(x'_i - (1 - d_i^{(t)})x_i) + \gamma n\alpha^*$$

$$= \sum_{i=1}^{n} \alpha_i^{(t)}(x_i' - (1 - d_i^{(t)})x_i) + (b^*)^{-1}\beta\lambda_1$$

$$\leq \sum_{i=1}^{n} \alpha_i^{(t)}(x_i' - (1 - d_i^{(t)})x_i) + ey \leq 1.$$

Combined with (6.182) and (6.165), the relation above implies that $x' + \gamma e \in a_t(x)$. Therefore (ii) holds and Proposition 6.28 holds too.

For every $x = (x_1, \ldots, x_n) \in R^n$, put

$$\|x\|_1 = \sum_{i=1}^{n} |x_i|, \quad \|x\|_\infty = \max\{|x_i| : i = 1, \ldots, n\}.$$

It is easy to see that for every nonnegative integer t, for every pair of vectors $x, y \in R_+^n$, and for $\|\cdot\| = \|\cdot\|_p$, where $p = 1, 2, \infty$,

$$H(a_t(x), a_t(y)) \leq \|((1 - d_i^{(t)})x_i)_{i=1}^n - ((1 - d_i^{(t)})y_i)_{i=1}^n\| \leq (1 - d_*)\|x - y\|.$$

$$(6.183)$$

Thus we have defined the mappings a_t and the cost functions u_t, $t = 0, 1, \ldots$. The control system considered in this section is a special case of the control system studied in Section 6.9. As we have already mentioned before, this control system corresponds to the nonstationary Robinson–Solow–Srinivasan model. Note that this control system satisfies the assumptions posed in Section 6.9 and therefore all the results stated there hold for this system. Indeed, fix

$$M_0 > (\alpha_* d_*)^{-1},$$

and set

$$K = \{z \in R_+^n : z \leq M_0 e\}.$$

In view of Lemma 6.25, $a_t(K) \subset K$, $t = 0, 1, \ldots$. Relation (6.160) follows from (6.183). Evidently, (6.161) is true by (6.164). Lemma 6.26 implies that u_t is upper semicontinuous for every nonnegative integer t. By Proposition 6.38, assumption (A5) holds. (A3) follows from Proposition 6.27.

6.11 Auxiliary Results for Theorems 6.23

Lemma 6.29 *Let δ be a positive number. Then there exists an integer $T_0 \geq 4$ such that for every integer $\tau_1 \geq 0$, every integer $\tau_2 \geq T_0 + \tau_1$, every program $\{x(t)\}_{t=\tau_1}^{\tau_2}$ for which there exists $z \in R^n$ satisfying*

$$z \in a_{\tau_2-1}(x(\tau_2 - 1)) \text{ and } z \geq x(\tau_2) + \delta e, \tag{6.184}$$

and every $\tilde{x}_0 \in K$, there exists a program $\{\tilde{x}(t)\}_{t=\tau_1}^{\tau_2}$ such that

$$\tilde{x}(\tau_1) = \tilde{x}_0, \ \tilde{x}(\tau_2) \geq x(\tau_2).$$

Proof Fix a positive number D_0 for which

$$\|z\| \leq D_0 \text{ for all } z \in K. \tag{6.185}$$

There exists a positive number c_0 such that

$$\|z\|_2 \leq c_0\|z\| \text{ for all } z \in K. \tag{6.186}$$

Fix an integer $T_0 \geq 4$ such that

$$8D_0 c_0 \kappa^{T_0} < \delta. \tag{6.187}$$

Assume that integers $\tau_1 \geq 0$, $\tau_2 \geq T_0 + \tau_1$, a that program $\{x(t)\}_{t=\tau_1}^{\tau_2}$ and $z \in R_+^n$ satisfy (6.184), and that $\tilde{x}_0 \in K$. In view of (6.160), there exists a program $\{\tilde{x}(t)\}_{t=\tau_1}^{\tau_2-1}$ such that

$$\tilde{x}(\tau_1) = \tilde{x}_0,$$

$$\|\tilde{x}(t+1) - x(t+1)\| \leq \kappa\|\tilde{x}(t) - x(t)\|, \ t = \tau_1, \ldots, \tau_2 - 2. \tag{6.188}$$

By (6.160) and (6.184), there exists

$$\tilde{x}(\tau_2) \in a_{\tau_2-1}(\tilde{x}(\tau_2 - 1)) \tag{6.189}$$

such that

$$\|\tilde{x}(\tau_2) - z\| \leq \kappa\|x(\tau_2 - 1) - \tilde{x}(\tau_2 - 1)\|. \tag{6.190}$$

Evidently, $\{\tilde{x}(t)\}_{t=\tau_1}^{\tau_2}$ is a program. It follows from (6.185), (6.189), and (6.190) that

$$\|\tilde{x}(\tau_2) - z\| \leq \kappa^{\tau_2-\tau_1}\|\tilde{x}(\tau_1) - x(\tau_1)\| \leq \kappa^{\tau_2-\tau_1}(2D_0) \leq \kappa^{T_0}(2D_0)$$

and by (6.186)

$$\|\tilde{x}(\tau_2) - z\|_2 \leq 2D_0 c_0 \kappa^{T_0}.$$

This implies that for all integers $i = 1, \ldots, n$,

$$|\tilde{x}_i(\tau_2) - z_i| \leq 2D_0 c_0 \kappa^{T_0}.$$

Combined with (6.184) and (6.187), this implies that

$$\tilde{x}(\tau_2) \geq z - 2D_0 c_0 \kappa^{T_0} e \geq x(\tau_2).$$

This completes the proof of Lemma 6.29.

Lemma 6.30 *Let δ be a positive number. Then there exists an integer $\bar{L} \geq 1$ such that for every integer $L \geq \bar{L}$, there exists an integer $\tau \geq L$ for which the following assertion holds:*

For every integer $T \geq \tau$ and every program $\{x(t)\}_{t=0}^{T}$ satisfying

$$\sum_{t=0}^{T-1} u_t(x(t), x(t+1)) = U(x(0), T), \qquad (6.191)$$

the inequality

$$\sum_{t=0}^{L-1} u_t(x(t), x(t+1)) \geq U(x(0), L) - \delta$$

is valid.

Proof Lemma 6.29 implies that there exists an integer $L_0 \geq 4$ such that the following property holds:

(P6) If integers $\tau_1 \geq 0$ and $\tau_2 \geq L_0 + \tau_1$, if for a program $\{x(t)\}_{t=\tau_1}^{\tau_2}$ there exists $z \in R_+^n$ which satisfies

$$z \in a_{\tau_2-1}(x(\tau_2 - 1)) \text{ and } z \geq x(\tau_2) + \gamma e,$$

and if $\tilde{x}_0 \in K$, then there exists a program $\{\tilde{x}(t)\}_{t=\tau_1}^{\tau_2}$ such that

$$\tilde{x}(\tau_1) = \tilde{x}_0, \ \tilde{x}(\tau_2) \geq x(\tau_2).$$

In view of (6.161), there exists an integer $\bar{L} \geq 1$ such that for every natural number $t \geq \bar{L}$, we have

$$\sup\{u_t(x, x') : (x, x') \in \text{graph}(a_t)\} \leq \delta(32L_0)^{-1}. \qquad (6.192)$$

Assume that an integer $L \geq \bar{L}$, and fix a natural number

$$\tau \geq L + L_0 + 2. \tag{6.193}$$

Assume that an integer $T \geq \tau$ and that a program $\{x(t)\}_{t=0}^{T}$ satisfies (6.191). We claim that

$$\sum_{t=0}^{L-1} u_t(x(t), x(t+1)) \geq U(x(0), L) - \delta.$$

Proposition 6.22 implies that there exists a program $\{\tilde{x}(t)\}_{t=0}^{L}$ satisfying

$$\tilde{x}(0) = x(0), \ \sum_{t=0}^{L-1} u_t(\tilde{x}(t), \tilde{x}(t+1)) = U(x(0), L). \tag{6.194}$$

There are two cases:

$$u_t(x(t), x(t+1)) < \Delta_t, \ t = L + L_0, \dots, T - 1; \tag{6.195}$$

$$\max\{u_t(x(t), x(t+1)) - \Delta_t : \ t = L + L_0, \dots, T - 1\} \geq 0. \tag{6.196}$$

Assume that (6.195) is valid. Assumption (A5) implies that there exists a program $\{x^{(1)}(t)\}_{t=0}^{T}$ which satisfies

$$x^{(1)}(t) = \tilde{x}(t), \ t = 0, \dots, L,$$

$$u_t(x^{(1)}(t), x^{(1)}(t+1)) \geq \Delta_t, \ t = L + 1, \dots, T - 1. \tag{6.197}$$

It follows from (6.191), (6.192), (6.194), (6.195), and (6.197) that

$$0 \geq \sum_{t=0}^{T-1} u_t(x^{(1)}(t), x^{(1)}(t+1)) - \sum_{t=0}^{T-1} u_t(x(t), x(t+1))$$

$$= \sum_{t=0}^{L-1} u_t(\tilde{x}(t), \tilde{x}(t+1)) + \sum_{t=L}^{T-1} u_t(x^{(1)}(t), x^{(1)}(t+1))$$

$$- \sum_{t=0}^{T-1} u_t(x(t), x(t+1))$$

$$\geq U(x(0), L) + \sum_{t=L+1}^{T-1} \Delta_t - \sum_{t=0}^{T-1} u_t(x(t), x(t+1))$$

$$\geq U(x(0), L) - \sum_{t=0}^{L-1} u_t(x(t), x(t+1))$$

$$+ \sum_{t=L+1}^{T-1} \Delta_t - \sum_{t=L}^{L+L_0-1} u_t(x(t), x(t+1)) - \sum_{t=L+L_0}^{T-1} \Delta_t$$

$$\geq U(x(0), L) - \sum_{t=0}^{L-1} u_t(x(t), x(t+1)) - \delta$$

and

$$\sum_{t=0}^{L-1} u_t(x(t), x(t+1)) \geq U(x(0), L) - \delta.$$

Assume that (6.196) is true. Then there exists an integer S_0 for which

$$L + L_0 \leq S_0 - 1 \leq T - 1, \tag{6.198}$$

$$u_{S_0-1}(x(S_0 - 1), x(S_0)) \geq \Delta_{S_0-1},$$

$u_t(x(t), x(t+1)) < \Delta_t$ for every integer t satisfying $L_0 + L \leq t < S_0 - 1$.

$$\tag{6.199}$$

Assumption (A5) implies that there exists a program $\{x^{(2)}(t)\}_{t=0}^{S_0-L_0}$ which satisfies

$$x^{(2)}(t) = \tilde{x}(t), \ t = 0, \dots, L, \tag{6.200}$$

$u_t(x^{(2)}(t), x^{(2)}(t+1)) \geq \Delta_t$ for all integers t satisfying $L + 1 \leq t \leq S_0 - L_0 - 1$.
$$\tag{6.201}$$

It follows from (6.198) and (A5) that there exists $y \in R^n$ such that

$$y \in a_{S_0-1}(x(S_0 - 1)) \text{ and } y \geq x(S_0) + \gamma e. \tag{6.202}$$

Property (P6) and (6.202) imply that there exists a program $\{x^{(2)}(t)\}_{t=S_0-L_0}^{S_0}$ such that

$$x^{(2)}(S_0) \geq x(S_0). \tag{6.203}$$

Evidently, $\{x^{(2)}(t)\}_{t=0}^{S_0}$ is a program. In view of (6.203) and (A3), there exists a program $\{x^{(2)}(t)\}_{t=0}^{T}$ such that

$$u_t(x^{(2)}(t), x^{(2)}(t+1)) \geq u_t(x(t), x(t+1)), \ t = S_0, \dots, T - 1. \tag{6.204}$$

It follows from (6.204), (6.191), (6.192), (6.194), and (6.199)–(6.201) that

$$
0 \geq \sum_{t=0}^{T-1} u_t(x^{(2)}(t), x^{(2)}(t+1)) - U(x(0), T)
$$

$$
= \sum_{t=0}^{T-1} u_t(x^{(2)}(t), x^{(2)}(t+1)) - \sum_{t=0}^{T-1} u_t(x(t), x(t+1))
$$

$$
\geq \sum_{t=0}^{L-1} u_t(\tilde{x}(t), \tilde{x}(t+1)) - \sum_{t=0}^{L-1} u_t(x(t), x(t+1))
$$

$$
+ \sum_{t=L}^{S_0-1} u_t(x^{(2)}(t), x^{(2)}(t+1)) - \sum_{t=L}^{S_0-1} u_t(x(t), x(t+1))
$$

$$
\geq U(x(0), L) - \sum_{t=0}^{L-1} u_t(x(t), x(t+1)) + \sum_{t=L}^{S_0-L_0-1} u_t(x^{(2)}(t), x^{(2)}(t+1))
$$

$$
- \sum_{t=L}^{S_0-L_0-1} u_t(x(t), x(t+1)) - \sum_{t=S_0-L_0}^{S_0-1} u_t(x(t), x(t+1))
$$

$$
\geq U(x(0), L) - \sum_{t=0}^{L-1} u_t(x(t), x(t+1))
$$

$$
- \sum_{t=L}^{L_0+L-1} u_t(x(t), x(t+1)) - \sum_{t=S_0-L_0}^{S_0-1} u_t(x(t), x(t+1))
$$

$$
\geq U(x(0), L) - \sum_{t=0}^{L-1} u_t(x(t), x(t+1)) - \delta,
$$

$$
\sum_{t=0}^{L-1} u_t(x(t), x(t+1)) \geq U(x(0), L) - \delta.
$$

Thus in both cases, the inequality above holds. This completes the proof of Lemma 6.30.

6.12   Proof of Theorem 6.23

Proposition 6.22 implies that for every $z \in K$ and every integer $T \geq 1$, there exists a program $\{x_{z,T}(t)\}_{t=0}^{T}$ such that

$$x_{z,T}(0) = z, \quad \sum_{t=0}^{T-1} u_t(x_{z,T}(t), x_{z,T}(t+1)) = U(z, T). \tag{6.205}$$

Let δ be a positive number. Lemma 6.30 implies that there exists an integer $L_\delta \geq 1$ such that the following property holds:

(P7) For every integer $L \geq L_\delta$, there exists an integer $\tau_L \geq L$ such that for every natural number $T \geq \tau_L$ and every $z \in K$,

$$\sum_{t=0}^{L-1} u_t(x_{z,T}(t), x_{z,T}(t+1)) \geq U(z, L) - \delta/4.$$

Let $z \in K$. There exist a strictly increasing sequence of natural numbers $\{T_k\}_{k=1}^{\infty}$ and a program $\{x_z(t)\}_{t=0}^{\infty}$ such that for each integer $t \geq 0$, we have

$$x_{z,T_k}(t) \to x_z(t) \text{ as } k \to \infty. \tag{6.206}$$

Evidently,

$$x_z(0) = z. \tag{6.207}$$

Let an integer L satisfy $L \geq L_\delta$, and let an integer $\tau_L \geq L$ be as guaranteed by the property (P7). By (6.206) and upper semicontinuity of the functions u_t, $t = 0, 1, \ldots$, there exists an integer $k \geq 1$ such that

$$T_k \geq \tau_L, \quad \sum_{t=0}^{L-1} u_t(x_z(t), x_z(t+1)) \geq \sum_{t=0}^{L-1} u_t(x_{z,T_k}(t), x_{z,T_k}(t+1)) - \delta/4. \tag{6.208}$$

In view of (P7), (6.208), and the choice of τ_L,

$$\sum_{t=0}^{L-1} u_t(x_{z,T_k}(t), x_{z,T_k}(t+1)) \geq U(z, L) - \delta/4.$$

Combined with (6.208) this implies that

$$\sum_{t=0}^{L-1} u_t(x_z(t), x_z(t+1)) \geq U(z, L) - \delta.$$

Theorem 6.23 is proved.

Chapter 7
One-Dimensional Robinson–Solow–Srinivasan Model

In this chapter we prove turnpike results for a class of discrete-time optimal control problems. These control problems arise in economic dynamics and describe the nonstationary Robinson–Solow–Srinivasan model. We study the structure of approximate solutions which is independent of the length of the interval, for all sufficiently large intervals.

7.1 Preliminaries and Main Results

Denote by $\text{Card}(E)$ the cardinality of a set E. Let R^1 (R^1_+) be the set of real (nonnegative) numbers. For each mapping $a : X \to 2^Y \setminus \{\emptyset\}$, where X, Y are nonempty sets, put $\text{graph}(a) = \{(x, y) \in X \times Y : y \in a(x)\}$. For each integer $t \geq 0$, let

$$\alpha_t > 0, \quad d_t \in (0, 1] \tag{7.1}$$

and for each integer $t \geq 0$, let $w_t : [0, \infty) \to [0, \infty)$ be a strictly increasing continuous function such that

$$w_t(0) = 0 \text{ and } \inf\{w_t(z) : \text{ an integer } t \geq 0\} > 0 \text{ for all } z > 0. \tag{7.2}$$

We suppose that the following assumption holds:

(A1) For each $\epsilon > 0$ there exists $\delta > 0$ such that $w_t(\delta) \leq \epsilon$ for each integer $t \geq 0$.

We now give a formal description of the model.

Let $t \geq 0$ be an integer. For each $x \in R_+^1$ set

$$a_t(x) = \left\{ y \in R_+^1 : \ y \geq (1 - d_t)x \text{ and } \alpha_t(y - (1 - d_t)x) \leq 1 \right\}. \tag{7.3}$$

It is clear that for each $x \in R_+^1$,

$$a_t(x) = \left[(1 - d_t)x, \ \alpha_t^{-1} + (1 - d_t)x \right]$$

and that $\mathrm{graph}(a_t)$ is a closed subset of $R_+^1 \times R_+^1$. Suppose that

$$\inf\{d_t : \ t = 0, 1, \ldots\} > 0, \tag{7.4}$$

$$\inf\{\alpha_t : \ t = 0, 1, \ldots\} > 0, \tag{7.5}$$

$$\sup\{\alpha_t : \ t = 0, 1, \ldots\} < \infty, \tag{7.6}$$

$$\sup\{w_t(M) : \ t = 0, 1, \ldots\} < \infty \text{ for each } M > 0. \tag{7.7}$$

The constraint mappings $a_t, t = 0, 1, \ldots$ have already been defined. Now we define the cost functions $u_t, \ t = 0, 1, \ldots$. For each integer $t \geq 0$ and each $(x, x') \in \mathrm{graph}(a_t)$, set

$$u_t\left(x, x'\right) = \sup\{w_t(y) : \ 0 \leq y \leq x \text{ and } y + \alpha_t\left(x' - (1 - d_t)x\right) \leq 1\}. \tag{7.8}$$

Clearly, for each integer $t \geq 0$ and each $(x, x') \in \mathrm{graph}(a_t)$,

$$u_t\left(x, x'\right) = w_t\left(\min\{x, 1 - \alpha_t(x' - (1 - d_t)x)\}\right).$$

Choose $\alpha^*, \alpha_*, d_* > 0$ such that

$$\alpha_* < \alpha_t < \alpha^*, \ d_* < d_t \text{ for all integers } t \geq 0. \tag{7.9}$$

Clearly, for each integer $t \geq 0$, the function $u_t : \mathrm{graph}(a_t) \to [0, \infty)$ is upper semicontinuous.

A sequence $\{x(t)\}_{t=0}^{\infty} \subset R_+^1$ is called a program if $x(t + 1) \in a_t(x(t))$ for all integers $t \geq 0$. Let T_1, T_2 be integers such that $T_1 < T_2$. A sequence $\{x(t)\}_{t=T_1}^{T_2} \subset R_+^1$ is called a program if $x(t+1) \in a_t(x(t))$ for all integers t satisfying $T_1 \leq t < T_2$. In the sequel we assume that the supremum over an empty set is $-\infty$.

For each $x_0 \in R_+^1$ and each pair of integers $T_1 < T_2$, set

$$U(x_0, T_1, T_2) = \sup \left\{ \sum_{t=T_1}^{T_2-1} u_t(x(t), x(t+1)) : \{x(t)\}_{t=T_1}^{T_2} \right.$$

$$\left. \text{is a program and } x(T_1) = x_0 \right\}. \tag{7.10}$$

Let $x_0, \tilde{x}_0 \in R_+^1$ and let $T_1 < T_2$ be integers. Set

$$U(x_0, \tilde{x}_0, T_1, T_2) = \sup \left\{ \sum_{t=T_1}^{T_2-1} u_t(x(t), x(t+1)) : \{x(t)\}_{t=T_1}^{T_2} \right.$$

$$\left. \text{is a program such that } x(T_1) = x_0, \ x(T_2) \geq \tilde{x}_0 \right\}. \tag{7.11}$$

Let T_1, T_2 be integers such that $T_1 < T_2$. Set

$$\widehat{U}_M(T_1, T_2) = \sup \left\{ \sum_{t=T_1}^{T_2-1} u_t(x(t), x(t+1)) \right.$$

$$\left. : \{x(t)\}_{t=T_1}^{T_2} \text{ is a program and } x(T_1) \leq M \right\}. \tag{7.12}$$

Upper semicontinuity of u_t, $t = 0, 1, \ldots$, compactness of sets of admissible programs, and the optimization theorem of Weierstrass imply the following results.

Proposition 7.1 *For each $x_0 \in R_+^1$ and each pair of integers $T_1 < T_2$, there exists a program $\{x(t)\}_{t=T_1}^{T_2}$ such that $x(T_1) = x_0$ and*

$$\sum_{t=T_1}^{T_2-1} u_t(x(t), x(t+1)) = U(x_0, T_1, T_2).$$

Proposition 7.2 *For each natural number T and each $M > 0$, there exists a program $\{x(t)\}_{t=0}^{T}$ such that*

$$\sum_{t=0}^{T-1} u_t(x(t), x(t+1)) = \widehat{U}_M(0, T).$$

and $x(0) \leq M$.

Fix

$$M_* > (\alpha_* d_*)^{-1} + 1. \tag{7.13}$$

It is clear that the model considered here is a particular case of the model discussed in Chapter 6 with $n = 1$ (see Section 6.3). Therefore all the results of Chapter 6 can be applied.

Theorem 6.4 and Lemma 6.9 imply the following result.

Theorem 7.3 *There exists $\bar{M} > 0$ such that for each $x_0 \in [0, M_*]$, there exists a program $\{\bar{x}(t)\}_{t=0}^\infty$ such that $\bar{x}(0) = x_0$, for each pair of integers $T_1, T_2 \geq 0$ satisfying $T_1 < T_2$,*

$$\left| \sum_{t=T_1}^{T_2-1} u_t(\bar{x}(t), \bar{x}(t+1)) - \widehat{U}_{M_*}(T_1, T_2) \right| \leq \bar{M}$$

and that for each integer $T > 0$,

$$\sum_{t=0}^{T-1} u_t(\bar{x}(t), \bar{x}(t+1)) = U(\bar{x}(0), \bar{x}(T), 0, T).$$

Lemma 6.9 and Proposition 6.6 imply the following result.

Theorem 7.4 *Let $x_0 \in [0, M_*]$ and let a program $\{\bar{x}(t)\}_{t=0}^\infty$ be as guaranteed by Theorem 7.3. Assume that $\{x(t)\}_{t=0}^\infty$ is a program. Then either the sequence $\{\sum_{t=0}^{T-1} u_t(x(t), x(t+1)) - \sum_{t=0}^{T-1} u_t(\bar{x}(t), \bar{x}(t+1))\}_{T=1}^\infty$ is bounded or*

$$\sum_{t=0}^{T-1} u_t(x(t), x(t+1)) - \sum_{t=0}^{T-1} u_t(\bar{x}(t), \bar{x}(t+1)) \to -\infty \text{ as } T \to \infty.$$

Other results of this chapter were obtained in [100].

Let $\bar{M} > 0$ be as guaranteed by Theorem 7.3. Fix $x_{*0} \in [0, M_*]$ and let a program $\{x^*(t)\}_{t=0}^\infty$ be as guaranteed by Theorem 7.3. Namely,

$$x^*(0) = x_{*0},$$

$$\sum_{t=0}^{T-1} u_t(x^*(t), x^*(t+1)) = U(x^*(0), x^*(T), 0, T) \tag{7.14}$$

for each integer $T > 0$ and

$$\left| \sum_{t=T_1}^{T_2-1} u_t(x^*(t), x^*(t+1)) - \widehat{U}_{M_*}(T_1, T_2) \right| \leq \bar{M} \tag{7.15}$$

for each pair of integers T_1, T_2 satisfying $0 \leq T_1 < T_2$.

For each integer $t \geq 0$, set

$$y^*(t) = \min \left\{ x^*(t), \ 1 - \alpha_t(x^*(t+1) - (1 - d_t)x^*(t)) \right\}. \tag{7.16}$$

We will show that the program $\{x^*(t)\}_{t=0}^{\infty}$ is the turnpike for the model.

A function $w : [0, \infty) \to R^1$ is called strictly concave if for each $x, y \in [0, \infty)$ satisfying $x \neq y$ and each $\alpha \in (0, 1)$,

$$w(\alpha x + (1 - \alpha)y) > \alpha w(x) + (1 - \alpha)w(y).$$

The following two results are consequences of the optimization theorem of Weierstrass.

Proposition 7.5 *Assume that* $w : [0, \infty) \to [0, \infty)$ *is continuous strictly concave function. Let* $\epsilon, M > 0$. *Then there exists* $\delta_0 > 0$ *such that for each* $x, y \in [0, M]$ *satisfying* $|x - y| \geq \epsilon$, $w(2^{-1}(x + y)) - 2^{-1}w(x) - 2^{-1}w(y) \geq \delta_0$.

Proposition 7.6 *Assume that* $w : [0, \infty) \to [0, \infty)$ *is a strictly increasing continuous function,* $M > 0$, *and* $\epsilon \in (0, M)$. *Then* $\inf\{w(x) - w(y) : x, y \in [0, M]$ *and* $x \geq y + \epsilon\} > 0$.

We suppose that the following assumptions hold:

(A2) For each $\epsilon, M > 0$ there exists $\epsilon_0 > 0$ such that for each $x, y \in (0, M]$ satisfying $|x - y| \geq \epsilon$ and each integer $t \geq 0$,

$$w_t \left(2^{-1}(x + y) \right) - 2^{-1}w_t(x) - 2^{-1}w_t(y) \geq \epsilon_0.$$

(A3) For each $M > 0$ and each $\epsilon \in (0, M]$, there is $\epsilon_1 > 0$ such that for each integer $t \geq 0$ and each $x, y \in [0, M]$ satisfying $x \geq y + \epsilon$,

$$w_t(x) - w_t(y) \geq \epsilon_1.$$

(A4) For each $M > 0$ and each $\epsilon > 0$, there exists $\delta > 0$ such that for each integer $t \geq 0$ and each $x, y \in [0, M]$ satisfying $|x - y| \leq \delta$, the inequality $|w_t(x) - w_t(y)| \leq \epsilon$ holds.

Note that (A2) is an assumption of uniform concavity of the functions w_t, $t = 0, 1, \ldots$, (A3) is an assumption of uniform strict monotonicity of the functions w_t, $t = 0, 1, \ldots$, and (A4) is an assumption of uniform equicontinuity of the functions w_t, $t = 0, 1, \ldots$.

It is easy to see that (A1) follows from (A4) and (7.2).
We assume that

$$d^* := \sup\{d_t : t = 0, 1, \dots\} < 1. \qquad (7.17)$$

The following theorems describe the structure of optimal program of the model.

Theorem 7.7 *Let $M > 0$ and $\epsilon > 0$. Then there exists a natural number Q such that for each pair of integers $T_1 \geq 0$ and $T_2 \geq Q + T_1$ and each program $\{x(t)\}_{t=T_1}^{T_2}$ which satisfies $x(T_1) \leq M_*$, $\sum_{t=T_1}^{T_2-1} u_t(x(t), x(t+1)) \geq U(x(T_1), T_1, T_2) - M$, the following inequality holds:*

$$Card\left(\{t \in \{T_1, \dots, T_2\} : |x(t) - x^*(t)| > \epsilon\}\right) \leq Q.$$

Theorem 7.7 is proved in Section 7.5.

Theorem 7.8 *Let $M, \epsilon > 0$. Then there exist a natural number p and $\delta > 0$ such that for each pair of integers $T_1 \geq 0$, $T_2 \geq 2p + T_1$ and each program $\{x(t)\}_{t=T_1}^{T_2}$ satisfying $x(T_1) \leq M_*$, $\sum_{t=T_1}^{T_2-1} u_t(x(t), x(t+1)) \geq U(x(T_1), x(T_2), T_1, T_2) - \delta$,*

$$U(x(T_1), x(T_2), T_1, T_2) \geq U(x(T_1), T_1, T_2) - M$$

the inequality $|x(t) - x^*(t)| \leq \epsilon$ holds for all integers $t \in [T_1 + p, T_2 - p]$.

Theorem 7.9 *Let $M > 0$ and $\epsilon > 0$. Then there exist a natural number p and $\delta > 0$ such that for each pair of integers $T_1 \geq 0$, $T_2 \geq p + T_1$ and each program $\{x(t)\}_{t=T_1}^{T_2}$ which satisfies $x(T_1) \leq M_*$, $|x(T_1) - x^*(T_1)| \leq \delta$,*

$$\sum_{t=T_1}^{T_2-1} u_t(x(t), x(t+1)) \geq U(x(T_1), x(T_2), T_1, T_2) - \delta,$$

$$U(x(T_1), x(T_2), T_1, T_2) \geq U(x(T_1), T_1, T_2) - M$$

the inequality $|x(t) - x^*(t)| \leq \epsilon$ holds for all integers $t \in [T_1, T_2 - p]$.

Theorem 7.8 is proved in Section 7.3, while Theorem 7.9 is proved in Section 7.4.

Theorem 7.10 *Let $\epsilon > 0$. Then there exist a natural number p and $\delta > 0$ such that for each pair of integers $T_1 \geq 0$, $T_2 \geq 2p + T_1$ and each program $\{x(t)\}_{t=T_1}^{T_2}$ which satisfies $x(T_1) \leq M_*$, $\sum_{t=T_1}^{T_2-1} u_t(x(t), x(t+1)) \geq U(x(T_1), T_1, T_2) - \delta$, the inequality $|x(t) - x^*(t)| \leq \epsilon$ holds for all integers $t \in [T_1 + p, T_2 - p]$.*

Theorem 7.11 *Let $\epsilon > 0$. Then there exist a natural number p and $\delta > 0$ such that for each pair of integers $T_1 \geq 0$, $T_2 \geq p + T_1$ and each program $\{x(t)\}_{t=T_1}^{T_2}$ which satisfies $x(T_1) \leq M_*$, $|x(T_1) - x^*(T_1)| \leq \delta$,*

$$\sum_{t=T_1}^{T_2-1} u_t(x(t), x(t+1)) \geq U(x(T_1), T_1, T_2) - \delta$$

the inequality $|x(t) - x^*(t)| \leq \epsilon$ *holds for all integers* $t \in [T_1, T_2 - p]$.

Theorems 7.10 and 7.11 easily follow from Theorems 7.8 and 7.9, respectively. A program $\{x(t)\}_{t=0}^{\infty}$ is called good if the sequence

$$\left\{ \sum_{t=0}^{T-1} u_t(x(t), x(t+1)) - \sum_{t=0}^{T-1} u_t(x^*(t), x^*(t+1)) \right\}_{T=1}^{\infty}$$

is bounded. In view of Theorem 7.4, if the sequence $\{x(t)\}_{t=0}^{\infty}$ is not good, then

$$\lim_{T \to \infty} \left[\sum_{t=0}^{T-1} u_t(x(t), x(t+1)) - \sum_{t=0}^{T-1} u_t(x^*(t), x^*(t+1)) \right] = -\infty.$$

In Section 7.6 we prove the following result.

Theorem 7.12 *Assume that a program* $\{x(t)\}_{t=0}^{\infty}$ *is good. Then* $x(t) - x^*(t) \to 0$ *as* $t \to \infty$.

A program $\{x(t)\}_{t=0}^{\infty}$ is called overtaking optimal if for each program $\{x'(t)\}_{t=0}^{\infty}$ satisfying $x'(0) = x(0)$,

$$\limsup_{T \to \infty} \left[\sum_{t=0}^{T-1} u_t(x'(t), x'(t+1)) - \sum_{t=0}^{T-1} u_t(x(t), x(t+1)) \right] \leq 0.$$

In Section 7.6 we prove the following result.

Theorem 7.13 *Let* $x_0 \in [0, M_*]$ *and let a program* $\{\bar{x}(t)\}_{t=0}^{\infty}$ *be as guaranteed by Theorem 7.3. Then* $\{\bar{x}(t)\}_{t=0}^{\infty}$ *is a unique overtaking optimal program with the initial state* x_0.

7.2 Auxiliary Results

We begin with the result which follows from Lemma 6.9 and Corollary 6.20.

Proposition 7.14 *There exist* $M_1 > 0$ *and a natural number* L *such that for each pair of integers* $T_1 \geq 0$, $T_2 \geq T_1 + L$ *and each* $x_0 \in [0, M_*]$, *the inequality* $|U(x_0, T_1, T_2) - \widehat{U}_{M_*}(T_1, T_2)| \leq M_1$ *holds.*

The following lemma shows the uniform equicontinuity of the functions u_t, $t = 0, 1, \ldots$.

Lemma 7.15 *Let $\epsilon > 0$ and $M \geq M_*$. Then there exists $\delta > 0$ such that for each integer $t \geq 0$ and each*

$$(x, x'), \ (y, y') \in \text{graph}(a_t) \tag{7.18}$$

satisfying

$$x, y \leq M, \ |x - y|, |x' - y'| \leq \delta \tag{7.19}$$

the inequality $|u_t(x, x') - u_t(y, y')| \leq \epsilon$ holds.

Proof By (A4) there is $\delta > 0$ such that for each integer $t \geq 0$ and each $z, z' \in [0, M]$ satisfying $|z - z'| \leq 2\delta(1 + \alpha^*)$, the following inequality holds:

$$|w_t(z) - w_t(z')| \leq \epsilon/2. \tag{7.20}$$

Let an integer $t \geq 0$ and let (x, x'), (y, y') satisfy (7.18) and (7.19). In order to prove the lemma, it is sufficient to show that $u_t(y, y') \geq u_t(x, x') - \epsilon$. By (7.8) there is $z \geq 0$ such that

$$z \leq x, \ z \leq 1 - \alpha_t \left(x' - (1 - d_t)x\right), \ u_t\left(x, x'\right) = w_t(z). \tag{7.21}$$

By (7.9), (7.19), and (7.21),

$$y \geq x - \delta \geq z - \delta, \tag{7.22}$$

$$1 - \alpha_t \left(y' - (1 - d_t)y\right) \geq 1 - \alpha_t \left(x' + \delta - (1 - d_t)(x - \delta)\right) \geq z - 2\alpha^*\delta. \tag{7.23}$$

Put

$$z' = \max\left\{0, \ z - \delta(1 + 2\alpha^*)\right\}. \tag{7.24}$$

In view of (7.22) and (7.24),

$$0 \leq z' \leq y. \tag{7.25}$$

It follows from (7.3), (7.18), (7.23), and (7.24) that

$$1 - \alpha_t \left(y' - (1 - d_t)y\right) \geq z'. \tag{7.26}$$

Relations (7.8), (7.18), (7.24), (7.25), and (7.26) imply that

$$w_t\left(z'\right) \leq u_t\left(y, y'\right). \tag{7.27}$$

By (7.19), (7.21), (7.24), (7.25), and the choice of δ (see (7.20)),

$$\left| w_t(z') - w_t(z) \right| \leq \epsilon/2.$$

Together with (7.21) and (7.27), this implies that

$$u_t(y, y') \geq w_t(z) - \epsilon/2 \geq u_t(x, x') - \epsilon/2.$$

Lemma 7.15 is proved.

The next result easily follows from (7.8) and the strict monotonicity of w_t, $t = 0, 1, \ldots$.

Lemma 7.16 *Let $t \geq 0$ be an integer, $(x, x') \in graph(a_t)$, and let $y \in [0, x]$ satisfy $y + \alpha_t(x' - (1 - d_t)x) \leq 1$. Then $w_t(y) = u_t(x, x')$ if and only if $y = \min\{x, 1 - \alpha_t(x' - (1 - d_t)x)\}$.*

By Proposition 6.12, there exist $\widehat{\Delta} > 0$ and a program $\{\widehat{x}(t)\}_{t=0}^{\infty}$ such that

$$\widehat{x}(0) < 1$$

and that

$$u_t(\widehat{x}(t), \widehat{x}(t+1)) \geq \widehat{\Delta}$$

for all integers $t \geq 0$.

Let a positive number γ satisfy

$$\gamma < 1/2 \text{ and } \gamma < 4^{-1}\widehat{\Delta}. \tag{7.28}$$

Lemmas 6.9 and 6.16 imply the following result.

Lemma 7.17 *Let $M_1 > 0$. Then there exist integers $L_1, L_2 \geq 4$ such that for each pair of integers $T_1 \geq 0$, $T_2 \geq L_1 + L_2 + T_1$, each program $\{x(t)\}_{t=T_1}^{T_2}$ satisfying*

$$x(T_1) \leq M_*, \quad \sum_{t=T_1}^{T_2-1} u_t(x(t), x(t+1)) \geq U(x(T_1), T_1, T_2) - M_1$$

and each integer $\tau \in [T_1 + L_1, T_2 - L_2]$, the following inequality holds:

$$\max\{u_t(x(t), x(t+1)) : t = \tau, \ldots, \tau + L_2 - 1\} \geq \gamma.$$

Lemmas 6.9 and 6.17 imply the following result.

Lemma 7.18 *Let $M_1 > 0$. Then there exist natural numbers $\bar{L}_1, \bar{L}_2$ and $M_2 > 0$ such that for each pair of integers $\tau_1 \geq 0$, $\tau_2 \geq \bar{L}_1 + \bar{L}_2 + \tau_1$ and each program $\{x(t)\}_{t=\tau_1}^{\tau_2}$ satisfying*

$$x(\tau_1) \le M_*, \quad \sum_{t=\tau_1}^{\tau_2-1} u_t(x(t), x(t+1)) \ge U(x(\tau_1), \tau_1, \tau_2) - M_1$$

the following assertion holds:
   If integers $T_1, T_2 \in [\tau_1, \tau_2 - \bar{L}_2]$ satisfy $\bar{L}_1 \le T_2 - T_1$, then

$$\sum_{t=T_1}^{T_2-1} u_t(x(t), x(t+1)) \ge U(x(T_1), T_1, T_2) - M_2.$$

Recall that the positive constant $\bar{M}$ was fixed in Section 7.1 after the statement of Theorem 7.4.

Lemma 7.19 *Let $\epsilon > 0$. Then there exist a natural number $L_0 \ge 4$ and $\delta > 0$ such that for each pair of integers $\tau_1 \ge 0$, $\tau_2 \ge \tau_1 + L_0$ and each $z_1, z_2, z_1', z_2' \in [0, M_*]$ satisfying*

$$\left| z_i - z_i' \right| \le \delta, \ i = 1, 2, \tag{7.29}$$

$$U(z_1, z_2, \tau_1, \tau_2) \ge \widehat{U}_{M_*}(\tau_1, \tau_2) - \bar{M} - 2 \tag{7.30}$$

the inequality

$$U\left(z_1', z_2', \tau_1, \tau_2\right) \ge U(z_1, z_2, \tau_1, \tau_2) - \epsilon$$

holds.

Proof By Lemma 7.17 there exist natural numbers $L_1, L_2 \ge 4$ such that the following property holds:
   (P1) for each pair of integers $T_1 \ge 0$, $T_2 \ge L_1 + L_2 + T_1$, each program $\{x(t)\}_{t=T_1}^{T_2}$ which satisfies

$$x(0) \le M_*, \quad \sum_{t=T_1}^{T_2-1} u_t(x(t), x(t+1)) \ge U(x(T_1), T_1, T_2) - \bar{M} - 4$$

and each integer $\tau \in [T_1 + L_1, T_2 - L_2]$, the following inequality holds:

$$\max\{u_t(x(t), x(t+1)) : \ t = \tau, \ldots, \tau + L_2 - 1\} \ge \gamma.$$

In view of (A1), there is $\gamma_0 > 0$ such that

$$w_t(\gamma_0) < \gamma/4 \text{ for each integer } t \ge 0. \tag{7.31}$$

(A4) implies that there is $\delta_1 > 0$ such that

$$\delta_1 < \gamma_0/4; \tag{7.32}$$

for each integer $t \geq 0$ and each $y, y' \in [0, M_*]$ satisfying $|y - y'| \leq 2\delta_1$,

$$\left| w_t(y) - w_t(y') \right| \leq 8^{-1}\epsilon. \tag{7.33}$$

Choose natural numbers

$$L_3 > L_1 + L_2 + 4, \quad L_0 > 8 + 4(L_1 + L_2 + 2L_3). \tag{7.34}$$

By Lemma 7.15, there exists $\delta_2 > 0$ such that

$$\delta_2\alpha^* < \gamma_0/4 \tag{7.35}$$

and that for each integer $t \geq 0$ and each (x, x'), $(y, y') \in \text{graph}(a_t)$ satisfying $x, y \leq M_*$, $|x - x'|$, $|y - y'| \leq 2\delta_2$, the following inequality holds:

$$|u_t\left(x, x'\right) - u_t\left(y, y'\right)| \leq \left(8^{-1}\epsilon\right)(L_3 + L_2 + 1)^{-1}. \tag{7.36}$$

Set

$$\delta = \min\left\{\delta_1, \delta_2, (4^{-1}\gamma_0)(1 + \alpha^*)^{-1}, 8^{-1}(1 - d^*)^{L_3 + L_2}\delta_1(1 + \alpha^*)^{-1}\right\}. \tag{7.37}$$

Assume that integers

$$\tau_1 \geq 0, \quad \tau_2 \geq \tau_1 + L_0, \tag{7.38}$$

$$z_1, z_2, z_1', z_2' \in [0, M_*], \quad |z_i - z_i'| \leq \delta, \quad i = 1, 2, \tag{7.39}$$

$$U(z_1, z_2, \tau_1, \tau_2) \geq \widehat{U}_{M_*}(\tau_1, \tau_2) - \bar{M} - 2. \tag{7.40}$$

By the continuity of w_t, $t = 0, 1, \ldots$, there is a program $\{x(t)\}_{t=\tau_1}^{\tau_2}$ such that

$$x(\tau_1) = z_1, \quad x(\tau_2) \geq z_2, \tag{7.41}$$

$$\sum_{t=\tau_1}^{\tau_2-1} u_t(x(t), x(t+1)) = U(z_1, z_2, \tau_1, \tau_2) \geq \widehat{U}_{M_*}(\tau_1, \tau_2) - \bar{M} - 2.$$

It follows from (P1), (7.34), (7.38), (7.39), and (7.41) that

$$\max\{u_t(x(t), x(t+1)): \ t = L_3 + \tau_1, \ldots, L_3 + \tau_1 + L_2 - 1\} \geq \gamma, \tag{7.42}$$

$$\max\{u_t(x(t), x(t+1)): \ t = \tau_2 - L_3 - L_2, \ldots, \tau_2 - L_3 - 1\} \geq \gamma. \tag{7.43}$$

By (7.42) and (7.43), there are integers

$$t_1 \in [L_3 + \tau_1, L_3 + \tau_1 + L_2 - 1], \quad t_2 \in [\tau_2 - L_3 - L_2, \tau_2 - L_3 - 1] \qquad (7.44)$$

such that

$$u_{t_1}(x(t_1), x(t_1 + 1)) \geq \gamma, \quad u_{t_2}(x(t_2), x(t_2 + 1)) \geq \gamma. \qquad (7.45)$$

For each integer $t \in [\tau_1, \tau_2 - 1]$, set

$$y(t) = \min\{x(t), \ 1 - \alpha_t(x(t + 1) - (1 - d_t)x_t)\}. \qquad (7.46)$$

In view of (7.46) and Lemma 7.16 for all integers $t \in [\tau_1, \tau_2 - 1]$,

$$u_t(x(t), x(t + 1)) = w_t(y(t)). \qquad (7.47)$$

It follows from (7.31), (7.45), (7.46), and the monotonicity of $w_t, t = 0, 1, \ldots$ that

$$y(t_1), \ y(t_2) > \gamma_0. \qquad (7.48)$$

Now we construct a program $\{\tilde{x}(t)\}_{t=\tau_1}^{\tau_2}$. Set

$$\tilde{x}(\tau_1) = z_1' \qquad (7.49)$$

and for all integers $t = \tau_1, \ldots, t_1 - 1$, put

$$\tilde{x}(t + 1) = (1 - d_t)\tilde{x}(t) + x(t + 1) - (1 - d_t)x(t). \qquad (7.50)$$

Clearly, $\{\tilde{x}(t)\}_{t=\tau_1}^{t_1}$ is a program. By (7.39), (7.41), (7.49), and (7.50), for all $t = \tau_1, , \ldots, t_1$,

$$|\tilde{x}(t) - x(t)| \leq |\tilde{x}(\tau_1) - x(\tau_1)| \leq \delta. \qquad (7.51)$$

Put

$$\tilde{x}(t_1 + 1) = x(t_1 + 1) + \delta. \qquad (7.52)$$

In view of (7.46) and (7.48),

$$\alpha_{t_1}(x(t_1 + 1) - (1 - d_{t_1})x(t_1)) \leq 1 - y(t_1) \leq 1 - \gamma_0. \qquad (7.53)$$

By (7.37) and (7.51)–(7.53),

$$\alpha_{t_1}(\tilde{x}(t_1 + 1) - (1 - d_{t_1})\tilde{x}(t_1))$$

$$\leq \alpha_{t_1}(x(t_1 + 1) - (1 - d_{t_1})x(t_1) + 2\delta) \leq 1 - \gamma_0 + 2\delta\alpha^* < 1. \qquad (7.54)$$

By (7.51) and (7.52),

$$\tilde{x}(t_1 + 1) - (1 - d_{t_1})\tilde{x}(t_1) \geq x(t_1 + 1) - (1 - d_{t_1})x(t_1) \geq 0.$$

Therefore $\{\tilde{x}(t)\}_{t=\tau_1}^{t_1+1}$ is a program. It follows from (7.37), (7.51), (7.52), the choice of δ_2 (see (7.35) and (7.36)), (7.39)–(7.41), (7.49), and Lemma 6.9 that

$$|u_t(x(t), x(t+1)) - u_t(\tilde{x}(t), \tilde{x}(t+1))| \leq (8^{-1}\epsilon)(L_3 + L_2 + 1)^{-1}, \ t = \tau_1, \ldots, t_1.$$

$$(7.55)$$

In view of (7.44) and (7.55),

$$\left| \sum_{t=\tau_1}^{t_1} u_t(x(t), x(t+1)) - \sum_{t=\tau_1}^{t_1} u_t(\tilde{x}(t), \tilde{x}(t+1)) \right| \qquad (7.56)$$

$$\leq (8^{-1}\epsilon)(L_3 + L_2 + 1)^{-1}(t_1 - \tau_1 + 1) \leq 8^{-1}\epsilon.$$

For all $t = t_1 + 2, \ldots, t_2$ set

$$\tilde{x}(t) = x(t) + (1 - d_{t-1})\tilde{x}(t-1) - (1 - d_{t-1})x(t-1). \qquad (7.57)$$

Relations (7.52) and (7.57) imply that $\tilde{x}(t) \geq x(t)$, $t = t_1 + 1, \ldots, t_2$. Clearly, $\{\tilde{x}(t)\}_{t=\tau_1}^{t_2}$ is a program. Set

$$\tilde{x}(t_2 + 1) = x(t_2 + 1) + (\alpha_{t_2})^{-1}\delta_1 + (1 - d_{t_2})\tilde{x}(t_2) - (1 - d_{t_2})x(t_2). \qquad (7.58)$$

By (7.32), (7.46), (7.48), and (7.58),

$$\alpha_{t_2}(\tilde{x}(t_2 + 1) - (1 - d_{t_2})\tilde{x}(t_2)) = \alpha_{t_2}(x(t_2 + 1) - (1 - d_{t_2})x(t_2) + (\alpha_{t_2})^{-1}\delta_1)$$
$$\leq 1 - y(t_2) + \delta_1 \leq 1 - \gamma_0 + \delta_1 < 1 - \gamma_0/2. \qquad (7.59)$$

By (7.58) and (7.59),

$$\tilde{x}(t_2 + 1) \in a_{t_2}(\tilde{x}(t_2)). \qquad (7.60)$$

In view of (7.32), (7.48), (7.52), and (7.57),

$$\tilde{x}(t_2) > y(t_2) - \delta_1 > \gamma_0/2. \qquad (7.61)$$

Relation (7.59) implies that

$$y(t_2) - \delta_1 \leq 1 - \alpha_{t_2}(\tilde{x}(t_2 + 1) - (1 - d_{t_2})\tilde{x}(t_2)). \tag{7.62}$$

It follows from (7.8), (7.61), (7.62), the choice of δ_1 (see (7.33)), (7.39), (7.41), (7.47), and (7.61) that

$$u_{t_2}(\tilde{x}(t_2), \tilde{x}(t_2 + 1)) \geq w_{t_2}(y(t_2) - \delta_1)$$

$$\geq w_{t_2}(y(t_2)) - \epsilon/8 = u_{t_2}(x(t_2), x(t_2 + 1)) - \epsilon/8. \tag{7.63}$$

For all $t = t_2 + 1 \ldots, \tau_2 - 1$ put

$$\tilde{x}(t + 1) = (1 - d_t)\tilde{x}(t) + x(t + 1) - (1 - d_t)x(t). \tag{7.64}$$

By (7.57), (7.58), and (7.64),

$$\tilde{x}(t) \geq x(t), \ t = t_2 + 1, \ldots, \tau_2. \tag{7.65}$$

Now it is easy to see that $\{\tilde{x}(t)\}_{t=\tau_1}^{\tau_2}$ is a program. In view of (7.8), (7.64), and (7.65),

$$u_t(\tilde{x}(t), \tilde{x}(t + 1)) \geq u_t(x(t), x(t + 1)), \ t = t_2 + 1, \ldots, \tau_2 - 1. \tag{7.66}$$

Relations (7.17) and (7.64) imply that for all $t = t_2 + 1, \ldots, \tau_2 - 1$,

$$\tilde{x}(t + 1) - x(t + 1) = (1 - d_t)(\tilde{x}(t) - x(t)) \geq (1 - d^*)(\tilde{x}(t) - x(t)).$$

Combined with (7.9), (7.17), (7.47), (7.57), and (7.58), this implies that

$$\tilde{x}(\tau_2) - x(\tau_2) \geq (1 - d^*)^{\tau_2 - t_2 - 1}(\tilde{x}(t_2 + 1) - x(t_2 + 1)) \geq (1 - d^*)^{L_3 + L_2}\delta_1(\alpha^*)^{-1}. \tag{7.67}$$

It follows from (7.37), (7.39), (7.41), and (7.67) that

$$\tilde{x}(\tau_2) \geq x(\tau_2) + (1 - d^*)^{L_3 + L_2}\delta_1(\alpha^*)^{-1} \geq z_2 + (1 - d^*)^{L_3 + L_2}\delta_1(\alpha^*)^{-1}$$

$$\geq z_2' - \delta + (1 - d^*)^{L_3 + L_2}\delta_1(\alpha^*)^{-1} > z_2'.$$

By the relation above, (7.41), (7.49), (7.57), the relation $\tilde{x}(t) \geq x(t), \ t = t_1 + 1, \ldots, t_2$, (7.8), (7.56), and (7.63), and (7.66),

$$U(z_1', z_2', \tau_1, \tau_2) \geq \sum_{t=\tau_1}^{\tau_2 - 1} u_t(\tilde{x}(t), \tilde{x}(t + 1)) \geq \sum_{t=\tau_1}^{\tau_2 - 1} u_t(x(t), x(t + 1))$$

$$+ \left[\sum_{t=\tau_1}^{\tau_2 - 1} u_t(\tilde{x}(t), \tilde{x}(t + 1)) - \sum_{t=\tau_1}^{\tau_2 - 1} u_t(x(t), x(t + 1)) \right]$$

$$\geq U(z_1, z_2, \tau_1, \tau_2) + \sum_{t=\tau_1}^{t_1} u_t(\tilde{x}(t), \tilde{x}(t+1)) - \sum_{t=\tau_1}^{t_1} u_t(x(t), x(t+1))$$

$$+u_{t_2}(\tilde{x}(t_2), \tilde{x}(t_2+1)) - u_{t_2}(x(t_2), x(t_2+1)) \geq U(z_1, z_2, \tau_1, \tau_2) - 8^{-1}\epsilon - \epsilon/8.$$

Lemma 7.19 is proved.

Lemma 7.20 *Let $\epsilon \in (0, 1)$. Then there exist a natural number $L_0 \geq 4$ and $\delta > 0$ such that for each pair of integers $\tau_1 \geq 0$, $\tau_2 \geq \tau_1 + L_0$ and each $z_1, z_2 \in [0, M_*]$ satisfying*

$$|z_1 - x^*(\tau_1)|, \ |z_2 - x^*(\tau_2)| \leq \delta \tag{7.68}$$

the inequality $|U(x^*(\tau_1), x^*(\tau_2), \tau_1, \tau_2) - U(z_1, z_2, \tau_1, \tau_2)| \leq \epsilon$ holds.

Proof By Lemma 7.19 there exist a natural number $L_0 \geq 4$ and $\delta > 0$ such that the following property holds:

(P2) for each pair of integers $\tau_1 \geq 0$, $\tau_2 \geq \tau_1 + L_0$ and each $z_1, z_2, z_1', z_2' \in [0, M_*]$ satisfying $|z_i - z_i'| \leq \delta$, $i = 1, 2$ and

$$U(z_1, z_2, \tau_1, \tau_2) \geq \widehat{U}_{M_*}(\tau_1, \tau_2) - \bar{M} - 2$$

the inequality

$$U(z_1', z_2', \tau_1, \tau_2) \geq U(z_1, z_2, \tau_1, \tau_2) - \epsilon/4$$

holds.

Assume that integers $\tau_1 \geq 0$, $\tau_2 \geq \tau_1 + L_0$ and that $z_1, z_2 \in [0, M_*]$ satisfy (7.68). By (7.68), (P2), (7.15), (7.13)

$$U(z_1, z_2, \tau_1, \tau_2) \geq U(x^*(\tau_1), x^*(\tau_2), \tau_1, \tau_2) - \epsilon/4. \tag{7.69}$$

Together with (7.14) and (7.15), this implies that

$$U(z_1, z_2, \tau_1, \tau_2) \geq \widehat{U}_{M_*}(\tau_1, \tau_2) - \bar{M} - 2. \tag{7.70}$$

Property (P2), (7.68), and (7.70) imply that

$$U(x^*(\tau_1), x^*(\tau_2), \tau_1, \tau_2) \geq U(z_1, z_2, \tau_1, \tau_2) - \epsilon/4.$$

Combined with (7.69) this implies that

$$|U(z_1, z_2, \tau_1, \tau_2) - U(x^*(\tau_1), x^*(\tau_2), \tau_1, \tau_2)| \leq \epsilon/2.$$

Lemma 7.20 is proved.

Lemma 7.21 *Let* $\epsilon > 0$. *Then there exist a natural number* $L_0 \geq 4$ *and* $\delta > 0$ *such that for each pair of integers* $\tau_1 \geq 0$, $\tau_2 \geq \tau_1 + L_0$, *each* $z_1, z_2 \in [0, M_*]$, *and each program* $\{x(t)\}_{t=\tau_1}^{\tau_2}$ *satisfying*

$$|z_1 - x^*(\tau_1)|, \ |z_2 - x^*(\tau_2)| \leq \delta, \tag{7.71}$$

$$x(\tau_1) = z_1, \ x(\tau_2) \geq z_2, \tag{7.72}$$

$$\sum_{t=\tau_1}^{\tau_2-1} u_t(x(t), x(t+1)) \geq U(z_1, z_2, \tau_1, \tau_2) - \delta \tag{7.73}$$

the following inequality holds for all $t = \tau_1, \ldots, \tau_2 - 1$:

$$|y^*(t) - \min\{x(t), 1 - \alpha_t(x(t+1) - (1 - d_t)x(t))\}| \leq \epsilon. \tag{7.74}$$

Proof By (A2) there is a positive number ϵ_1 such that for each integer $t \geq 0$ and each $x, y \in [0, M_*]$ satisfying $|x - y| \geq \epsilon/2$,

$$w_t\left(2^{-1}(x+y)\right) - 2^{-1}w_t(x) - 2^{-1}w_t(y) \geq 2\epsilon_1. \tag{7.75}$$

Lemma 7.20 implies that there is $\delta > 0$ and a natural number $L_0 \geq 4$ such that

$$\delta < \epsilon_1/2;$$

for each pair of integers $\tau_1 \geq 0$, $\tau_2 \geq \tau_1 + L_0$ and each $z_1, z_2 \in [0, M_*]$ satisfying $|z_1 - x^*(\tau_1)|$, $|z_2 - x^*(\tau_2)| \leq \delta$, the following inequality holds:

$$\left|U(x^*(\tau_1), x^*(\tau_2), \tau_1, \tau_2) - U(z_1, z_2, \tau_1, \tau_2)\right| \leq \epsilon_1/4. \tag{7.76}$$

Assume that pair of integers $\tau_1 \geq 0$, $\tau_2 \geq \tau_1 + L_0$, $z_1, z_2 \in [0, M_*]$ and that a program $\{x(t)\}_{t=\tau_1}^{\tau_2}$ satisfies (7.71)–(7.73). We show that for all $t = \tau_1, \ldots, \tau_2 - 1$ (7.74) holds. Assume the contrary. Then there is an integer $s \in [\tau_1, \tau_2 - 1]$ such that

$$\left|y^*(s) - \min\{x(s), 1 - \alpha_s(x(s+1) - (1 - d_s)x(s))\}\right| > \epsilon. \tag{7.77}$$

For all $t = \tau_1, \ldots, \tau_2 - 1$ set

$$y(t) = \min\{x(t), \ 1 - \alpha_t(x(t+1) - (1 - d_t)x(t))\}. \tag{7.78}$$

By (7.78) and Lemma 7.16,

$$u_t(x(t), x(t+1)) = w_t(y(t)), \ t = \tau_1 \ldots, \tau_2 - 1. \tag{7.79}$$

It is not difficult to see that $\{2^{-1}x(t) + 2^{-1}x^*(t)\}_{t=\tau_1}^{\tau_2}$ is a program and that

$$u_t\left(2^{-1}(x(t) + x^*(t)), 2^{-1}(x(t+1) + x^*(t+1))\right)$$

$$\geq w_t\left(2^{-1}y(t) + 2^{-1}y^*(t)\right), \ t = \tau_1, \ldots, \tau_2 - 1. \tag{7.80}$$

By (7.73) and the choice of δ (see (7.76)),

$$\sum_{t=\tau_1}^{\tau_2-1} u_t(x(t), x(t+1)) \geq U(z_1, z_2, \tau_1, \tau_2) - \delta \geq U(x^*(\tau_1), x^*(\tau_2), \tau_1, \tau_2)$$

$$-\delta - \epsilon_1/4. \tag{7.81}$$

In view of (7.72),

$$2^{-1}x(\tau_1) + 2^{-1}x^*(\tau_1)$$

$$= (z_1 + x^*(\tau_1))/2, \ 2^{-1}x(\tau_2) + 2^{-1}x^*(\tau_2)$$

$$\geq 2^{-1}(x^*(\tau_2) + z_2)). \tag{7.82}$$

In view of (7.71) and the choice of δ (see (7.76)),

$$\left| U(2^{-1}z_1 + 2^{-1}x^*(\tau_1), 2^{-1}z_2 + 2^{-1}x^*(\tau_2), \tau_1, \tau_2) \right.$$

$$\left. -U(x^*(\tau_1), x^*(\tau_2), \tau_1, \tau_2) \right| \leq \epsilon_1/4. \tag{7.83}$$

Relations (7.16), (7.79), (7.80), and Lemma 7.16 imply that for all $t = \tau_1, \ldots, \tau_2 - 1$,

$$u_t\left(2^{-1}(x(t)+x^*(t)), 2^{-1}(x(t+1)+x^*(t+1))\right) \geq 2^{-1}w_t(y(t)) + 2^{-1}w_t(y^*(t))$$

$$= 2^{-1}u_t(x(t), x(t+1)) + 2^{-1}u_t\left(x^*(t), x^*(t+1)\right). \tag{7.84}$$

By the choice of ϵ_1 (see (7.75)), (7.77), and (7.78),

$$w_s(2^{-1}(y^*(s) + y(s))) - w_s(y^*(s))/2 - w_s(y(s))/2 \geq 2\epsilon_1. \tag{7.85}$$

It follows from (7.16), (7.79), (7.80), (7.85), and Lemma 7.16 that

$$u_s\left(2^{-1}(x(s) + x^*(s)), 2^{-1}(x(s+1) + x^*(s+1))\right)$$

$$\geq 2^{-1}u_s(x(s), s(s+1)) + 2^{-1}u_s\left(x^*(s), x^*(s+1)\right) + 2\epsilon_1. \tag{7.86}$$

In view of (7.14), (7.72), (7.81), (7.82), (7.84), (7.86), and the relation $\delta < \epsilon_1/2$,

$$U(2^{-1}z_1 + 2^{-1}x^*(\tau_1), 2^{-1}z_2 + 2^{-1}x^*(\tau_2), \tau_1, \tau_2)$$

$$\geq \sum_{t=\tau_1}^{\tau_2-1} u_t(2^{-1}(x(t)+x^*(t)), 2^{-1}(x(t+1)+x^*(t+1)))$$

$$\geq 2^{-1}\sum_{t=\tau_1}^{\tau_2-1} u_t(x(t), x(t+1)) + 2^{-1}\sum_{t=\tau_1}^{\tau_2-1} u_t(x^*(t), x^*(t+1)) + 2\epsilon_1$$

$$\geq 2^{-1}U(x^*(\tau_1), x^*(\tau_2), \tau_1, \tau_2) + 2^{-1}U(x^*(\tau_1), x^*(\tau_2), \tau_1, \tau_2)$$

$$- \delta - \epsilon_1/4 + 2\epsilon_1$$

$$\geq U(x^*(\tau_1), x^*(\tau_2), \tau_1, \tau_2) + \epsilon_1.$$

This contradicts (7.83). The contradiction we have reached proves (7.74) for all $t = \tau_1, \ldots, \tau_2 - 1$. Lemma 7.21 is proved.

Lemma 7.22 *Let $\epsilon > 0$. Then there exist a natural number $L_0 \geq 4$ and $\delta > 0$ such that for each pair of integers $\tau_1 \geq 0$, $\tau_2 \geq \tau_1 + L_0$ and each program $\{x(t)\}_{t=\tau_1}^{\tau_2}$ which satisfies*

$$x(\tau_1) \leq M_*, \quad \sum_{t=\tau_1}^{\tau_2-1} u_t(x(t), x(t+1)) \geq U(x(\tau_1), x(\tau_2), \tau_1, \tau_2) - \delta \qquad (7.87)$$

and

$$U(x(\tau_1), x(\tau_2), \tau_1, \tau_2) \geq \widehat{U}_{M_*}(\tau_1, \tau_2) - \bar{M} - 2, \qquad (7.88)$$

for each integer $t \in [\tau_1, \tau_2 - L_0]$, the following inequality holds:

$$x(t) + \epsilon \geq 1 - \alpha_t(x(t+1) - (1 - d_t)x(t)). \qquad (7.89)$$

Proof By Lemma 7.17, there exist natural numbers $L_1, L_2 \geq 4$ such that the following property holds:

(P3) for each pair of integers $T_1 \geq 0$, $T_2 \geq L_1 + L_2 + T_1$ and each program $\{x(t)\}_{t=T_1}^{T_2}$ which satisfies $x(0) \leq M_*$ and

$$\sum_{t=T_1}^{T_2-1} u_t(x(t), x(t+1)) \geq U(x(T_1), T_1, T_2) - \bar{M} - 4$$

and each integer $\tau \in [T_1 + L_1, T_2 - L_2]$, the following inequality holds:

$$\max\{u_t(x(t), x(t+1)) : t = \tau, \ldots, \tau + L_2 - 1\} \geq \gamma.$$

By (A1) there is $\gamma_0 > 0$ such that

$$w_t(\gamma_0) < \gamma/4 \text{ for each integer } t \geq 0. \tag{7.90}$$

Choose $\gamma_1 > 0$ such that

$$\gamma_1 (d_*)^{-1} < \gamma_0/4 \tag{7.91}$$

and natural numbers L_3, L_0 such that

$$L_3 > 4(L_1 + L_2), \quad (M_* + 1)(1 - d_*)^{L_3} < \gamma_0/4, \tag{7.92}$$

$$L_0 \geq 4(L_1 + L_2 + L_3 + 4). \tag{7.93}$$

Choose a positive number γ_2 such that

$$\gamma_2 < \gamma_1/4, \quad \gamma_2 < (1 - d^*)^{L_3+L_2}(\alpha^*)^{-1}\epsilon/8. \tag{7.94}$$

By (A3) there is $\delta \in (0, 1)$ such that the following property holds:
is valid (P4) for each integer $t \geq 0$ and each $z_1, z_2 \in [0, M_*]$ satisfying $z_2 \geq z_1 + \gamma_2 \min\{1, \alpha_*\}$, the inequality $w_t(z_2) - w_t(z_1) \geq 4\delta$.
   Assume that integers $\tau_1 \geq 0$, $\tau_2 \geq \tau_1 + L_0$, a program $\{x(t)\}_{t=\tau_1}^{\tau_2}$ satisfies (7.87), (7.88), and an integer

$$t_0 \in [\tau_1, \tau_2 - L_0]. \tag{7.95}$$

We show that (7.89) holds with $t = t_0$. Assume the contrary. Then

$$x(t_0) + \epsilon < 1 - \alpha_{t_0}(x(t_0 + 1) - (1 - d_{t_0})x(t_0)). \tag{7.96}$$

For all $t = \tau_1, \ldots, \tau_2 - 1$ set

$$y(t) = \min\{x(t), \ 1 - \alpha_t(x(t+1) - (1 - d_t)x(t))\}. \tag{7.97}$$

Now we define a program $\{\tilde{x}(t)\}_{t=\tau_1}^{\tau_2}$. Set

$$\tilde{x}(t) = x(t), \ t = \tau_1, \ldots, t_0, \ \tilde{x}(t_0 + 1) = x(t_0 + 1) + \alpha_{t_0}^{-1}\epsilon. \tag{7.98}$$

Clearly,

$$\tilde{x}(t_0 + 1) \geq (1 - d_{t_0})\tilde{x}(t_0).$$

By (7.96)–(7.98),

$$\alpha_{t_0}(\tilde{x}(t_0+1) - (1-d_{t_0})\tilde{x}(t_0))$$

$$= \alpha_{t_0}(x(t_0+1) - (1-d_{t_0})x(t_0)) + \epsilon < 1 - x(t_0) \le 1 - y(t_0). \qquad (7.99)$$

Therefore $\{\tilde{x}(t)\}_{t=\tau_1}^{t_0+1}$ is a program. In view of (7.8) and (7.97)–(7.99) and Lemma 7.16,

$$u_{t_0}(\tilde{x}(t_0), \tilde{x}(t_0+1)) \ge w_{t_0}(y(t_0)) = u_{t_0}(x(t_0), x(t_0+1)). \qquad (7.100)$$

Property (P3) and relations (7.92), (7.93), and (7.95) imply that

$$\max\{u_t(x(t), x(t+1)) : t = t_0 + 1 + L_3, \ldots, t_0 + L_3 + L_2\} \ge \gamma. \qquad (7.101)$$

It follows from (7.101) that there exists an integer t_1 such that

$$t_0 + 1 + L_3 \le t_1 \le t_0 + L_3 + L_2, \ u_{t_1}(x(t_1), x(t_1+1)) \ge \gamma. \qquad (7.102)$$

By (7.97), Lemma 7.16, (7.90), and (7.102),

$$\gamma \le u_{t_1}(x(t_1), x(t_1+1)) = w_{t_1}(y(t_1)) \text{ and } x(t_1) \ge y(t_1) \ge \gamma_0. \qquad (7.103)$$

We show that there is an integer t_2 such that

$$t_0 + 1 \le t_2 \le t_1 - 1, \qquad (7.104)$$

$$x(t_2+1) - (1-d_{t_2})x(t_2) \ge \gamma_1. \qquad (7.105)$$

Let us assume the contrary. Then for all integers $t \in [t_0 + 1, t_1 - 1]$,

$$x(t+1) - (1-d_t)x(t) \le \gamma_1. \qquad (7.106)$$

By (7.13), (7.87), and Lemma 6.9, $x(t_0+1) \le M_*$, and in view of (7.106) and (7.9) for all integers $t \in [t_0 + 1, t_1 - 1]$,

$$x(t+1) \le (1-d_t)x(t) + \gamma_1 \le (1-d_*)x(t) + \gamma_1.$$

Combined with (7.91), (7.92), and (7.102), this implies that

$$x(t_1) \le M_*(1-d_*)^{t_1-t_0-1} + \gamma_1 \sum_{i=0}^{\infty}(1-d_*)^i \le M_*(1-d_*)^{L_3} + \gamma_1 d_*^{-1} < \gamma_0/2.$$

This contradicts (7.103). Therefore there is an integer t_2 for which (7.104) and (7.105) hold. For each integer t satisfying $t_0 + 1 \le t < t_2$, set

$$\tilde{x}(t+1) = (1-d_t)\tilde{x}(t) + x(t+1) - (1-d_t)x(t). \tag{7.107}$$

By (7.98) and (7.107),

$$\tilde{x}(t) \ge x(t), \quad t = t_0 + 1, \ldots, t_2, \tag{7.108}$$

$\{\tilde{x}(t)\}_{t=\tau_1}^{t_2}$ is a program, and in view of (7.8) for any integer t satisfying $t_0 + 1 \le t < t_2$,

$$u_t(\tilde{x}(t), \tilde{x}(t+1)) \ge u_t(x(t), x(t+1)). \tag{7.109}$$

It follows from (7.17) and (7.107) that for any integer t satisfying $t_0 + 1 \le t < t_2$,

$$\tilde{x}(t+1) - x(t+1) = (1-d_t)(\tilde{x}(t) - x(t)) \ge (1-d^*)(\tilde{x}(t) - x(t)). \tag{7.110}$$

By (7.92), (7.98), (7.102), (7.104), and (7.110),

$$\tilde{x}(t_2) - x(t_2) \ge (1-d^*)^{t_2-t_0-1}(\tilde{x}(t_0+1) - x(t_0+1))$$
$$\ge (1-d^*)^{L_3+L_2-1}\alpha_{t_0}^{-1}\epsilon \ge (1-d^*)^{L_3+L_2-1}(\alpha^*)^{-1}\epsilon. \tag{7.111}$$

Set

$$\tilde{x}(t_2+1) = (1-d_{t_2})\tilde{x}(t_2) + x(t_2+1) - (1-d_{t_2})x(t_2) - \gamma_2. \tag{7.112}$$

In view of (7.94), (7.105), and (7.112),

$$\tilde{x}(t_2+1) \ge (1-d_{t_2})\tilde{x}(t_2). \tag{7.113}$$

It follows from (7.17), (7.94), (7.111), and (7.112) that

$$\tilde{x}(t_2+1) = x(t_2+1) + (1-d_{t_2})(\tilde{x}(t_2) - x(t_2)) - \gamma_2$$
$$\ge x(t_2+1) + (1-d^*)^{L_3+L_2}(\alpha^*)^{-1}\epsilon - \gamma_2 > x(t_2+1).$$

Thus

$$\tilde{x}(t_2+1) > x(t_2+1). \tag{7.114}$$

Relations (7.9), (7.97), and (7.112) imply that

$$1 - \alpha_{t_2}(\tilde{x}(t_2+1) - (1-d_{t_2})\tilde{x}(t_2))$$
$$= 1 - \alpha_{t_2}(x(t_2+1) - (1-d_{t_2})x(t_2)) + \alpha_{t_2}\gamma_2$$
$$\ge y(t_2) + \alpha_{t_2}\gamma_2 \ge y(t_2) + \alpha_*\gamma_2. \tag{7.115}$$

Relations (7.94), (7.97), and (7.111) imply that

$$\tilde{x}(t_2) \geq x(t_2) + \gamma_2 \geq y(t_2) + \gamma_2. \tag{7.116}$$

By (7.97), (7.115), (7.116), (P4), and Lemma 7.16,

$$u_{t_2}(\tilde{x}(t_2), \tilde{x}(t_2 + 1)) \geq w_{t_2}(y(t_2) + \gamma_2 \min\{1, \alpha_*\})$$
$$\geq w_{t_2}(y(t_2)) + 4\delta = u_{t_2}(x(t_2), x(t_2 + 1)) + 4\delta. \tag{7.117}$$

For any integer t satisfying $t_2 + 1 \leq t \leq \tau_2 - 1$, put

$$\tilde{x}(t + 1) = (1 - d_t)\tilde{x}(t) + x(t + 1) - (1 - d_t)x(t). \tag{7.118}$$

Relations (7.114) and (7.118) imply that,

$$\tilde{x}(t) \geq x(t), \ t = t_2 + 1, \ \ldots, \tau_2, \tag{7.119}$$

$\{\tilde{x}(t)\}_{t=\tau_1}^{\tau_2}$ is a program, and in view of (7.8) for all integers t satisfying $t_2 + 1 \leq t \leq \tau_2 - 1$,

$$u_t(\tilde{x}(t), \tilde{x}(t + 1)) \geq u_t(x(t), x(t + 1)). \tag{7.120}$$

It follows from (7.87), (7.98), (7.100), (7.109), (7.117), (7.119), and (7.120) that

$$U(x(\tau_1), x(\tau_2), \tau_1, \tau_2) \geq \sum_{t=\tau_1}^{\tau_2-1} u_t(\tilde{x}(t), \tilde{x}(t + 1))$$

$$\geq \sum_{t=\tau_1}^{\tau_2-1} u_t(x(t), x(t + 1)) + 4\delta \geq U(x(\tau_1), x(\tau_2), \tau_1, \tau_2) - \delta + 4\delta.$$

The contradiction we have reached proves (7.89) with $t = t_0$ and Lemma 7.22 itself.

Corollary 7.23 *For each integer $t \geq 0$,*

$$x^*(t) \geq 1 - \alpha_t(x^*(t + 1) - (1 - d_t)\dot{x}^*(t)).$$

We chose $\bar{M} > 0$ in Section 7.1 being as guaranteed by Theorem 7.3. Clearly this constant can be replaced by any other number larger than $\bar{M}$. Taking into account this remark, we obtain the following corollary of Lemma 7.22.

Corollary 7.24 *Let $\epsilon, M > 0$. Then there exist a natural number $L_0 \geq 4$ and $\delta > 0$ such that for each pair of integers $\tau_1 \geq 0$, $\tau_2 \geq \tau_1 + L_0$ and each program $\{x(t)\}_{t=\tau_1}^{\tau_2}$ which satisfies (7.87) and $U(x(\tau_1), x(\tau_2), \tau_1, \tau_2) \geq \widehat{U}_{M_*}(\tau_1, \tau_2) - \bar{M} - M$ and for each integer $t \in [\tau_1, \tau_2 - L_0]$, the inequality (7.89) holds.*

Lemma 7.25 *Let $\epsilon > 0$. Then there exist $\delta > 0$ and a natural number L_0 such that for each pair of integers $\tau_1 \geq 0$, $\tau_2 \geq \tau_1 + L_0$ and each program $\{x(t)\}_{t=\tau_1}^{\tau_2}$ satisfying*

$$x(\tau_1) \leq M_*, \ |x(\tau_i) - x^*(\tau_i)| \leq \delta, \ i = 1, 2, \tag{7.121}$$

$$\sum_{t=\tau_1}^{\tau_2-1} u_t(x(t), x(t+1)) \geq U(x(\tau_1), x(\tau_2), \tau_1, \tau_2) - \delta \tag{7.122}$$

the following inequality holds:

$$|x(t) - x^*(t)| \leq \epsilon, \ t = \tau_1, \ldots, \tau_2 - L_0. \tag{7.123}$$

Proof Choose a positive number ϵ_1 such that

$$(\alpha_*^{-1} + 1)(8\epsilon_1)(d_*)^{-1} < \epsilon/8. \tag{7.124}$$

By Lemma 7.21 there exist a natural number $L_1 \geq 4$ and $\delta_1 > 0$ such that the following property holds:
(P5) For each pair of integers $\tau_1 \geq 0$, $\tau_2 \geq \tau_1 + L_1$, each $z_1, z_2 \in [0, M_*]$, and each program $\{x(t)\}_{t=\tau_1}^{\tau_2}$ satisfying

$$|z_1 - x^*(\tau_1)|, \ |z_2 - x^*(\tau_2)| \leq \delta_1, \ x(\tau_1) = z_1, \ x(\tau_2) \geq z_2$$

and

$$\sum_{t=\tau_1}^{\tau_2-1} u_t(x(t), x(t+1)) \geq U(z_1, z_2, \tau_1, \tau_2)) - \delta_1$$

the following inequality holds for all $t = \tau_1, \ldots, \tau_2 - 1$:

$$\left| y^*(t) - \min\{x(t), \ 1 - \alpha_t(x(t+1) - (1-d_t)x(t))\} \right| \leq \epsilon_1.$$

By Lemma 7.22 there exist a natural number $L_2 \geq 4$ and $\delta_2 > 0$ such that

$$M_*(1 - d_*)^{L_2} < \epsilon/8 \tag{7.125}$$

and the following property holds:
(P6) For each pair of integers $\tau_1 \geq 0$, $\tau_2 \geq \tau_1 + L_2$ and each program $\{x(t)\}_{t=\tau_1}^{\tau_2}$ satisfying $x(\tau_1) \leq M_*$,

$$\sum_{t=\tau_1}^{\tau_2-1} u_t(x(t), x(t+1)) \geq U(x(\tau_1), x(\tau_2), \tau_1, \tau_2) - \delta_2$$

and

$$U(x(\tau_1), x(\tau_2), \tau_1, \tau_2) \geq \widehat{U}_{M_*}(\tau_1, \tau_2) - \bar{M} - 2$$

we have that for each integer $t \in [\tau_1, \tau_2 - L_2]$, the inequality $x(t) + \epsilon_1 \geq 1 - \alpha_t(x(t+1) - (1 - d_t)x(t))$ holds.

By Lemma 7.20 there exist a natural number $L_3 \geq 4$ and $\delta_3 > 0$ such that the following property holds:

(P7) For each pair of integers $\tau_1 \geq 0$, $\tau_2 \geq \tau_1 + L_3$ and each $z_1, z_2 \in [0, M_*]$ satisfying $|z_1 - x^*(\tau_1)|$, $|z_2 - x^*(\tau_2)| \leq \delta_3$, the following inequality holds:

$$|U(x^*(\tau_1), x^*(\tau_2), \tau_1, \tau_2) - U(z_1, z_2, \tau_1, \tau_2)| \leq 4^{-1}.$$

Set

$$\delta = \min\{\delta_1, \delta_2, \delta_3, \epsilon/8\} \tag{7.126}$$

and choose a natural number

$$L_0 > 4(L_1 + L_2 + L_3). \tag{7.127}$$

Assume that integers $\tau_1 \geq 0$, $\tau_2 \geq \tau_1 + L_0$ and a program $\{x(t)\}_{t=\tau_1}^{t_2}$ satisfies (7.121) and (7.122). For all $t = \tau_1, \ldots, \tau_2 - 1$ set

$$y(t) = \min\{x(t), \ 1 - \alpha_t(x(t+1) - (1 - d_t)x(t))\}. \tag{7.128}$$

By (P5), (7.121), (7.122), and (7.126)–(7.128),

$$|y^*(t) - y(t)| \leq \epsilon_1, \ t = \tau_1, \ldots, \tau_2 - 1. \tag{7.129}$$

In view of (P7), (7.14), (7.15), (7.121), (7.126), and (7.127),

$$U(x(\tau_1), x(\tau_2), \tau_1, \tau_2) \geq U(x^*(\tau_1), x^*(\tau_2), \tau_1, \tau_2) - 4^{-1} \geq \widehat{U}_{M_*}(\tau_1, \tau_2) - \bar{M} - 4^{-1}. \tag{7.130}$$

By (P6), (7.121), (7.122), (7.126), (7.127), and (7.130) for each integer $t \in [\tau_1, \tau_2 - L_2]$,

$$x(t) + \epsilon_1 \geq 1 - \alpha_t(x(t+1) - (1 - d_t)x(t)). \tag{7.131}$$

By (7.128) and (7.131) for each integer $t \in [\tau_1, \tau_2 - L_2]$,

$$|y(t) - (1 - \alpha_t(x(t+1) - (1 - d_t)x(t)))| \leq \epsilon_1. \tag{7.132}$$

Corollary 7.23 and (7.16) imply that for all integers $t \in [\tau_1, \tau_2 - 1]$,

$$y^*(t) = 1 - \alpha_t(x^*(t+1) - (1 - d_t)x^*(t)). \tag{7.133}$$

By (7.129), (7.132), and (7.133) for all integers $t = \tau_1, \ldots, \tau_2 - L_2$,

$$|\alpha_t(x^*(t+1) - (1 - d_t)x^*(t)) - (\alpha_t(x(t+1) - (1 - d_t)x(t)))|$$
$$\leq |y^*(t) - y(t)| + |y(t) - (1 - \alpha_t(x(t+1) - (1 - d_t)x(t)))| \leq \epsilon_1 + \epsilon_1. \tag{7.134}$$

Assume that an integer τ satisfies

$$\tau_2 - L_2 \geq \tau > \tau_1 + L_2. \tag{7.135}$$

Then

$$s := \tau - L_2 - 1 \geq \tau_1.$$

By (7.9), (7.134), and (7.135) for all integers $t \in [s, s + L_2]$,

$$\left|x(t+1) - x^*(t+1)\right| \leq (1 - d_t)|x(t) - x^*(t)|$$
$$+ \alpha_t^{-1} \left|\alpha_t(x^*(t+1) - (1 - d_t)x^*(t)) - (\alpha_t(x(t+1) - (1 - d_t)x(t)))\right|$$
$$\leq (1 - d_*)\left|x(t) - x^*(t)\right| + \alpha_*^{-1}(2\epsilon_1). \tag{7.136}$$

In view of (7.124)–(7.126), (7.136), and the choice of δ,

$$|x(\tau) - x^*(\tau)| \leq M_*(1 - d_*)^{L_2+1} + \alpha_*^{-1}(2\epsilon_1) \sum_{i=0}^{\infty} (1 - d_*)^i$$

$$= M_*(1 - d_*)^{L_2+1} + \alpha_*^{-1}(2\epsilon_1)d_*^{-1} < \epsilon/4.$$

Thus

$$|x(\tau) - x^*(\tau)| \leq \epsilon/4 \tag{7.137}$$

for all integers τ satisfying $\tau_1 + L_2 < \tau \leq \tau_2 - L_2$.

Assume that an integer

$$\tau \leq \tau_1 + L_2 \text{ and } \tau \geq \tau_1. \tag{7.138}$$

By (7.9), (7.127), (7.134), and (7.138), for all integers t satisfying $\tau_1 \leq t < \tau$,

$$\left|x(t+1) - x^*(t+1)\right| \leq (1 - d_t)|x(t) - x^*(t)|$$
$$+ \alpha_t^{-1} \left|\alpha_t(x^*(t+1) - (1 - d_t)(x^*(t)) - \alpha_t(x(t+1) - (1 - d_t)x(t))\right|$$

$$\leq (1 - d_*) \left| x(t) - x^*(t) \right| + \alpha_*^{-1}(2\epsilon_1).$$

In view of the relation above, (7.121), (7.124), and (7.126),

$$|x(\tau) - x^*(\tau)| \leq |x(\tau_1) - x^*(\tau_1)| + \alpha_*^{-1}(2\epsilon_1) \sum_{i=0}^{\infty} (1 - d_*)^i \leq \delta + \alpha_*^{-1}(2\epsilon_1)d_*^{-1} < \epsilon.$$

Thus $|x(\tau) - x^*(\tau)| < \epsilon$ for all integers $\tau = \tau_1, \ldots, \tau_1 + L_2$. Together with (7.137) this implies that $|x(\tau) - x^*(\tau)| < \epsilon$ for $\tau = \tau_1, \ldots, \tau_2 - L_2$. Lemma 7.25 is proved.

Lemma 7.26 *Let $\epsilon, M > 0$. Then there exists a natural number $L_0 \geq 4$ such that for each pair of integers $\tau_1 \geq 0$, $\tau_2 \geq \tau_1 + L_0$ and each program $\{x(t)\}_{t=\tau_1}^{\tau_2}$ satisfying*

$$x(\tau_1) \leq M_*, \tag{7.139}$$

$$\sum_{t=\tau_1}^{\tau_2 - 1} u_t(x(t), x(t+1)) \geq U(x(\tau_1), \tau_1, \tau_2) - M \tag{7.140}$$

the following inequality holds:

$$Card(\{t \in \{\tau_1, \ldots, \tau_2 - 1\} :$$

$$|y^*(t) - \min\{x(t),\ 1 - \alpha_t(x(t+1) - (1 - d_t)x(t))\}| > \epsilon\}) \leq L_0. \tag{7.141}$$

Proof By Proposition 7.14 there exist $M_1 > 0$ and a natural number L_1 such that for each pair of integers $T_1 \geq 0$, $T_2 \geq T_1 + L_1$ and each $x_0 \in [0, M_*]$,

$$|U(x_0, T_1, T_2) - \widehat{U}_{M_*}(T_1, T_2)| \leq M_1. \tag{7.142}$$

By (A2) there is $\epsilon_1 \in (0, 1)$ such that the following property holds:
(P8) for each integer $t \geq 0$ and each $z_1, z_2 \in [0, M_*]$ satisfying $|z_1 - z_2| \geq \epsilon$,

$$w_t(2^{-1}z_1 + 2^{-1}z_2) - 2^{-1}w_i(z_1) - 2^{-1}w_t(z_2) \geq \epsilon_1.$$

Choose a natural number

$$L_0 > \epsilon_1^{-1}(\bar{M} + M + M_1) + 4 + L_1. \tag{7.143}$$

Assume that integers $\tau_1 \geq 0$, $\tau_2 \geq \tau_1 + L_0$ and a program $\{x(t)\}_{t=\tau_1}^{\tau_2}$ satisfies (7.139) and (7.140). By (7.139) and (7.140)–(7.143),

$$\sum_{t=\tau_1}^{\tau_2-1} u_t(x(t), x(t+1)) \geq \widehat{U}_{M_*}(\tau_1, \tau_2) - M_1 - M.$$

(7.144)

For $t = \tau_1, \ldots, \tau_2 - 1$ set

$$y(t) = \min\{x(t), 1 - \alpha_t(x(t+1) - (1-d_t)x(t))\}.$$

(7.145)

In view of (7.145) and Lemma 7.16,

$$w_t(y(t)) = u_t(x(t), x(t+1)), \ t = \tau_1, \ldots, \tau_2 - 1.$$

(7.146)

Clearly, $\{2^{-1}(x(t) + x^*(t)) : t = \tau_1, \ldots, \tau_2\}$ is a program, and for all $t = \tau_1, \ldots, \tau_2 - 1$,

$$u_t\left(2^{-1}(x(t) + x^*(t)), 2^{-1}(x(t+1) + x^*(t+1))\right) \geq w_t(2^{-1}(y(t) + y_*(t))).$$

(7.147)

Set

$$E = \{t \in \{\tau_1, \ldots, \tau_2 - 1\} : |y(t) - y^*(t)| > \epsilon\}.$$

(7.148)

It follows from (7.139), (P8), (7.148), and Lemma 6.9 that for each $t \in E$,

$$w_t(2^{-1}y(t) + 2^{-1}y^*(t)) \geq 2^{-1}w_t(y(t)) + 2^{-1}w_t(y^*(t)) + \epsilon_1.$$

(7.149)

By (7.15), (7.16), (7.139), (7.144), (7.146), (7.147), (7.149), and Lemma 7.16,

$$\widehat{U}_{M_*}(\tau_1, \tau_2) \geq \sum_{t=\tau_1}^{\tau_2-1} u_t(2^{-1}(x(t) + x^*(t)), 2^{-1}(x(t+1) + x^*(t+1)))$$

$$\geq \sum_{t=\tau_1}^{\tau_2-1} w_t(2^{-1}(y(t) + y^*(t)))$$

$$\geq 2^{-1}\sum_{t=\tau_1}^{\tau_2-1} w_t(y(t)) + 2^{-1}\sum_{t=\tau_1}^{\tau_2-1} w_t(y^*(t)) + \epsilon_1 \text{Card}(E)$$

$$\geq 2^{-1}\left(\widehat{U}_{M_*}(\tau_1, \tau_2) - \bar{M}\right) + 2^{-1}\left(\widehat{U}_{M_*}(\tau_1, \tau_2) - M_1 - M\right)$$

$$+ \epsilon_1 \text{Card}(E), \ \epsilon_1 \text{Card}(E) \leq \bar{M} + M + M_1$$

and in view of (7.143),

$$\text{Card}(E) \leq \epsilon_1^{-1}(\bar{M} + M + M_1) < L_0.$$

Lemma 7.26 is proved.

Lemma 7.27 *Let $M, \epsilon > 0$. Then there exist $\delta > 0$ and a natural number $L_0 \geq 4$ such that for each pair of integers $\tau_1 \geq 0$, $\tau_2 \geq \tau_1 + L_0$ and each program $\{x(t)\}_{t=\tau_1}^{\tau_2}$ satisfying*

$$x(\tau_1) \leq M_*, \quad \sum_{t=\tau_1}^{\tau_2-1} u_t(x(t), x(t+1)) \geq U(x(\tau_1), x(\tau_2), \tau_1, \tau_2) - \delta, \qquad (7.150)$$

$$U(x(\tau_1), x(\tau_2), \tau_1, \tau_2) \geq U(x(\tau_1), \tau_1, \tau_2) - M \qquad (7.151)$$

the following inequality holds:

$$Card(\{t \in \{\tau_1, \ldots, \tau_2 - 1\} :$$

$$|x(t+1) - (1 - d_t)x(t) - (x^*(t+1) - (1 - d_t)x^*(t))| > \epsilon\}) \leq L_0.$$

Proof Choose a positive number ϵ_1 such that

$$\left(1 + \alpha_*^{-1}\right)(2\epsilon_1) < \epsilon. \qquad (7.152)$$

By Lemma 7.26 there is an integer $L_1 \geq 4$ such that the following property holds:
(P9) for each pair of integers $\tau_1 \geq 0$, $\tau_2 \geq \tau_1 + L_1$, each program $\{x(t)\}_{t=\tau_1}^{\tau_2}$ satisfying $x(\tau_1) \leq M_*$, $\sum_{t=\tau_1}^{\tau_2-1} u_t(x(t), x(t+1)) \geq U(x(\tau_1), \tau_1, \tau_2) - 2 - M$, we have

$$Card(\{t \in \{\tau_1, \ldots, \tau_2 - 1\} :$$

$$|y^*(t) - \min\{x(t), \ 1 - \alpha_t(x(t+1) - (1 - d_t)x(t))\}| > \epsilon_1\}) \leq L_1.$$

By Proposition 7.14 there exist $M_1 > 0$ and a positive number L_2 such that for each pair of integers $T_1 \geq 0$, $T_2 \geq T_1 + L_2$ and each $x_0 \in [0, M_*]$,

$$|U(x_0, T_1, T_2) - \widehat{U}_{M_*}(T_1, T_2)| \leq M_1. \qquad (7.153)$$

By Corollary 7.24 there exist a natural number $L_3 \geq 4$ and $\delta \in (0, 1/4)$ such that the following property holds:
(P10) for each pair of integers $\tau_1 \geq 0$, $\tau_2 \geq \tau_1 + L_3$, each program $\{x(t)\}_{t=\tau_1}^{\tau_2}$ satisfying $x(\tau_1) \leq M_*$, $\sum_{t=\tau_1}^{\tau_2-1} u_t(x(t), x(t+1)) \geq U(x(\tau_1), x(\tau_2), \tau_1, \tau_2) - \delta$,

$$U(x(\tau_1), x(\tau_2), \tau_1, \tau_2) \geq \widehat{U}_{M_*}(\tau_1, \tau_2) - \bar{M} - 2 - M_1 - M$$

and for each integer $t \in [\tau_1, \tau_2 - L_3]$, we have

$$x(t) + \epsilon_1 \geq 1 - \alpha_t(x(t+1) - (1 - d_t)(x(t))).$$

Choose a natural number

$$L_0 > 4(L_1 + L_2 + L_3). \tag{7.154}$$

Assume that integers $\tau_1 \geq 0$, $\tau_2 \geq \tau_1 + L_0$ and a program $\{x(t)\}_{t=\tau_1}^{\tau_2}$ satisfies (7.150) and (7.151). By (7.150)–(7.154),

$$\sum_{t=\tau_1}^{\tau_2-1} u_t(x(t), x(t+1)) \geq U(x(\tau_1), \tau_1, \tau_2) - M - 1 \geq \widehat{U}_{M_*}(\tau_1, \tau_2) - M - 1 - M_1. \tag{7.155}$$

By (P10), (7.150), (7.154), and (7.155) for each integer $t \in [\tau_1, \tau_2 - L_3]$,

$$x(t) + \epsilon_1 \geq 1 - \alpha_t(x(t+1) - (1 - d_t)x(t)). \tag{7.156}$$

For each integer $t \in \{\tau_1, \ldots, \tau_2 - 1\}$, set

$$y(t) = \min\{x(t), \ 1 - \alpha_t(x(t+1) - (1 - d_t)x(t))\}. \tag{7.157}$$

Set

$$E_1 = \{t \in \{\tau_1, \ldots, \tau_2 - 1\}: \ |y^*(t) - y(t)| > \epsilon_1\}. \tag{7.158}$$

It follows from (P9), (7.150), (7.151), (7.154), (7.155), (7.157), and (7.158) that

$$\mathrm{Card}(E_1) \leq L_1. \tag{7.159}$$

By Corollary 7.23, (7.12), and (7.156)–(7.158), for each $t \in [\tau_1, \tau_2 - L_3] \setminus E_1$,

$$\left| \alpha_t(x(t+1) - (1 - d_t)x(t)) - \alpha_t(x^*(t+1) - (1 - d_t)x^*(t)) \right|$$
$$= \left| y^*(t) - (1 - \alpha_t(x(t+1) - (1 - d_t)x(t))) \right| \leq \left| y^*(t) - y(t) \right|$$
$$+ |y(t) - (1 - \alpha_t(x(t+1) - (1 - d_t)x(t)))| \leq \epsilon_1 + \epsilon_1. \tag{7.160}$$

and (7.160) for all integers $t \in [\tau_1, \tau_2 - L_3] \setminus E_1$,

$$\left| x(t+1) - (1 - d_t)x(t) - (x^*(t+1) - (1 - d_t)x^*(t)) \right| \leq \alpha_*^{-1}(2\epsilon_1) < \epsilon. \tag{7.161}$$

By (7.154), (7.159), and (7.161),

$$\mathrm{Card}(\{t \in \{\tau_1, \ldots, \tau_2 - 1\}:$$

$$|x(t+1) - (1 - d_t)x(t) - (x^*(t+1) - (1 - d_t)x^*(t))| > \epsilon\})$$

$$\leq \mathrm{Card}(\{\tau_1, \ldots, \tau_2 - 1\} \setminus (\{\tau_1, \ldots, \tau_2 - L_3\} \setminus E_1))$$

$$\leq L_3 + 1 + \mathrm{Card}(E_1) \leq L_3 + 1 + L_1 < L_0.$$

Lemma 7.27 is proved.

7.3   Proof of Theorem 7.8

By Lemma 7.25 there are $\delta_1 > 0$ and a natural number L_1 such that the following property holds:

(P11) for each pair of integers $\tau_1 \geq 0$, $\tau_2 \geq \tau_1 + L_1$ and each program $\{x(t)\}_{t=\tau_2}^{\tau_2}$ satisfying $|x(\tau_i) - x^*(\tau_i)| \leq \delta_1$, $i = 1, 2$, $x(\tau_1) \leq M_*$,

$$\sum_{t=\tau_1}^{\tau_2 - 1} u_t(x(t), x(t + 1)) \geq U(x(\tau_1), x(\tau_2), \tau_1, \tau_2) - \delta_1$$

the inequality $|x(t) - x^*(t)| \leq \epsilon$ holds for all $t = \tau_1, \ldots, \tau_2 - L_1$.

Choose a positive number

$$\delta_2 < \delta_1 d_* / 8. \tag{7.162}$$

By Lemma 7.27 there exist

$$\delta \in (0, \delta_2)$$

and a natural number $L_2 \geq 4$ such that the following property holds:

(P12) for each pair of integers $\tau_1 \geq 0$, $\tau_2 \geq \tau_1 + L_2$ and each program $\{x(t)\}_{t=\tau_1}^{\tau_2}$ satisfying

$$x(\tau_1) \leq M_*, \quad \sum_{t=\tau_1}^{\tau_2 - 1} u_t(x(t), x(t + 1)) \geq U(x(\tau_1), x(\tau_2), \tau_1, \tau_2) - \delta,$$

$$U(x(\tau_1), x(\tau_2), \tau_1, \tau_2) \geq U(x(\tau_1), \tau_2, \tau_2) - M - 2$$

we have

$$\mathrm{Card}(\{t \in \{\tau_1, \ldots, \tau_2 - 1\} :$$

$$|x(t + 1) - (1 - d_t)x(t) - (x^*(t + 1) - (1 - d_t)x^*(t))| > \delta_2\}) \leq L_2.$$

Choose natural numbers $L_3 \geq 4$ and p such that

$$M_*(1 - d_*)^{L_3} < \delta_1/8, \tag{7.163}$$

$$p > 2(L_1 + L_2 + 4)(4L_3 + 8). \tag{7.164}$$

Assume that integers $\tau_1 \geq 0$, $\tau_2 \geq \tau_1 + 2p$ and that a program $\{x(t)\}_{t=\tau_1}^{\tau_2}$ satisfies

$$x(\tau_1) \leq M_*, \quad \sum_{t=\tau_1}^{\tau_2-1} u_t(x(t), x(t+1)) \geq U(x(\tau_1), x(\tau_2), \tau_1, \tau_2) - \delta, \tag{7.165}$$

$$U(x(\tau_1), x(\tau_2), \tau_1, \tau_2) \geq U(x(\tau_1), \tau_1, \tau_2) - M. \tag{7.166}$$

Set

$$E_1 = \{t \in \{\tau_1, \ldots, \tau_2 - 1\} :$$

$$|x(t+1) - (1 - d_t)x(t) - (x^*(t+1) - (1 - d_t)x^*(t))| > \delta/2\}. \tag{7.167}$$

By (P12) and (6.164)–(6.167),

$$\text{Card}(E_1) \leq L_2. \tag{7.168}$$

Set

$$E_2 = \{t \in \{\tau_1, \ldots, \tau_2\} : \text{ there is } s \in E_1 \text{ such that } |t - s| \leq 2L_3 + 2\}. \tag{7.169}$$

In view of (7.164), (7.168), and (7.169),

$$\text{Card}(E_2) \leq L_2(4L_3 + 5) < p/2. \tag{7.170}$$

Relations (7.164) and (7.170) imply that there are $S_1, S_2 \in [\tau_1, \tau_2]$ such that

$$S_1 \leq \tau_1 + p/2, \quad S_2 \geq \tau_2 - p/2, \quad S_1, S_2 \notin E_2. \tag{7.171}$$

It follows from (7.164), (7.169), and (7.171) that

$$\{S_1, \ldots, S_1 + 2L_3 + 2\} \cap E_1 = \emptyset, \quad \{S_2 - 2L_3 - 2, \ldots, S_2\} \cap E_1 = \emptyset. \tag{7.172}$$

By (7.167) and (7.172) for all integers

$$t \in \{S_1, \ldots, S_1 + 2L_3 + 2\} \cup \{S_2 - 2L_3 - 2, \ldots, S_2\},$$

we have

$$|x(t+1) - (1 - d_t)x(t) - (x^*(t+1) - (1 - d_t)x^*(t))| \leq \delta_2$$

and in view of (7.9),

$$|x(t+1)-x^*(t+1)| \le (1-d_t)|x(t)-x^*(t)| + \delta_2 \le (1-d_*)|x(t)-x^*(t)| + \delta_2.$$

(7.173)

By (7.162), (7.163), (7.165), and (7.173),

$$|x(S_1 + 2L_3 + 2) - x^*(S_1 + 2L_3 + 2)| \le |x(S_1) - x^*(S_1)|(1-d_*)^{2L_3}$$

$$+ \delta_2 \sum_{i=0}^{\infty} (1-d_*)^i \le M_*(1-d_*)^{2L_3} + \delta_2 d_*^{-1} < \delta_1,$$

$$|x(S_2) - x^*(S_2)| \le |x(S_2 - 2L_3 - 2) - x^*(S_2 - 2L_3 - 2)|(1-d_*)^{2L_3}$$

$$+ \delta_2 \sum_{i=0}^{\infty} (1-d_*)^i \le M_*(1-d_*)^{2L_3} + \delta_2 d_*^{-1} < \delta_1.$$

Thus

$$|x(S_2) - x^*(S_2)| < \delta_1, \ |x(S_1 + 2L_3 + 2) - x^*(S_1 + 2L_3 + 2)| < \delta_1. \quad (7.174)$$

In view of (7.165),

$$\sum_{t=S_1+2L_3+2}^{S_2-1} u_t(x(t), x(t+1)) \ge U(x(S_1+2L_3+2), x(S_2), S_1+2L_3+2, S_2) - \delta.$$

(7.175)

By (7.164) and (7.171),

$$S_2 - (S_1 + 2L_3 + 2) \ge p > L_1. \quad (7.176)$$

It follows from (7.174), (7.175), (7.176), and (P11) that $|x(t) - x^*(t)| \le \epsilon$ for all $t = S_1 + 2L_3 + 2, \ldots, S_2 - L_1$. Theorem 7.8 is proved.

7.4 Proof of Theorem 7.9

By Lemma 7.25 there exist $\delta_1 \in (0, \epsilon)$ and a natural number L_1 such that the following property holds:

(P13) for each pair of integers $\tau_1 \ge 0$, $\tau_2 \ge \tau_1 + L_1$ and each program $\{x(t)\}_{t=\tau_1}^{\tau_2}$ satisfying $x(\tau_1) \le M_*$, $|x(\tau_i) - x^*(\tau_i)| \le \delta_1$, $i = 1, 2$, and

$$\sum_{t=\tau_1}^{\tau_2-1} u_t(x(t), x(t+1)) \geq U(x(\tau_1), x(\tau_2), \tau_1, \tau_2) - \delta_1$$

the inequality $|x(t) - x^*(t)| \leq \epsilon$ holds for all $t = \tau_1, \ldots, \tau_2 - L_1$.

By Theorem 7.8 there exist a natural number L_2 and $\delta \in (0, \delta_1)$ such that the following property holds:

(P14) for each pair of integers $T_1 \geq 0$, $T_2 \geq T_1 + 2L_2$ and each program $\{x(t)\}_{t=T_1}^{T_2}$ which satisfies $x(T_1) \leq M_*$, $\sum_{t=T_1}^{T_2-1} u_t(x(t), x(t+1)) \geq U(x(T_1), x(T_2), T_1, T_2) - \delta$,

$$U(x(T_1), x(T_2), T_1, T_2) \geq U(x(T_1), T_1, T_2) - M$$

we have

$$|x(t) - x^*(t)| \leq \delta_1, \ t \in [T_1 + L_2, T_2 - L_2]. \tag{7.177}$$

Choose a natural number

$$p > 2(2L_2 + 2L_1). \tag{7.178}$$

Assume that integers $T_1 \geq 0$, $T_2 \geq T_1 + p$ and that a program $\{x(t)\}_{t=T_1}^{T_2}$ satisfies

$$x(T_1) \leq M_*, \ |x(T_1) - x^*(T_1)| \leq \delta, \tag{7.179}$$

$$\sum_{t=T_1}^{T_2-1} u_t(x(t), x(t+1)) \geq U(x(T_1), x(T_2), T_1, T_2) - \delta, \tag{7.180}$$

$$U(x(T_1), x(T_2), T_1, T_2) \geq U(x(T_1), T_1, T_2) - M. \tag{7.181}$$

By (P14) and (7.178)–(7.181), the inequality (7.177) is true. By (7.177),

$$|x(T_1 + L_2 + L_1) - x^*(T_1 + L_2 + L_1)| \leq \delta_1. \tag{7.182}$$

In view of (7.179), (7.180), (7.182), and (P13),

$$|x(t) - x^*(t)| \leq \epsilon, \ t \in \{T_1, \ldots, T_1 + L_2\}. \tag{7.183}$$

By (7.177) and (7.183), $|x(t) - x^*(t)| \leq \epsilon$, $t = T_1, \ldots, T_2 - L_2$. Theorem 7.9 is proved.

7.5 Proof of Theorem 7.7

By Lemma 7.18 there are $M_0 > 0$ and natural numbers $\bar{L}_1, \bar{L}_2$ such that the following property holds:

(P15) for each pair of integers $\tau_1 \geq 0$, $\tau_2 \geq \tau_1 + \bar{L}_1 + \bar{L}_2$, each program $\{x(t)\}_{t=\tau_1}^{\tau_2}$ which satisfies $x(\tau_1) \leq M_*$, $\sum_{t=\tau_1}^{\tau_2-1} u_t(x(t), x(t+1)) \geq U(x(\tau_1), \tau_1, \tau_2) - M - 1$, and each pair of integers $T_1, T_2 \in [\tau_1, \tau_2 - \bar{L}_2]$ satisfying $\bar{L}_1 \leq T_2 - T_1$, the inequality $\sum_{t=T_1}^{T_2-1} u_t(x(t), x(t+1)) \geq U(x(T_1), T_1, T_2) - M_0$ holds.

By Theorem 7.8 there exist a natural number p and $\delta > 0$ such that the following property holds:

(P16) for each pair of integers $T_1 \geq 0$, $T_2 \geq T_1 + 2p$ and each program $\{x(t)\}_{t=T_1}^{T_2}$ which satisfies $x(T_1) \leq M_*$, $\sum_{t=T_1}^{T_2-1} u_t(x(t), x(t+1)) \geq U(x(T_1), x(T_2), T_1, T_2) - \delta$,

$$U(x(T_1), x(T_2), T_1, T_2) \geq U(x(T_1), T_1, T_2) - M_0 - 1$$

the inequality $|x(t) - x^*(t)| \leq \epsilon$ holds for all integers $t \in [T_1 + p, T_2 - p]$.

Choose a natural number Q such that

$$Q > 8(\bar{L}_1 + \bar{L}_2) + 2(2 + M\delta^{-1})[6p + \bar{L}_1 + 2\bar{L}_2 + 8]. \tag{7.184}$$

Assume that integers

$$T_1 \geq 0, \ T_2 \geq Q + T_1 \tag{7.185}$$

and a program $\{x(t)\}_{t=T_1}^{T_2}$ satisfies

$$x(T_1) \leq M_*, \ \sum_{t=T_1}^{T_2-1} u_t(x(t), x(t+1)) \geq U(x(T_1), T_1, T_2) - M. \tag{7.186}$$

Define by induction a sequence of nonnegative integers $T_1 = \tau_0 < \tau_1 < \ldots, \ldots < \tau_q = T_2$ with an integer $q \geq 1$. Put $\tau_0 = T_1$. Assume that an integer $j \geq 0$, τ_i, $i = 0, \ldots, j$ has been defined and define τ_{j+1}. If $\tau_j = T_2$, then $j = q$ and the construction is completed. Assume that $\tau_j < T_2$. There are two cases:

$$\sum_{t=\tau_j}^{T_2-1} u_t(x(t), x(t+1)) \geq U(x(\tau_j), x(T_2), \tau_j, T_2) - \delta; \tag{7.187}$$

$$\sum_{t=\tau_j}^{T_2-1} u_t(x(t), x(t+1)) < U(x(\tau_j), x(T_2), \tau_j, T_2) - \delta. \tag{7.188}$$

If (7.187) holds, then set $\tau_{j+1} = T_2, q = j + 1$ and the construction is completed. If (7.188) holds, then there is a natural number $\tau_{j+1} > \tau_j$ such that

$$\sum_{t=\tau_j}^{\tau_{j+1}-1} u_t(x(t), x(t+1)) < U(x(\tau_j), x(\tau_{j+1}), \tau_j, \tau_{j+1}) - \delta; \qquad (7.189)$$

if an integer s satisfies $\tau_j < s < \tau_{j+1}$, then

$$\sum_{t=\tau_j}^{s-1} u_t(x(t), x(t+1)) \geq U(x(\tau_j), x(s), \tau_j, s) - \delta. \qquad (7.190)$$

Therefore by induction we defined a sequence of τ_i. It is not difficult to see that this sequence is finite. Let τ_q be its last element.

We construct by induction a sequence of programs. First we define a program $\{x^{(q)}\}_{t=T_1}^{T_2}$. If

$$\sum_{t=\tau_q-1}^{\tau_q-1} u_t(x(t), x(t+1)) \geq U(x(\tau_{q-1}), x(\tau_q), \tau_{q-1}, \tau_q) - \delta, \qquad (7.191)$$

then set

$$x^{(q)}(t) = x(t), \ t = T_1, \ldots, T_2. \qquad (7.192)$$

If

$$\sum_{t=\tau_q-1}^{\tau_q-1} u_t(x(t), x(t+1)) < U(x(\tau_{q-1}), x(\tau_q), \tau_{q-1}, \tau_q) - \delta, \qquad (7.193)$$

then there exists a program $\{x^{(q)}(t)\}_{t=T_1}^{T_2}$ such that

$$x^{(q)}(t) = x(t), \ t = T_1, \ldots, \tau_{q-1}, \ x^{(q)}(T_2) \geq x(T_2),$$

$$\sum_{t=\tau_q-1}^{\tau_q-1} u_t(x^{(q)}(t), x^{(q)}(t+1)) \geq \sum_{t=\tau_q-1}^{\tau_q-1} u_t(x(t), x(t+1)) + \delta. \qquad (7.194)$$

Assume that an integer $j \geq 1$ satisfies $j \leq q$ and that we have already defined a program $\{x^{(j)}(t)\}_{t=T_1}^{T_2}$ such that

$$x^{(j)}(t) = x(t), \ t = T_1, \ldots, \tau_{j-1}, \ x^{(j)}(T_2) \geq x(T_2), \qquad (7.195)$$

$$\sum_{t=\tau_{j-1}}^{T_2-1} u_t(x^{(j)}(t), x^{(j)}(t+1)) \geq \sum_{t=\tau_{j-1}}^{T_2-1} u_t(x(t), x(t+1)) + \delta(q-j). \qquad (7.196)$$

Clearly for $j = q$ the assumption is true.

Assume that an integer $j \geq 2$ and define a program $\{x^{(j-1)}(t)\}_{t=T_1}^{T_2}$. Set

$$x^{(j-1)}(t) = x(t), \ t = T_1, \ldots, \tau_{j-2}. \qquad (7.197)$$

By (7.189),

$$\sum_{t=\tau_{j-2}}^{\tau_{j-1}-1} u_t(x(t), x(t+1)) < U(x(\tau_{j-2}), x(\tau_{j-1}), \tau_{j-2}, \tau_{j-1}) - \delta. \qquad (7.198)$$

By (7.198) there is a program $\{x^{(j-1)}(t)\}_{t=\tau_{j-2}}^{\tau_j-1}\}$ such that

$$x^{(j-1)}(\tau_{j-1}) \geq x(\tau_{j-1}), \qquad (7.199)$$

$$\sum_{t=\tau_{j-2}}^{\tau_{j-1}-1} u_t(x^{(j-1)}(t), x^{(j-1)}(t+1)) \geq \sum_{t=\tau_{j-2}}^{\tau_{j-1}-1} u_t(x(t), x(t+1)) + \delta. \qquad (7.200)$$

Clearly, $\{x^{(j-1)}(t)\}_{t=T_1}^{\tau_j-1}$ is a program. For all integers t satisfying $\tau_{j-1} \leq t < T_2$, put

$$x^{(j-1)}(t+1) = (1-d_t)x^{(j-1)}(t) + x^{(j)}(t+1) - (1-d_t)x^{(j)}(t). \qquad (7.201)$$

By (7.195), (7.197), (7.199), and (7.201), $\{x^{(j-1)}(t)\}_{t=T_1}^{T_2}$ is a program such that

$$x^{(j-1)}(T_2) \geq x^{(j)}(T_2) \geq x(T_2). \qquad (7.202)$$

By (7.8), (7.195), (7.197), and (7.201),

$$\sum_{t=\tau_{j-1}}^{T_2-1} u_t(x^{(j-1)}(t), x^{(j-1)}(t+1)) - \sum_{t=\tau_{j-1}}^{T_2-1} u_t(x^{(j)}(t), x^{(j)}(t+1)) \geq 0. \qquad (7.203)$$

Therefore $\{x^{j-1}(t)\}_{t=T_1}^{T_2}$ is a program, (7.197) and (7.202) hold, and in view of (7.196), (7.200), and (7.203),

$$\sum_{t=\tau_{j-2}}^{T_2-1} u_t(x^{(j-1)}(t), x^{(j-1)}(t+1)) - \sum_{t=\tau_{j-2}}^{T_2-1} u_t(x(t), x(t+1))$$

$$\geq \delta + \sum_{t=\tau_{j-1}}^{T_2-1} u_t(x^{(j)}(t), x^{(j)}(t+1)) - \sum_{t=\tau_{j-1}}^{T_2-1} u_t(x(t), x(t+1)) \geq \delta(q-j+1).$$

Therefore the assumptions posed for j hold also for $j-1$, and by induction we construct a sequence of programs $\{x^{(j)}(t)\}_{t=T_1}^{T_2}$, $j = q, \ldots, 1$ satisfying (7.195) and (7.196) for $j = q, \ldots, 1$. In particular, $x^{(1)}(T_1) = x(T_1)$, $x^{(1)}(T_2) \geq x(T_2)$,

$$\sum_{t=T_1}^{T_2-1} u_t(x^{(1)}(t), x^{(1)}(t+1)) - \sum_{t=T_1}^{T_2-1} u_t(x(t), x(t+1)) \geq \delta(q-1)$$

and in view of (7.186),

$$\delta(q-1) \leq U(x(T_1), T_1, T_2) - \sum_{t=T_1}^{T_2-1} u_t(x(t), x(t+1)) \leq M$$

and

$$q \leq M\delta^{-1} + 1. \tag{7.204}$$

Assume that an integer j satisfies

$$0 \leq j \leq q-1, \ \tau_{j+1} - \tau_j > \bar{L}_2 + 4 + 2p + \bar{L}_1. \tag{7.205}$$

If $j < q-1$, then (7.190) holds for all integers s satisfying $\tau_j < s < \tau_{j+1}$ and in particular

$$\sum_{t=\tau_j}^{\tau_{j+1}-2} u_t(x(t), x(t+1)) \geq U(x(\tau_j), x(\tau_{j+1}-1), \tau_j, \tau_{j+1}-1) - \delta. \tag{7.206}$$

If $j = q-1$, then (7.206) holds or (7.187) is valid. It is easy to see that (7.206) holds in all the cases. By (7.184)–(7.186), (7.205), and (P15),

$$\sum_{t=\tau_j}^{\tau_{j+1}-2-\bar{L}_2} u_t(x(t), x(t+1)) \geq U(x(\tau_j), \tau_j, \tau_{j+1}-\bar{L}_2-1) - M_0. \tag{7.207}$$

Relation (7.206) implies that

$$\sum_{t=\tau_j}^{\tau_{j+1}-2-\bar{L}_2} u_t(x(t), x(t+1)) \geq U(x(\tau_j), x(\tau_{j+1} - 1 - \bar{L}_2), \tau_j, \tau_{j+1} - \bar{L}_2 - 1) - \delta.$$

$$(7.208)$$

By (P16), (7.186), (7.205), (7.207), and (7.208),

$$|x(t) - x^*(t)| \leq \epsilon \text{ for all integers } t \in [\tau_j + p, \tau_{j+1} - 1 - p - \bar{L}_2]. \qquad (7.209)$$

Thus (7.209) holds for all integers j satisfying (7.205) and

$$\{t \in \{T_1, \ldots, T_2\} : |x(t) - x^*(t)| > \epsilon\}$$

$$\subset \{T_1, \ldots, T_2\} \setminus \cup\{\{\tau_j + p, \ldots \tau_{j+1} - 1 - p - \bar{L}_2\} :$$

$$j \text{ is an integer such that (7.205) holds}\} \subset \cup\{\{\tau_j, \ldots, \tau_{j+1}\} :$$

$$j \text{ is an integer such that } 0 \leq j \leq q - 1, \ \tau_{j+1} - \tau_j \leq \bar{L}_2 + 4 + 2p + \bar{L}_1\}$$

$$\cup\{\{\tau_j, \ldots, \tau_j + p - 1\} \cup \{\tau_{j+1} - p - \bar{L}_2, \ldots, \tau_{j+1}\} :$$

$$j \text{ is an integer satisfying (7.205)}\}.$$

Together with (7.184) and (7.204), this implies that

$$\text{Card}(\{t \in \{T_1, \ldots, T_2\} : |x(t) - x^*(t)| > \epsilon\})$$

$$\leq q(\bar{L}_2 + 6 + 2p + \bar{L}_1) + 2q(p + 2 + \bar{L}_2 + 1)$$

$$\leq 2q(8 + 2\bar{L}_2 + \bar{L}_1 + 4p) < 2(M\delta^{-1} + 1)(8 + 2\bar{L}_2 + 4p + \bar{L}_1) < Q.$$

Theorem 7.7 is proved.

7.6   Proofs of Theorems 7.12 and 7.13

Proof of Theorem 7.12 We may assume that $M_* > x(0)$. There exist $M_0 > 0$ such that

$$\left| \sum_{t=0}^{T-1} u_t(x(t), x(t+1)) - \sum_{t=0}^{T-1} u_t(x^*(t), x^*(t+1)) \right| \leq M_0$$

for all integers $T \geq 1$. Combined with (7.15) this implies that for each integer $T \geq 0$, the inequality

$$\left| \sum_{t=0}^{T-1} u_t(x(t), x(t+1)) - \widehat{U}_{M_*}(0, T) \right| \leq \bar{M} + M_0$$

holds. Now Theorem 7.12 follows from Theorem 7.7.

Proof of Theorem 7.13 Since M_* is an arbitrary positive number satisfying (7.13) and $x_{*,0}$ is an arbitrary element of $[0, M_*]$, we may assume without loss of generality that $x_0 = x_*$. Then $\{\bar{x}(t)\}_{t=0}^{\infty} = \{x^*(t)\}_{t=0}^{\infty}$. Let a program $\{x(t)\}_{t=0}^{\infty}$ satisfy $x(0) = x_{*,0} = x^*(0)$. We show that $\limsup_{T \to \infty}[\sum_{t=0}^{T-1} u_t(x(t), x(t+1)) - \sum_{t=0}^{T-1} u_t(x^*(t), x^*(t+1))] \leq 0$. By Theorem 7.4, we may assume without loss of generality that the sequence

$$\left\{ \sum_{t=0}^{T-1} u_t(x(t), x(t+1)) - \sum_{t=0}^{T-1} u_t(x^*(t), x^*(t+1)) \right\}$$

is bounded. By Theorem 7.12,

$$\lim_{t \to \infty} |x(t) - x^*| = 0. \tag{7.210}$$

In view of Lemma 6.9,

$$x^*(t), \ x(t) \leq M_*, \ t = 0, 1, \ldots \tag{7.211}$$

By (7.15), (7.210), (7.211), and Lemma 7.19,

$$\lim_{T \to \infty} |U(0, T, x^*(0), x^*(T)) - U(0, T, zx(0), x(T))| = 0. \tag{7.212}$$

Relations (7.14) and (7.212) imply that

$$\limsup_{T \to \infty} \left[\sum_{t=0}^{T-1} u_t(x(t), x(t+1)) - \sum_{t=0}^{T-1} u_t(x^*(t), x^*(t+1)) \right]$$

$$\leq \limsup_{T \to \infty} \left[U(0, T, x(0), x(T)) - U(0, T, x^*(0), x^*(T)) \right] = 0$$

and $\{x^*(t)\}_{t=0}^{\infty}$ is an overtaking optimal program.

Assume now that $\{x(t)\}_{t=0}^{\infty}$ is an overtaking optimal program satisfying $x(0) = x^*(0)$. Clearly,

$$\lim_{T \to \infty} \left[\sum_{t=0}^{T-1} u_t(x(t), x(t+1)) - \sum_{t=0}^{T-1} u_t(x^*(t), x^*(t+1)) \right] = 0.$$

Together with (7.15) and Theorem 7.9, this implies the validity of Theorem 7.13.

Chapter 8
Optimal Programs

In this chapter we continue to study the Robinson–Solow–Srinivasan model and compare different optimality criterions. In particular, we are interested in good programs, agreeable and weakly maximal programs.

8.1 Preliminaries

Let R^1 (R^1_+) be the set of real (nonnegative) numbers, and let R^n be the n-dimensional Euclidean space with nonnegative orthant

$$R^n_+ = \{x = (x_1, \ldots, x_n) \in R^n : x_i \geq 0, \ i = 1, \ldots, n\}.$$

For every pair of vectors $x = (x_1, \ldots, x_n)$, $y = (y_1, \ldots, y_n) \in R^n$, define their inner product by

$$xy = \sum_{i=1}^{n} x_i y_i$$

and let $x >> y, x > y, x \geq y$ have their usual meaning.

Let $e(i)$, $i = 1, \ldots, n$ be the ith unit vector in R^n and e be an element of R^n_+ all of whose coordinates are unity. For every $x \in R^n$, denote by $\|x\|$ its Euclidean norm in R^n.

Let $a = (a_1, \ldots, a_n) >> 0, b = (b_1, \ldots, b_n) >> 0, d \in (0, 1)$,

$$c_i = b_i/(1 + da_i), \ i = 1, \ldots, n.$$

© The Author(s), under exclusive license to Springer Nature Switzerland AG 2020
A. J. Zaslavski, *Turnpike Theory for the Robinson–Solow–Srinivasan Model*,
Springer Optimization and Its Applications 166,
https://doi.org/10.1007/978-3-030-60307-6_8

Clearly, there exists $\sigma \in \{1, \ldots, n\}$ such that

$$c_\sigma \geq c_i \text{ for all } i = 1, \ldots, n. \tag{8.1}$$

We may assume without loss of generality that for each $i \in \{1, \ldots, n\}$

$$c_\sigma = c_i \text{ if and only if } i \geq \sigma. \tag{8.2}$$

(Note that in Chapters 2 and 3, we assumed that $\sigma = n$.) The planner's preferences are formalized by a continuous, strictly increasing, concave, and differentiable function $w : [0, \infty) \to R^1$.
Let

$$\Omega = \{(x, x') \in R_+^n \times R_+^n : x' - (1-d)x \geq 0 \text{ and } a(x' - (1-d)x) \leq 1\}, \tag{8.3}$$

for every $(x, x') \in \Omega$,

$$\Lambda(x, x') = \{y \in R_+^n : 0 \leq y \leq x \text{ and } ey \leq 1 - a(x' - (1-d)x)\}. \tag{8.4}$$

and

$$u\left(x, x'\right) = \max\left\{w(by) : y \in \Lambda(x, x')\right\}. \tag{8.5}$$

Recall that a sequence $\{x(t), y(t)\}_{t=0}^{\infty}$ is called a program if for each integer $t \geq 0$

$$(x(t), y(t)) \in R_+^n \times R_+^n, \ x(t+1) \geq (1-d)x(t),$$

$$0 \leq y(t) \leq x(t), \ a(x(t+1) - (1-d)x(t)) + ey(t) \leq 1. \tag{8.6}$$

Let T_1, T_2 be integers such that $0 \leq T_1 < T_2$. A pair of sequences $(\{x(t)\}_{t=T_1}^{T_2}, \{y(t)\}_{t=T_1}^{T_2-1})$ is called a *program* if $x(T_2) \in R_+^n$ and for each integer t satisfying $T_1 \leq t < T_2$ relations (8.6) hold.
For $i = 1, \ldots, n$ set

$$\widehat{q}_i = a_i b_i / (1 + da_i), \ \widehat{p}_i = w'(c_\sigma)\widehat{q}_i. \tag{8.7}$$

We have the following important auxiliary result.

Lemma 8.1 $w(c_\sigma) \geq w(by) + \widehat{p}x' - \widehat{p}x$ *for any* $(x, x') \in \Omega$ *and for any* $y \in \Lambda(x, x')$.

For the proof of this result, see the proof of Lemma 2.2, and note that its proof remains valid without the assumption that $c_\sigma > c_i$ for all $i \in \{1, \ldots, n\} \setminus \{\sigma\}$.
For any $(x, x') \in \Omega$ and any $y \in \Lambda(x, x')$, set

$$\delta(x, y, x') = \widehat{p}(x - x') - (w(by) - w(c_\sigma)). \tag{8.8}$$

By Lemma 8.1,

$$\delta\left(x, y, x'\right) \geq 0 \text{ for each } \left(x, x'\right) \in \Omega \text{ and each } y \in \Lambda\left(x, x'\right). \tag{8.9}$$

It is easy to see that the following lemma holds.

Lemma 8.2 *Let $T > 0$ be an integer and $(\{x(t)\}_{t=0}^{T}, \{y(t)\}_{t=0}^{T-1})$ be a program. Then*

$$\sum_{t=0}^{T-1}(w(by(t)) - w(c_\sigma)) = \widehat{p}(x(0) - x(T)) - \sum_{t=0}^{T-1}\delta(x(t), y(t), x(t+1)).$$

The model considered here is a particular case of the model studied in Chapter 5. Therefore we can use all the results obtained there. A program $\{x(t), y(t)\}_{t=0}^{\infty}$ is called good if there exists $M \in R^1$ such that

$$\sum_{t=0}^{T}(w(by(t)) - w(c_\sigma)) \geq M \text{ for all integers } T \geq 0.$$

A program is called bad if $\lim_{T\to\infty} \sum_{t=0}^{T}(w(by(t)) - w(c_\sigma)) = -\infty$.

By Theorem 5.8, for any initial state $x_0 \in R_+^n$, there exists good program $\{x(t), y(t)\}_{t=0}^{\infty}$ such that $x(0) = x_0$.

Lemmas 5.3 and 8.2 and (8.9) imply the following result.

Proposition 8.3 *A program $\{x(t), y(t)\}_{t=0}^{\infty}$ is good if and only if*

$$\sum_{t=0}^{\infty}\delta(x(t), y(t), x(t+1)) < \infty.$$

A program $\{x(t), y(t)\}_{t=0}^{\infty}$ is bad if and only if

$$\sum_{t=0}^{\infty}\delta(x(t), y(t), x(t+1)) = \infty.$$

Corollary 8.4 *Any program that is not good is bad.*

Let $x_0 \in R_+^n$, and define

$$\Delta(x_0) = \inf\left\{\sum_{t=0}^{\infty}\delta(x(t), y(t), x(t+1))\right\}, \tag{8.10}$$

where the infimum is taken over all programs $\{x(t), y(t)\}_{t=0}^{\infty}$ with $x(0) = x_0$.

By Theorem 5.8 and Proposition 8.3, $\Delta(x_0) < \infty$.

We now conclude this section by the following result.

Proposition 8.5 *Let $x_0 \in R_+^n$. Then there exists a program $\{x(t), y(t)\}_{t=0}^{\infty}$ from x_0 such that*

$$\sum_{t=0}^{\infty} \delta\left(x(t), y(t), x(t+1)\right) = \Delta(x_0).$$

For its proof, see Proposition 2.10.

8.2   Optimality Criteria

A program $\{x^*(t), y^*(t)\}_{t=0}^{\infty}$ is called finitely optimal if for each integer $T > 0$ and each program $\{x(t), y(t)\}_{t=0}^{\infty}$ satisfying $x(0) = x^*(0)$ and $x(t) = x^*(t)$ for all $t \geq T$, the following inequality holds:

$$\sum_{t=0}^{T-1}\left[w(by(t)) - w(by^*(t))\right] \leq 0.$$

A program $\{x^*(t), y^*(t)\}_{t=0}^{\infty}$ is called weakly optimal if for each program

$$\{x(t), y(t)\}_{t=0}^{\infty}$$

satisfying $x(0) = x^*(0)$, the following inequality holds:

$$\liminf_{T \to \infty} \sum_{t=0}^{T}\left[w(by(t)) - w(by^*(t))\right] \leq 0.$$

A program $\{x^*(t), y^*(t)\}_{t=0}^{\infty}$ is called overtaking optimal if

$$\limsup_{T \to \infty} \sum_{t=0}^{T}\left[w(by(t)) - w(by^*(t))\right] \leq 0$$

for every program $\{x(t), y(t)\}_{t=0}^{\infty}$ satisfying $x(0) = x^*(0)$.

A program $\{x^*(t), y^*(t)\}_{t=0}^{\infty}$ is called weakly maximal if for each integer $T > 0$ and each program $(\{x(t)\}_{t=0}^{T}, \{y(t)\}_{t=0}^{T-1})$ satisfying $x(0) = x^*(0)$, $x(T) \geq x^*(T)$, the following inequality holds:

$$\sum_{t=0}^{T-1}\left[w(by(t)) - w(by^*(t))\right] \leq 0.$$

A program $\{x^*(t), y^*(t)\}_{t=0}^{\infty}$ is called agreeable if for all integers $t \geq 0$,

$$u(x^*(t), x^*(t+1)) = w(by^*(t))$$

and if for any natural number T_0 and any $\epsilon > 0$, there exists an integer $T_\epsilon > T_0$ such that for any integer $T \geq T_\epsilon$ and any program $(\{x(t)\}_{t=0}^{T}, \{y(t)\}_{t=0}^{T-1})$ satisfying $x(0) = x^*(0)$, there exists a program $(\{x'(t)\}_{t=0}^{T}, \{y'(t)\}_{t=0}^{T-1})$ such that

$$x'(0) = x(0), \; x'(t) = x^*(t), \; t = 0, \ldots, T_0,$$

$$\sum_{t=0}^{T-1} w(by'(t)) \geq \sum_{t=0}^{T-1} w(by(t)) - \epsilon.$$

For any natural number T and any $x_0 \in R_+^n$, let

$$U(x_0, T) = \sup \left\{ \sum_{t=0}^{T-1} w(by(t)) : (\{x(t)\}_{t=0}^{T}, \{y(t)\}_{t=0}^{T-1}) \text{ is a program from } x_0 \right\}.$$

We follow the convention that the supremum of an empty set is negative infinity.

8.3 Four Theorems

We now present the results which are proved in this section. All of them were obtained in [54].

Theorem 8.6 *Assume that* $\{x(t), y(t)\}_{t=0}^{\infty}$ *is a program such that*

$$\sum_{t=0}^{\infty} \delta(x(t), y(t), x(t+1)) = \Delta(x(0)). \tag{8.11}$$

Then $\{x(t), y(t)\}_{t=0}^{\infty}$ *is a weakly maximal program.*

By Theorem 5.9, if $\{x(t), y(t)\}_{t=0}^{\infty}$ is a weakly maximal program and

$$\limsup_{t \to \infty} \|y(t)\| > 0$$

then $\{x(t), y(t)\}_{t=0}^{\infty}$ is a good program.

Theorem 8.7 *Let* $\{x(t), y(t)\}_{t=0}^{\infty}$ *be weakly optimal. Then it is weakly maximal.*

Theorem 8.8 *Any agreeable program is weakly maximal.*

Theorem 8.9 *Assume that $c_\sigma > c_i$ for all $i \in \{1, \ldots, n\} \setminus \{\sigma\}$ and that for each good program $\{x(t), y(t)\}_{t=0}^\infty$,*

$$\lim_{t \to \infty} (x(t), y(t)) = (\widehat{x}, \widehat{x}),$$

where

$$\widehat{x} = (1/(1 + da_\sigma))e_\sigma.$$

Then for each weakly maximal program $\{x(t), y(t)\}_{t=0}^\infty$ which satisfies

$$\limsup_{t \to \infty} \|y(t)\| > 0$$

the following equality holds:

$$\sum_{t=0}^\infty \delta(x(t), y(t), x(t+1)) = \Delta(x(0)).$$

8.4   Proof of Theorem 8.6

Assume that $\{x(t), y(t)\}_{t=0}^\infty$ is not a weakly maximal program. Then there exist an integer $\tau > 0$ and a program $(\{\bar{x}(t)\}_{t=0}^\tau, \{\bar{y}(t)\}_{t=0}^{\tau-1})$ such that

$$\bar{x}(0) = x(0), \quad \bar{x}(\tau) \geq x(\tau), \tag{8.12}$$

$$\sum_{t=0}^{\tau-1} w(b\bar{y}(t)) > \sum_{t=0}^{\tau-1} w(by(t)). \tag{8.13}$$

Set

$$z = \bar{x}(\tau) - x(\tau). \tag{8.14}$$

Define

$$\bar{y}(t) = y(t) \text{ for all integers } t \geq \tau, \tag{8.15}$$

$$\bar{x}(t) = x(t) + (1 - d)^{t-\tau}z \text{ for all integers } t > \tau. \tag{8.16}$$

It is not difficult to see that $\{\bar{x}(t), \bar{y}(t)\}_{t=0}^\infty$ is a program. By (8.14) and (8.16),

$$\lim_{t \to \infty} (\bar{x}(t) - x(t)) = 0. \tag{8.17}$$

It follows from Lemma 8.2, (8.12), and (8.15) that for each integer $T > \tau$,

$$\sum_{t=0}^{T} \delta(\bar{x}(t), \bar{y}(t), \bar{x}(t+1)) - \sum_{t=0}^{T} \delta(x(t), y(t), x(t+1))$$

$$= \sum_{t=0}^{T} [w(by(t)) - w(b\bar{y}(t))] + \widehat{p}(\bar{x}(0) - \bar{x}(T+1)) - \widehat{p}(x(0) - x(T+1))$$

$$= \sum_{t=0}^{\tau-1} [w(by(t)) - w(b\bar{y}(t))] + \widehat{p}(x(T+1) - \bar{x}(T+1)).$$

Combined with (8.13) and (8.17), this relation implies that

$$\lim_{T \to \infty} \left[\sum_{t=0}^{T} \delta(\bar{x}(t), \bar{y}(t), \bar{x}(t+1)) - \sum_{t=0}^{T} \delta(x(t), y(t), x(t+1)) \right]$$

$$= \sum_{t=0}^{\tau-1} [w(by(t)) - w(b\bar{y}(t))] < 0.$$

In view of this equation and (8.11),

$$\sum_{t=0}^{\infty} \delta(\bar{x}(t), \bar{y}(t), \bar{x}(t+1)) < \sum_{t=0}^{\infty} \delta(x(t), y(t), x(t+1)) = \Delta(x(0)).$$

This contradicts (8.12). The contradiction we have reached proves Theorem 8.6.

8.5 Proof of Theorem 8.7

Assume that the program $\{x(t), y(t)\}_{t=0}^{\infty}$ is not weakly maximal. Then there exist a natural number τ and a program

$$(\{\bar{x}(t)\}_{t=0}^{\tau}, \{\bar{y}(t)\}_{t=0}^{\tau-1})$$

such that

$$\bar{x}(0) = x(0), \quad \bar{x}(\tau) \geq x(\tau), \quad \sum_{t=0}^{\tau-1} w(b\bar{y}(t)) > \sum_{t=0}^{\tau-1} w(by(t)). \tag{8.18}$$

Define for all integers $t \geq \tau$,

$$\bar{y}(t) = y(t), \quad \bar{x}(t+1) = (1-d)\bar{x}(t) + [x(t+1) - (1-d)x(t)]. \quad (8.19)$$

It is not difficult to see that $\{\bar{x}(t), \bar{y}(t)\}_{t=0}^{\infty}$ is a program. By (8.19) for all natural numbers $T > \tau$,

$$\sum_{t=0}^{T} w(b\bar{y}(t)) - \sum_{t=0}^{T} w(by(t)) = \sum_{t=0}^{\tau-1} w(b\bar{y}(t)) - \sum_{t=0}^{\tau-1} w(by(t)).$$

This equality and (8.18) imply that

$$\lim_{T \to \infty} \left[\sum_{t=0}^{T} w(b\bar{y}(t)) - \sum_{t=0}^{T} w(by(t)) \right]$$

$$= \sum_{t=0}^{\tau-1} w(b\bar{y}(t)) - \sum_{t=0}^{\tau-1} w(by(t)) > 0.$$

Thus the program $\{x(t), y(t)\}_{t=0}^{\infty}$ is not weakly optimal. The contradiction we have reached proves the theorem.

8.6   Proof of Theorem 8.8

Assume the contrary. Then there is an agreeable program $\{x^*(t), y^*(t)\}_{t=0}^{\infty}$ which is not weakly maximal. Then there exist a natural number T_0 and a program $(\{\bar{x}(t)\}_{t=0}^{T_0}, \{\bar{y}(t)\}_{t=0}^{T_0-1})$ such that

$$\bar{x}(0) = x^*(0), \quad \bar{x}(T_0) \geq x^*(T_0), \quad (8.20)$$

$$\sum_{t=0}^{T_0-1} w(b\bar{y}(t)) > \sum_{t=0}^{T_0-1} w(by^*(t)) = \sum_{t=0}^{T_0-1} u(x^*(t), x^*(t+1)). \quad (8.21)$$

Put

$$\epsilon = 4^{-1} \left[\sum_{t=0}^{T_0-1} w(b\bar{y}(t)) - \sum_{t=0}^{T_0-1} w(by^*(t)) \right]. \quad (8.22)$$

Let a natural number $T_\epsilon > T_0$ be as guaranteed by the definition of an agreeable program. Clearly, there exists a program $(\{x(t)\}_{t=0}^{T_\epsilon}, \{y(t)\}_{t=0}^{T_\epsilon-1})$ such that

$$x(0) = x^*(0), \quad \sum_{t=0}^{T_\epsilon - 1} w(by(t)) = U(x^*(0), T_\epsilon). \tag{8.23}$$

By the choice of T_ϵ, (8.23) and the agreeability of $\{x^*(t), y^*(t)\}_{t=0}^\infty$, there is a program $(\{x'(t)\}_{t=0}^{T_\epsilon}, \{y'(t)\}_{t=0}^{T_\epsilon - 1})$ such that

$$x'(0) = x(0), \quad x'(t) = x^*(t), \quad t = 0, \ldots, T_0, \tag{8.24}$$

$$\sum_{t=0}^{T_\epsilon - 1} w(by'(t)) \geq \sum_{t=0}^{T_q - 1} w(by(t)) - \epsilon. \tag{8.25}$$

For all integers t satisfying $T_0 \leq t < T_\epsilon$, set

$$\bar{y}(t) = y'(t), \quad \bar{x}(t+1) = (1-d)\bar{x}(t) + x'(t+1) - (1-d)x'(t). \tag{8.26}$$

By (8.20), (8.24), and (8.26),

$$\bar{x}(t) \geq x'(t)), \quad t = T_0, \ldots, T_\epsilon$$

and $(\{\bar{x}(t)\}_{t=0}^{T_\epsilon}, \{\bar{y}(t)\}_{t=0}^{T_\epsilon - 1})$ is a program.
By (8.20)–(8.26),

$$U(x^*(0), T_\epsilon) \geq \sum_{t=0}^{T_\epsilon - 1} w(b\bar{y}(t)) = \sum_{t=0}^{T_0 - 1} w(b\bar{y}(t)) + \sum_{t=T_0}^{T_\epsilon - 1} w(b\bar{y}(t))$$

$$= \sum_{t=0}^{T_0 - 1} w(by^*(t)) + 4\epsilon + \sum_{t=T_0}^{T_\epsilon - 1} w(by'(t))$$

$$= \sum_{t=0}^{T_0 - 1} u(x^*(t), x^*(t+1)) + 4\epsilon + \sum_{t=T_0}^{T_\epsilon - 1} w(by'(t))$$

$$\geq \sum_{t=0}^{T_0 - 1} w(by'(t)) + \sum_{t=T_0}^{T_\epsilon - 1} w(by'(t)) + 4\epsilon$$

$$\geq \sum_{t=0}^{T_\epsilon - 1} w(by(t)) + 3\epsilon = U(x^*(0), T_\epsilon) + 3\epsilon,$$

a contradiction. The contradiction we have reached proves Theorem 8.8.

8.7 Proof of Theorem 8.9

Let $\{x(t), (t)\}_{t=0}^{\infty}$ be a weakly maximal program such that

$$\limsup_{t\to\infty} \|y(t)\| > 0. \tag{8.27}$$

We will show that

$$\sum_{t=0}^{\infty} \delta(x(t), y(t), x(t+1)) = \Delta(x(0)).$$

Let us assume the contrary. Then

$$\sum_{t=0}^{\infty} \delta(x(t), y(t), x(t+1)) > \Delta(x(0)). \tag{8.28}$$

By (8.27), the program $\{x(t), y(t)\}_{t=0}^{\infty}$ is good. Therefore

$$\lim_{t\to\infty} (x(t), y(t)) = (\widehat{x}, \widehat{x}). \tag{8.29}$$

In view of Proposition 8.5, there is a good program $\{\bar{x}(t), \bar{y}(t)\}_{t=0}^{\infty}$ such that

$$\bar{x}(0) = x(0), \ \sum_{t=0}^{\infty} \delta(\bar{x}(t), \bar{y}(t), \bar{x}(t+1)) = \Delta(x(0)). \tag{8.30}$$

Choose a positive number ϵ such that

$$8\epsilon < \sum_{t=0}^{\infty} \delta(x(t), y(t), x(t+1)) - \Delta(x(0)). \tag{8.31}$$

By Proposition 3.13 and the continuity of the function $\delta(\cdot, \cdot, \cdot)$, there exists

$$\delta \in (0, (1+\|\widehat{p}\|)^{-1}\epsilon 8^{-1})$$

such that the following property holds:
   (P1) For each $x, x' \in R_+^n$ satisfying

$$\|x - \widehat{x}\|, \ \|x' - \widehat{x}\| \le \delta$$

there exist $\bar{x} \ge x', y \in R_+^n$ such that

$$(x, \bar{x}) \in \Omega, \ y \in \Lambda(x, \bar{x}),$$

$$\|y - \widehat{x}\| \le (4(1 + \|\widehat{p}\|))^{-1}\epsilon, \ \|\bar{x} - \widehat{x}\| \le (4(1 + \|\widehat{p}\|))^{-1}\epsilon, \ \delta(x, y, \bar{x}) \le \epsilon/8.$$

Since $\{\bar{x}(t), \bar{y}(t)\}_{t=0}^{\infty}$ is a good program

$$\lim_{t \to \infty} (\bar{x}(t), \bar{y}(t)) = (\widehat{x}, \widehat{x}). \tag{8.32}$$

It follows from (8.29) and (8.32) that there is an integer $T_0 > 2$ such that

$$\|\bar{x}(t) - \widehat{x}\|, \ \|x(t) - \widehat{x}\| < \delta \text{ for all integers } t \ge T_0. \tag{8.33}$$

By (8.30) and (8.31), there exists an integer $\tau > T_0$ such that

$$\sum_{t=0}^{\tau-1} \delta(x(t), y(t), x(t+1)) - \sum_{t=0}^{\tau-1} \delta(\bar{x}(t), \bar{y}(t), \bar{x}(t+1))$$

$$\ge 2^{-1}\left[\sum_{t=0}^{\infty} \delta(x(t), y(t), x(t+1)) - \Delta(x(0))\right] > 4\epsilon. \tag{8.34}$$

In view of (8.33) and (P1), there exist

$$\bar{y} \in R_+^n, \ \bar{x} \ge x(\tau + 1) \tag{8.35}$$

such that

$$(\bar{x}(\tau), \bar{x}) \in \Omega, \ \bar{y} \in \Lambda(\bar{x}(\tau), \bar{x}),$$

$$\delta(\bar{x}(\tau), \bar{y}, \bar{x}) \le \epsilon/8,$$

$$\|\bar{x} - \widehat{x}\| \le \epsilon(4(1 + \|\widehat{p}\|))^{-1}. \tag{8.36}$$

Define

$$\tilde{x}(t) = \bar{x}(t), \ t = 0, \dots, \tau, \ \tilde{x}(\tau + 1) = \bar{x},$$

$$\tilde{y}(t) = \bar{y}(t), \ t = 0, \dots, \tau - 1, \ \tilde{y}(\tau) = \bar{y}. \tag{8.37}$$

It is not difficult to see that

$$(\{\tilde{x}(t)\}_{t=0}^{\tau+1}, \ \{\tilde{y}(t)\}_{t=0}^{\tau})$$

is a program. By (8.27) and (8.30),

$$\tilde{x}(0) = \bar{x}(0) = x(0). \tag{8.38}$$

Relations (8.35) and (8.27) imply that

$$\tilde{x}(\tau + 1) \geq x(\tau + 1). \tag{8.39}$$

It follows from (8.31), (8.34), and (8.36)–(8.38) that

$$\sum_{t=0}^{\tau} \delta(x(t), y(t), x(t+1)) - \sum_{t=0}^{\tau} \delta(\tilde{x}(t), \tilde{y}(t), \tilde{x}(t+1))$$

$$\geq \sum_{t=0}^{\tau-1} \delta(x(t), y(t), x(t+1)) - \sum_{t=0}^{\tau-1} \delta(\tilde{x}(t), \tilde{y}(t), \tilde{x}(t+1))$$

$$- \delta(\tilde{x}(\tau), \tilde{y}(\tau), \tilde{x}(\tau + 1))$$

$$\geq 2^{-1} \left[\sum_{t=0}^{\infty} \delta(x(t), y(t), x(t+1)) - \Delta x(0) \right] - \delta(\tilde{x}(\tau), \tilde{y}, \tilde{x})$$

$$\geq 2^{-1} \left[\sum_{t=0}^{\infty} \delta(x(t), y(t), x(t+1)) - \Delta(x(0)) \right] - \epsilon/8$$

$$\geq 4^{-1} \left[\sum_{t=0}^{\infty} \delta(x(t), y(t), x(t+1)) - \Delta(x(0)) \right]. \tag{8.40}$$

By Lemma 8.2, (8.31), (8.33), (8.36)–(8.38), (8.40), and the choice of δ,

$$\sum_{t=0}^{\tau} [w(by(t)) - w(b\tilde{y}(t))]$$

$$= \widehat{p}(x(0) - x(\tau + 1)) - \sum_{t=0}^{\tau} \delta(x(t), y(t), x(t+1))$$

$$- \left[\widehat{p}(\tilde{x}(0) - \tilde{x}(\tau + 1)) - \sum_{t=0}^{\tau} \delta(\tilde{x}(t), \tilde{y}(t), \tilde{x}(t+1)) \right]$$

$$\leq -4^{-1} \left[\sum_{t=0}^{\infty} \delta(x(t), y(t), x(t+1)) - \Delta(x(0)) \right] + \widehat{p} \left[\tilde{x}(\tau + 1) - x(\tau + 1) \right]$$

$$\leq -4^{-1} \left[\sum_{t=0}^{\infty} \delta(x(t), y(t), x(t+1)) - \Delta(x(0)) \right]$$

$$+ \|\widehat{p}\| \|\tilde{x} - \widehat{x}\| + \|\widehat{p}\| \|\widehat{x} - x(\tau + 1)\|$$

$$\leq -4^{-1} \left[\sum_{t=0}^{\infty} \delta(x(t), y(t), x(t+1)) - \Delta(x(0)) \right] + \epsilon/8 + \epsilon/4 \leq -2\epsilon + \epsilon < 0.$$

By the relation above, (8.38) and (8.39), $\{x(t), y(t)\}_{t=0}^{\infty}$ is not weakly maximal. The contradiction we have reached proves the theorem.

8.8 Maximal Programs

The following optimality criterion was introduced in [10].

A program $\{x(t), y(t)\}_{t=0}^{\infty}$ is called maximal if there exist no program

$$\left\{ x'(t), y'(t) \right\}_{t=0}^{\infty}$$

and a natural number S such that

$$x'(0) = x(0), \quad \sum_{t=0}^{S-1} [w(by'(t)) - w(by(t))] > 0,$$

$$w(by'(t)) \geq w(by(t)) \text{ for all integers } t \geq S+1.$$

Clearly any overtaking optimal program is maximal.

We assume that

$$\{i \in \{1, \ldots, n\} : c_\sigma = c_u\} = \{\sigma\}. \tag{8.41}$$

Set

$$\widehat{x} = (1 + da_\sigma)^{-1} e_\sigma. \tag{8.42}$$

We prove the following result which was obtained in [115].

Theorem 8.10 *Assume that for each good program* $\{x(t), y(t)\}_{t=0}^{\infty}$,

$$\lim_{t \to \infty} (x(t), y(t)) = (\widehat{x}, \widehat{x}).$$

Then any maximal program is overtaking optimal.

For its proof we use the following result which follows from Theorems 3.9 and 8.9.

Proposition 8.11 *Assume that for each good program* $\{u(t), v(t)\}_{t=0}^{\infty}$,

$$\lim_{t \to \infty} (u(t), v(t)) = (\widehat{x}, \widehat{x}).$$

Let a program $\{x(t), y(t)\}_{t=0}^{\infty}$ *be weakly maximal and satisfy*

$$\limsup_{t \to \infty} \|y(t)\| > 0.$$

Then the program $\{x(t), y(t)\}_{t=0}^{\infty}$ *is overtaking optimal.*

Proof of Theorem 8.10 Assume that a program $\{x(t), y(t)\}_{t=0}^{\infty}$ is maximal. We show that it is overtaking optimal.

Clearly, $\{x(t), y(t)\}_{t=0}^{\infty}$ is weakly maximal. By Proposition 8.11, in order to complete the proof, it is sufficient to show that

$$\limsup_{t \to \infty} \|y(t)\| > 0,$$

Assume the contrary. Then

$$\lim_{t \to \infty} \|y(t)\| = 0. \tag{8.43}$$

There exists a good program $\{\bar{x}(t), \bar{y}(t)\}_{t=0}^{\infty}$ such that

$$\bar{x}(0) = x(0). \tag{8.44}$$

Clearly,

$$\lim_{t \to \infty} \bar{y}(t) = \widehat{x}. \tag{8.45}$$

By (8.43) and (8.45), there are $\epsilon > 0$ and a natural number T_0 such that

$$w(b\bar{y}(t)) \ge w(by(t)) + \epsilon \text{ for all integers } t \ge T_0. \tag{8.46}$$

By (8.46), there is a natural number $T_1 > T_0$ such that

$$\sum_{t=0}^{T_1-1} w(b\bar{y}(t)) > \sum_{t=0}^{T_1-1} w(by(t)).$$

Together with (8.44) and (8.46), this implies that $\{x(t), y(t)\}_{t=0}^{\infty}$ is not maximal, a contradiction. The contradiction we have reached proves the theorem.

8.9 One-Dimensional Model

In this section we consider a particular case of the model considered in Section 8.1 with $n = 1$.

Let $a > 0$ and $d \in (0, 1]$.

Recall that a sequence $\{x(t), y(t)\}_{t=0}^{\infty}$ is called a program if for each integer $t \geq 0$

$$(x(t), y(t)) \in R_+^1 \times R_+^1,$$

$$x(t + 1) \geq (1 - d)x(t),$$

$$0 \leq y(t) \leq x(t),$$

$$a(x(t + 1) - (1 - d)x(t)) + y(t) \leq 1. \qquad (8.47)$$

Let T_1, T_2 be integers such that $0 \leq T_1 < T_2$. A pair of sequences

$$\left(\{x(t)\}_{t=T_1}^{T_2}, \{y(t)\}_{t=T_1}^{T_2-1} \right)$$

is called a program if $x(T_2) \in R_+^1$ and if for each integer t satisfying $T_1 \leq t < T_2$ relations (8.47) hold.

Let $w : [0, \infty) \to [0, \infty)$ be a continuous strictly increasing function which represents the preferences of the planner. Note that we do not assume the concavity of w.

In the sequel we assume that supremum of empty set is $-\infty$.

Let $x_0, \tilde{x}_0 \in R_+^1$ and let T be a natural number. Set

$$U(x_0, \tilde{x}_0, T) = \sup\{\sum_{t=0}^{T-1} w(y(t)) : (\{x(t)\}_{t=0}^{T}, \{y(t)\}_{t=0}^{T-1}) \qquad (8.48)$$

$$\text{is a program such that } x(0) = x_0, \ x(T) \geq \tilde{x}_0\}.$$

A program $\{x(t), y(t)\}_{t=0}^{\infty}$ is called weakly maximal if for all integers $T > 0$,

$$\sum_{t=0}^{T-1} w(y(t)) = U(x(0), x(T), T). \qquad (8.49)$$

Theorems 5.8 and 5.9 and relation (5.8) imply the following result.

Theorem 8.12 *For any $x_0 \in R_+^1$ there exists a weakly maximal program*

$$\{x(t), y(t)\}_{t=0}^{\infty}$$

such that

$$x(0) = x_0, \ \limsup_{t \to \infty} y(t) > 0$$

and

$$\liminf_{T \to \infty} T^{-1} \sum_{t=0}^{T-1} w(y(t)) > w(0).$$

Recall that a program $\{x_*(t), y_*(t)\}_{t=0}^{\infty}$ is called maximal if there is not a program

$$\{x(t), y(t)\}_{t=0}^{\infty}$$

for which there exists a natural number s such that

$$x(0) = x_*(0),$$

$$\sum_{t=0}^{s-1} w(y(t)) > \sum_{t=0}^{s-1} w(y_*(t))$$

and $w(y(t)) \geq w(y_*(t))$ for all integers $t \geq s$.

We prove the following theorem obtained in [105].

Theorem 8.13 *A program $\{x(t), y(t)\}_{t=0}^{\infty}$ is weakly maximal and satisfies*

$$\limsup_{t \to \infty} y(t) > 0 \qquad (8.50)$$

if and only if it is maximal.

Proof It is not difficult to see that if the program $\{x(t), y(t)\}_{t=0}^{\infty}$ is maximal, then it is weakly maximal and satisfies (8.50).

Assume that the program $\{x(t), y(t)\}_{t=0}^{\infty}$ is weakly maximal and satisfies (8.50). We show that it is maximal.

Assume the contrary. Then there exist a program $\{\bar{x}(t), \bar{y}(t)\}_{t=0}^{\infty}$ and an integer $s > 1$ such that

$$\bar{x}(0) = x(0), \qquad (8.51)$$

$$\sum_{t=0}^{s-1} w(\bar{y}(t)) > \sum_{t=0}^{s-1} w(y(t)) \qquad (8.52)$$

and

$$w(\bar{y}(t)) \geq w(y(t)) \text{ for all integers } t \geq s. \qquad (8.53)$$

Since the program $\{x(t), y(t)\}_{t=0}^{\infty}$ is weakly maximal, Equations (8.51)–(8.53) imply that

$$x(t) > \bar{x}(t) \text{ for all integers } t \geq s. \tag{8.54}$$

Since the function w is strictly increasing, Equation (8.53) implies that

$$\bar{y}(t) \geq y(t) \text{ for all integers } t \geq s. \tag{8.55}$$

We define a program $\{x^*(t), y^*(t)\}_{t=0}^{\infty}$. Set

$$x^*(t) = \bar{x}(t), \ t = 0, \ldots, s, \tag{8.56}$$

$$y^*(t) = \bar{y}(t), \ t = 0, \ldots, s-1$$

and for all integers $t \geq s$, put

$$y^*(t) = y(t), \tag{8.57}$$

$$x^*(t+1) = (1-d)x^*(t) + a^{-1}(1 - y^*(t)).$$

We show that $\{x^*(t), y^*(t)\}_{t=0}^{\infty}$ is a program. By (8.57) $y^*(t) \in [0, 1]$ for all integers $t \geq s$ and

$$x^*(t+1) \geq (1-d)x^*(t) \text{ for all integers } t \geq s. \tag{8.58}$$

This implies that

$$x^*(t) \geq 0 \text{ for all integers } t \geq s. \tag{8.59}$$

In view of (8.57) for all integers $t \geq s$,

$$a(x^*(t+1) - (1-d)x^*(t)) + y^*(t) = 1. \tag{8.60}$$

By definition, for all integers $t \geq 0$

$$x(t+1) \leq (1-d)x(t) + a^{-1}(1 - y(t)) \tag{8.61}$$

and

$$\bar{x}(t+1) \leq (1-d)\bar{x}(t) + a^{-1}(1 - \bar{y}(t)).$$

It follows from (8.57), (8.55), and (8.56)

$$y^*(s) = y(s) \leq \bar{y}(s) \leq \bar{x}(s) = x^*(s). \tag{8.62}$$

By induction we show that for all integers $t \geq s$,

$$x^*(t) \geq \bar{x}(t). \tag{8.63}$$

It follows from (8.56) that (8.63) is true for $t = s$.

Assume that an integer $t \geq s$ and (8.63) holds. By (8.57), (8.63), (8.55), and (8.61),

$$x^*(t + 1) \geq (1 - d)\bar{x}(t) + a^{-1}(1 - y(t))$$

$$\geq (1 - d)\bar{x}(t) + a^{-1}(1 - \bar{y}(t)) \geq \bar{x}(t + 1).$$

Therefore we have shown by induction that (8.63) holds for all integers $t \geq s$.

It follows from (8.55), (8.57), and (8.63) that for all integers $t \geq s$

$$y^*(t) = y(t) \leq \bar{y}(t) \leq \bar{x}(t) \leq x^*(t). \tag{8.64}$$

By (8.56)–(8.60) and (8.64), imply that $\{x^*(t), y^*(t)\}_{t=0}^{\infty}$ is a program.

In view of (8.61) and (8.57), for all integers $t \geq s$,

$$x^*(t + 1) - x(t + 1)$$

$$= (1 - d)x^*(t) + a^{-1}(1 - y^*(t)) - x(t + 1)$$

$$\geq (1 - d)x^*(t) + a^{-1}(1 - y(t)) - (1 - d)x(t) - a^{-1}(1 - y(t))$$

$$= (1 - d)(x^*(t) - x(t)).$$

This implies that

$$\liminf_{t \to \infty}(x^*(t) - x(t)) \geq 0. \tag{8.65}$$

Put

$$\Delta := \sum_{t=0}^{s-1} w(\bar{y}(t)) - \sum_{t=0}^{s-1} w(y(t)). \tag{8.66}$$

In view of (8.66) and (8.52),

$$\Delta > 0. \tag{8.67}$$

By Lemma 5.3, there is $M_0 > 0$ such that

$$x(t), \; \bar{x}(t), \; x^*(t) \leq M_0 \text{ for all integers } t \geq 0. \tag{8.68}$$

By (8.64) there is $\epsilon_0 > 0$ such that

$$\limsup_{t\to\infty} y(t) > 4\epsilon_0. \tag{8.69}$$

There is $\epsilon_1 > 0$ such that

$$|w(z_1) - w(z_2)| \le \Delta/16 \text{ for all } z_1, z_2 \in [0, 1] \tag{8.70}$$

$$\text{such that } |z_1 - z_2| \le 2\epsilon_1.$$

Put

$$\epsilon = \min\{\epsilon_0, \epsilon_1\}(1+a)^{-1}. \tag{8.71}$$

By (8.65) there is a natural number $p_0 > s + 2$ such that

$$x(t) - x^*(t) \le \epsilon/2 \text{ for all integers } t \ge p_0. \tag{8.72}$$

By (8.69), there is an integer $p > p_0 + 2$ such that

$$y(p) > 4\epsilon_0. \tag{8.73}$$

Put

$$\tilde{x}(t) = x^*(t), \ t = 0, \dots, p,$$
$$\tilde{y}(t) = y^*(t), \ t = 0, \dots, p - 1,$$
$$\tilde{x}(p+1) = x^*(p+1) + \epsilon, \ \tilde{y}(p) = y^*(p) - a\epsilon. \tag{8.74}$$

By (8.74),

$$\tilde{x}(p+1) - (1-d)\tilde{x}(p)$$
$$= x^*(p+1) + \epsilon - (1-d)x^*(p) \ge \epsilon. \tag{8.75}$$

It follows from (8.74), (8.57), (8.73), and (8.71) that

$$\tilde{y}(p) \ge 4\epsilon_0 - a\epsilon > 0. \tag{8.76}$$

By (8.74),

$$a(\tilde{x}(p+1) - (1-d)\tilde{x}(p)) + \tilde{y}(p)$$
$$\le a(x^*(p+1) - (1-d)x^*(p)) + a\epsilon + y^*(p) - a\epsilon \le 1. \tag{8.77}$$

In view of (8.74)–(8.77), $(\{\tilde{x}(t)\}_{t=0}^{p+1}, \{\tilde{y}(t)\}_{t=0}^{p})$ is a program. By (8.74) and (8.72),

$$\tilde{x}(p+1) = x^*(p+1) + \epsilon \geq x(p+1) - \epsilon/2 + \epsilon. \tag{8.78}$$

Equations (8.74), (8.56) and (8.51) imply that

$$\tilde{x}(0) = x^*(0) = x(0). \tag{8.79}$$

By (8.74), (8.76), (8.70), (8.71), and (8.66),

$$|w(\tilde{y}(p)) - w(y^*(p))| \leq \Delta/16. \tag{8.80}$$

It follows from (8.52), (8.56), (8.57), (8.66), (8.67), (8.74), and (8.80) that

$$\sum_{t=0}^{p} w(\tilde{y}(t)) - \sum_{t=0}^{p} w(y(t))$$

$$= \sum_{t=0}^{p-1} w(y^*(t)) - \sum_{t=0}^{p-1} w(y(t)) + w(\tilde{y}(p)) - w(y^*(p))$$

$$\geq \sum_{t=0}^{p-1} w(y^*(t)) - \sum_{t=0}^{p-1} w(y(t)) - \Delta/16$$

$$= \sum_{t=0}^{s-1} w(y^*(t)) - \sum_{t=0}^{s-1} w(y(t)) - \Delta/16$$

$$= \sum_{t=0}^{s-1} w(\tilde{y}(t)) - \sum_{t=0}^{s-1} w(y(t)) - \Delta/16 > \Delta/2 > 0. \tag{8.81}$$

Equations (8.78), (8.79), and (8.81) contradict the weak maximality of the program $\{x(t), y(t)\}_{t=0}^{\infty}$. The contradiction we have reached proves the theorem.

Chapter 9
Turnpike Phenomenon for the RSS Model with Nonconcave Utility Functions

In this chapter we study the turnpike properties for the Robinson–Solow–Srinivasan model. To have these properties means that the approximate solutions of the problems are essentially independent of the choice of an interval and endpoint conditions. The utility functions, which determine the optimality criterion, are nonconcave. We show that the turnpike properties hold and that they are stable under perturbations of an objective function. Moreover, we consider a class of RSS models which is identified with a complete metric space of utility functions. Using the Baire category approach, we show that the turnpike phenomenon holds for most of the models. All the results of this chapter are new.

9.1 Preliminaries and Main Results

Let R^1 (R^1_+) be the set of real (nonnegative) numbers, and let R^n be the n-dimensional Euclidean space with nonnegative orthant

$$R^n_+ = \{x = (x_1, \ldots, x_n) \in R^n : x_i \geq 0, \ i = 1, \ldots, n\}.$$

For every pair of vectors $x = (x_1, \ldots, x_n)$, $y = (y_1, \ldots, y_n) \in R^n$, define their inner product by

$$xy = \sum_{i=1}^{n} x_i y_i$$

and let $x >> y$, $x > y$, $x \geq y$ have their usual meaning.

Let $e(i)$, $i = 1, \ldots, n$ be the ith unit vector in R^n and e be an element of R^n_+ all of whose coordinates are unity. For every $x \in R^n$, denote by $\|x\|$ its Euclidean

© The Author(s), under exclusive license to Springer Nature Switzerland AG 2020
A. J. Zaslavski, *Turnpike Theory for the Robinson–Solow–Srinivasan Model*,
Springer Optimization and Its Applications 166,
https://doi.org/10.1007/978-3-030-60307-6_9

norm in R^n. For every $x \in R^n$ set

$$\|x\|_\infty = \max\{|x_i| : i = 1, \ldots, n\}.$$

Let $a = (a_1, \ldots, a_n) >> 0$, $b = (b_1, \ldots, b_n) >> 0$ be vectors of R^n, $d \in (0, 1)$,

$$c_i = b_i/(1 + da_i), \quad i = 1, \ldots, n.$$

We assume the following:

There exists $\sigma \in \{1, \ldots, n\}$ such that for all

$$i \in \{1, \ldots, n\} \setminus \{\sigma\}, \ c_\sigma > c_i. \tag{9.1}$$

Recall that a sequence $\{x(t), y(t)\}_{t=0}^\infty$ is called a program if for each integer $t \geq 0$

$$(x(t), y(t)) \in R_+^n \times R_+^n, \ x(t+1) \geq (1 - d)x(t),$$

$$0 \leq y(t) \leq x(t), \ a(x(t+1) - (1-d)x(t)) + ey(t) \leq 1. \tag{9.2}$$

Let T_1, T_2 be integers such that $0 \leq T_1 < T_2$. A pair of sequences

$$\left(\{x(t)\}_{t=T_1}^{T_2}, \{y(t)\}_{t=T_1}^{T_2-1}\right)$$

is called a program if $x(T_2) \in R_+^n$ and for each integer t satisfying $T_1 \leq t < T_2$ relations (9.2) are valid.

Define

$$\Omega = \{(x, x') \in R_+^n \times R_+^n : x' - (1-d)x \geq 0$$

$$\text{and } a(x' - (1-d)x) \leq 1\} \tag{9.3}$$

and a correspondence $\Lambda : \Omega \to R_+^n$ given by

$$\Lambda(x, x') = \{y \in R_+^n : 0 \leq y \leq x \text{ and } ey \leq 1 - a(x' - (1-d)x)\}, \ (x, x') \in \Omega. \tag{9.4}$$

Assume that $w : [0, \infty) \to R^1$ is a continuous strictly increasing function which represents the preferences of the planner.

For every $(x, x') \in \Omega$ set

$$u(x, x') = \max\{w(by) : y \in \Lambda(x, x')\}. \tag{9.5}$$

For each $x_0 \in R_+^n$ and each integer $T \geq 1$, set

$$U(x_0, T) = \sup \left\{ \sum_{t=0}^{T-1} w(by(t)) : \left(\{x(t)\}_{t=0}^{T}, \{y(t)\}_{t=0}^{T-1} \right) \right.$$

$$\left. \text{is a program such that } x(0) = x_0 \right\}. \tag{9.6}$$

Let $x_0, \tilde{x}_0 \in R_+^n$ and let T be a natural number. Define

$$U(x_0, \tilde{x}_0, T) = \sup \left\{ \sum_{t=0}^{T-1} w(by(t)) : \left(\{x(t)\}_{t=0}^{T}, \{y(t)\}_{t=0}^{T-1} \right) \right.$$

$$\left. \text{is a program such that } x(0) = x_0, \ x(T) \geq \tilde{x}_0 \right\}. \tag{9.7}$$

(Here we suppose that a supremum over empty set is $-\infty$.)

Let M_0 be a positive number and let $T \geq 1$ be an integer. Set

$$\widehat{U}(M_0, T) = \sup \left\{ \sum_{t=0}^{T-1} w(by(t)) : \right.$$

$$\left. \left(\{x(t)\}_{t=0}^{T}, \{y(t)\}_{t=0}^{T-1} \right) \text{ is a program such that } x(0) \leq M_0 e \right\}. \tag{9.8}$$

By Theorem 5.4, for each pair of numbers

$$M_1, M_2 > \max\{(da_i)^{-1} : i = 1, \ldots, n\}$$

we have

$$\mu := \mu(w) := \lim_{p \to \infty} \widehat{U}(M_i, p)/p, \ i = 1, 2. \tag{9.9}$$

Let $x \in R_+^n$ satisfy $(x, x) \in \Omega$. Then

$$x \geq (1 - d)x, \ a(x - (1 - d)x) \leq 1,$$

$$d \sum_{i=1}^{n} a_i x_i = dax \leq 1$$

and

$$x_i \leq (da_i)^{-1}, \ i = 1, \ldots, n.$$

Thus

$$\{x \in R_+^n : \ (x, x) \in \Omega\} \subset \{x \in R_+^n : \ x_i \leq (da_i)^{-1}, \ i = 1, \ldots, n\}. \qquad (9.10)$$

Clearly,

$$\sup\{w(by) : \ y \in \Lambda(x, x), \ x \in R_+^n, \ (x, x) \in \Omega\} \leq \mu(w). \qquad (9.11)$$

Suppose that

$$\sup\{w(by) : \ y \in \Lambda(x, x), \ x \in R_+^n, \ (x, x) \in \Omega\} = \mu(w). \qquad (9.12)$$

By (9.12), there exist $x(w), y(w) \in R_+^n$

$$(x(w), x(w)) \in \Omega, \ y(w) \in \Lambda(x(w), x(w))$$

such that

$$w(by(w)) = \mu(w).$$

Since the function w is strictly monotone, it is not difficult to see that the following lemma holds.

Lemma 9.1

$$\{y \in \cup\{\Lambda(x, x) : \ x \in R_+^n, \ (x, x) \in \Omega\}, \ w(by) = \mu(w)\}$$

is the set of all

$$y \in \cup\{\Lambda(x, x) : \ x \in R_+^n, \ (x, x) \in \Omega\}$$

such that

$$by \geq bz$$

for all

$$z \in \cup\{\Lambda(x, x) : \ x \in R_+^n, \ (x, x) \in \Omega\}\}.$$

Lemma 9.1 and the results of Chapter 2, applied with $w(t) = t, t \in R_+^1$, imply the following theorem.

Theorem 9.2

$$\{(x, y) \in R^n_+ \times R^n_+ : (x, x) \in \Omega, \ y \in \Lambda(x, x), \ w(by) = \mu(w)\}$$
$$= \{((1 + da_\sigma)^{-1}e(\sigma), (1 + da_\sigma)^{-1}e(\sigma))\}.$$

Set

$$\widehat{x} = (1 + da_\sigma)^{-1}e(\sigma), \ \widehat{y} = \widehat{x}. \tag{9.13}$$

Recall that a program $\{x(t), y(t)\}^\infty_{t=0}$ is (w)-good (or good if the function w is understood) if the sequence

$$\left\{ \sum_{t=0}^{T-1} [w(by(t)) - \mu] \right\}^\infty_{T=1}$$

is bounded. A program $\{x(t), y(t)\}^\infty_{t=0}$ is (w)-bad (or bad if the function w is understood) if

$$\lim_{T \to \infty} \sum_{t=0}^{T-1} [w(by(t)) - \mu] = -\infty.$$

By Proposition 5.7, any program that is not good is bad.

In this chapter we prove the following results.

Theorem 9.3 *Assume that for each good program* $\{x(t), y(t)\}^\infty_{t=0}$,

$$\lim_{t \to \infty} (x(t), y(t)) = (\widehat{x}, \widehat{x}). \tag{9.14}$$

Let

$$M > \max\{(a_i d)^{-1} : i = 1, \ldots, n\},$$

ϵ *be positive number and* $\Gamma \in (0, 1)$. *Then there exist a natural number* L *and a positive number* γ *such that for each integer* $T > 2L + 1$, *each* $z_0, z_1 \in R^n_+$ *satisfying* $z_0 \leq Me$ *and* $az_1 \leq \Gamma d^{-1}$, *and each program* $(\{x(t)\}^T_{t=0}, \{y(t)\}^{T-1}_{t=0})$ *which satisfies*

$$x(0) = z_0, \ x(T) \geq z_1, \ \sum_{t=0}^{T-1} w(by(t)) \geq U(z_0, z_1, T) - \gamma,$$

there are integers τ_1, τ_2 *such that*

$$\tau_1 \in [0, L], \ \tau_2 \in [T - L, T],$$

$$\|x(t) - \widehat{x}\|, \ \|y(t) - \widehat{x}\| \le \epsilon \text{ for all } t = \tau_1, \ldots, \tau_2 - 1 \text{ and } \|x(\tau_2) - \widehat{x}\| \le \epsilon.$$

Moreover if $\|x(0) - \widehat{x}\| \le \gamma$, *then* $\tau_1 = 0$, *and if* $\|x(T) - \widehat{x}\| \le \gamma$, *then* $\tau_2 = T$.

Theorem 9.4 *Assume that for each good program* $\{x(t), y(t)\}_{t=0}^{\infty}$, *Equation (9.14) holds. Let*

$$M > \max\{(a_i d)^{-1} : \ i = 1, \ldots, n\}$$

and ϵ *be a positive number. Then there exist a natural number* L *and a positive number* γ *such that for each integer* $T > 2L + 1$, *each* $z_0 \in R_+^n$ *satisfying* $z_0 \le Me$, *and each program* $\left(\{x(t)\}_{t=0}^T, \{y(t)\}_{t=0}^{T-1}\right)$ *which satisfies*

$$x(0) = z_0, \quad \sum_{t=0}^{T-1} w(by(t)) \ge U(z_0, T) - \gamma,$$

there are integers τ_1, τ_2 *such that*

$$\tau_1 \in [0, L], \ \tau_2 \in [T - L, T],$$

$$\|x(t) - \widehat{x}\|, \ \|y(t) - \widehat{x}\| \le \epsilon \text{ for all } t = \tau_1, \ldots, \tau_2 - 1 \text{ and } \|x(\tau_2) - \widehat{x}\| \le \epsilon.$$

Moreover if $\|x(0) - \widehat{x}\| \le \gamma$, *then* $\tau_1 = 0$, *and if* $\|x(T) - \widehat{x}\| \le \gamma$, *then* $\tau_2 = T$.

Clearly, Theorem 9.4 follows from Theorem 9.3 applied with $z_1 = 0$.

Theorem 9.5 *Assume that for each good program* $\{x(t), y(t)\}_{t=0}^{\infty}$, *Equation (9.14) holds and*

$$M_0 > \max\{(a_i d)^{-1} : \ i = 1, \ldots, n\}.$$

Let M_1, ϵ *be positive numbers and* $\Gamma \in (0, 1)$. *Then there exists a natural number* L *such that for each integer* $T > L$, *each* $z_0, z_1 \in R_+^n$ *satisfying* $z_0 \le Me$ *and* $a z_1 \le \Gamma d^{-1}$, *and each program* $(\{x(t)\}_{t=0}^T, \{y(t)\}_{t=0}^{T-1})$ *which satisfies*

$$x(0) = z_0, \ x(T) \ge z_1, \quad \sum_{t=0}^{T-1} w(by(t)) \ge U(z_0, z_1, T) - M_1,$$

the following inequality holds:

$$Card\{i \in \{0, \ldots, T - 1\} : \ \max\{\|x(t) - \widehat{x}\|, \ \|y(t) - \widehat{x}\|\} > \epsilon\} \le L.$$

$$\sum_{t=0}^{T-1}(w(bv(t)) - \mu(w)) \le M_2.$$

By property (i), (9.19), and (9.21), for each integer $k \ge 1$ and each integer $T \in \{0, \ldots, k\}$,

$$\sum_{t=0}^{T-1} w(by^{(k)}(t)) = \sum_{t=0}^{T-1} w(by^{(k)}(t)) - \sum\{w(by^{(k)}(t)) : t \in \{0, \ldots, k-1\}, \ t \ge T\}$$

$$\ge T\mu(w) - M_1 - M_2. \tag{9.22}$$

By extracting subsequence and using diagonalization process, we obtain that there exist a strictly increasing sequence of natural numbers $\{k_p\}_{p=1}^{\infty}$ such that for each integer $t \ge 0$, there exist

$$x(t) = \lim_{p \to \infty} x^{(k_p)}(t), \quad y(t) = \lim_{p \to \infty} y^{(k_p)}(t). \tag{9.23}$$

Clearly, $\{x(t), \ y(t)\}_{t=0}^{\infty}$ is a program. By (9.22) and (9.23), for all natural numbers s,

$$\sum_{t=0}^{T-1}(w(by(t)) - \mu(w)) \ge -M_1 - M_2$$

and $\{x(t), y(t)\}_{t=0}^{\infty}$ is a good program. By (9.15), there exists a natural number t_0 such that for all integers $t \ge t_0$,

$$\|x(t) - \widehat{x}\| \le \epsilon/8, \quad \|y(t) - \widehat{x}\| \le \epsilon/8. \tag{9.24}$$

By (9.23), there exists an integer $p \ge 1$ such that $k_p > t_0 + 4$,

$$\|x(t_0) - x^{(k_p)}(t_0)\| \le \epsilon/8, \quad \|y(t_0) - y^{(k_p)}(t_0)\| \le \epsilon/8. \tag{9.25}$$

In view of (9.24) and (9.25),

$$\|x^{(k_p)}(t_0) - \widehat{x}\| \le \epsilon/4, \quad \|y^{(k_p)}(t_0) - \widehat{y}\| \le \epsilon/4.$$

This contradicts (9.20). The contradiction we have reached proves Lemma 9.9.

Lemma 9.10 *Let $\epsilon \in (0, 1)$. Then there exists $\delta > 0$ such that for each integer $T \ge 2$ and each program $\left(\{x(t)\}_{t=0}^{T}, \{y(t)\}_{t=0}^{T-1}\right)$ satisfying*

$$\|x(0) - \widehat{x}\| \le \delta, \quad \|x(T) - \widehat{x}\| \le \delta,$$

$$U(x(0), x(T), T) \leq \sum_{t=0}^{T-1} w(by(t)) + \delta$$

the inequalities

$$\|x(t) - \widehat{x}\|_\infty \leq \epsilon, \ t = 0, \dots, T,$$

$$\|y(t) - \widehat{x}\|_\infty \leq \epsilon, \ t = 0, \dots, T - 1$$

hold.

Proof By Proposition 3.13 and continuity of w, for every natural number k, there exists

$$\delta_k \in (0, 4^{-k}\epsilon) \tag{9.26}$$

such that the following property holds:

(ii) for each $x, x' \in R_+^n$ satisfying

$$\|x - \widehat{x}\|, \|x' - \widehat{x}\| \leq \delta_k$$

there exist $\bar{x} \geq x', y \in R_+^n$ such that

$$(x, \bar{x}) \in \Omega, \ y \in \Lambda(x, \bar{x}),$$

$$\|y - \widehat{x}\| \leq 4^{-k}\epsilon, \ \|\bar{x} - \widehat{x}\| \leq 4^{-k}\epsilon,$$

$$|w(by) - w(b\widehat{x})| \leq 4^{-k}\epsilon.$$

We may assume without loss of generality that

$$\delta_{l+1} < \delta_k \text{ for all integers } k \geq 1.$$

Assume that the lemma does not hold. Then for each integer $k \geq 1$, there exist an integer $T_k \geq 2$ and a program $\left(\{x^{(k)}(t)\}_{t=0}^{T_k}, \{y^{(k)}(t)\}_{t=0}^{T_k-1}\right)$ such that

$$\|x^{(k)}(0) - \widehat{x}\| \leq \delta_k, \ \|x^{(k)}(T_k) - \widehat{x}\| \leq \delta_k, \tag{9.27}$$

$$\sum_{t=0}^{T_k-1} w(by^{(k)}(t)) \geq U(x^{(k)}(0), x^{(k)}(T_k), T_k) - \delta_k \tag{9.28}$$

and

$$\max\{\max\{\|x^{(k)}(t) - \widehat{x}\|_\infty : t = 0, \dots, T_k\},$$

$$\max\{\|y^{(k)}(t) - \widehat{y}\|_\infty : t = 0, \ldots, T_k - 1\}\} > \epsilon. \tag{9.29}$$

Let $k \geq 1$ be an integer. Property (ii) and (9.27) imply that there exist $\tilde{x}^{(k)}(1) \in R_+^n$ such that

$$\tilde{x}^{(k)}(1) \geq \widehat{x}, \tag{9.30}$$

$$(x^{(k)}(0), \tilde{x}^{(k)}(1)) \in \Omega \tag{9.31}$$

and

$$\tilde{y}^{(k)}(0) \in \Lambda(x^{(k)}(0), \tilde{x}^{(k)}(1)) \tag{9.32}$$

such that

$$\|\tilde{y}^{(k)} - \widehat{x}\| \leq 4^{-k}\epsilon, \quad \|\tilde{x}^{(k)}(1) - \widehat{x}\| \leq 4^{-k}\epsilon, \tag{9.33}$$

$$|w(b\tilde{y}^{(k)}(0)) - w(b\widehat{x})| \leq 4^{-k}\epsilon. \tag{9.34}$$

Property (ii) and (9.27) imply that there exist $\tilde{x}^{(k)}(T_k) \in R_+^n$ such that

$$\tilde{x}^{(k)}(T_k) \geq x^{(k)}(T_k), \tag{9.35}$$

such that

$$(\widehat{x}, \tilde{x}^{(k)}(T_k)) \in \Omega_k \tag{9.36}$$

and

$$\tilde{y}^{(k)}(T_k - 1) \in \Lambda(\widehat{x}, \tilde{x}^{(k)}(T_k)), \tag{9.37}$$

such that

$$\|\tilde{y}^{(k)}(T_k - 1) - \widehat{y}\| \leq 4^{-k}\epsilon, \quad \|\tilde{x}^{(k)}(T_k) - \widehat{x}\| \leq 4^{-k}\epsilon, \tag{9.38}$$

$$|w(b\tilde{y}^{(k)}(T_k - 1)) - w(b\widehat{x})| \leq 4^{-k}\epsilon. \tag{9.39}$$

Set

$$\bar{x}^{(k)}(0) = x^{(k)}(0), \quad \bar{x}^{(k)}(1) = \tilde{x}^{(k)}(1), \tag{9.40}$$

$$\bar{x}^{(k)}(t) = \widehat{x} + (1 - d)^t(\tilde{x}^{(k)}(1) - \bar{x}), \quad t = 1, \ldots, T_k - 1, \tag{9.41}$$

$$\bar{y}^{(k)}(0) = \tilde{y}^{(k)}(0), \quad \bar{y}^{(k)}(t) = \widehat{y},$$

$$t \in \{1, \ldots, T_k - 1\} \setminus \{T_k - 1\}, \quad \bar{y}^{(k)}(T_k - 1) = \tilde{y}(T_k - 1), \tag{9.42}$$

$$\bar{x}^{(k)}(T_k) = \tilde{x}^{(k)}(T_k) + (1-d)^{T_k}(\bar{x}^{(k)}(1) - \hat{x}). \tag{9.43}$$

In view of (9.30)–(9.32), (9.36), (9.37), and (9.40)–(9.43),

$$\left(\{\bar{x}^{(k)}(t)\}_{t=0}^{T_k}, \{\bar{y}^{(k)}(t)\}_{t=0}^{T_k-1}\right)$$

is a program. It follows from (9.28), (9.34), (9.35), (9.39), (9.40), (9.42), and (9.43) that

$$\sum_{t=0}^{T_k-1} w(by^{(k)}(t)) \geq U(x^{(k)}(0), x^{(k)}(T_k), T_k) - \delta_k$$

$$\geq -\delta_k + \sum_{t=0}^{T_k-1} w(b\bar{y}^{(k)}(t)) \geq T_k w(b\hat{y}) - 2 \cdot 4^{-k}\epsilon. \tag{9.44}$$

Property (ii) and (9.27) imply that there exist $z^{(k)}$, $\xi^{(k)} \in R_+^n$ such that

$$(x^{(k)}(T_k), x^{(k+1)}(0) + z^{(k)}) \in \Omega, \tag{9.45}$$

$$\|x^{(k+1)}(0) + z^{(k)} - \hat{x}\| \leq 4^{-k}\epsilon, \tag{9.46}$$

$$\xi^{(k)} \in \Lambda(x^{(k)}(T_k), x^{(k+1)}(0) + z^{(k)}), \tag{9.47}$$

$$\|\xi^{(k)} - \hat{y}\| \leq 4^{-k}\epsilon, \tag{9.48}$$

$$|w(b\xi^{(k)}) - w(b\hat{x})| \leq 4^{-k}\epsilon. \tag{9.49}$$

By induction we construct a program $(\{x(t), y(t)\}_{t=0}^{\infty})$. Set

$$x(t) = x^{(1)}(t), \ t = 0, \ldots, T_1, \ y(t) = y^{(1)}(t), \ t = 0, \ldots, T_1 - 1, \tag{9.50}$$

$$x(T_1 + 1) = x^{(2)}(0) + z^{(1)}, \ y(T_1) = \xi^{(1)}, \tag{9.51}$$

for all $t = 1, \ldots, T_2$,

$$x(T_1 + t + 1) = x^{(2)}(t) + (1-t)^t z^{(1)}$$

and for all $t = 0, \ldots, T_2 - 1$,

$$y(T_1 + t + 1) = y^{(2)}(t).$$

Clearly, $\left(\{x(t)\}_{t=0}^{T_2+T_1+1}, \{y(t)\}_{t=0}^{T_2+T_1}\right)$ is a program. Set

$$\tilde{z}^{(0)} = 0.$$

Assume that $k \geq 2$ is an integer and we defined a program

$$x(t), \ t = 0, \ldots, \sum_{i=1}^{k}(T_i + 1) - 1, \quad y(t), \ t = 0, \ldots, \sum_{i=1}^{k}(T_i + 1) - 2$$

such that (9.50) holds and for each integer $p \in [2, k]$ there exists $\tilde{z}^{(p-1)} \in R_+^n$ such that

$$\tilde{z}^{(p-1)} = z^{(p-1)} + (1 - d)^{T_{p-1}+1}\tilde{z}^{(p-2)}, \tag{9.52}$$

$$y\left(\sum_{i=1}^{p-1}(T_i + 1) - 1\right) = \xi^{(p-1)}, \tag{9.53}$$

$$x\left(\sum_{i=1}^{p-1}(T_i + 1)\right) = x^{(p)}(0) + \tilde{z}^{(p-1)}, \tag{9.54}$$

for all integers $t = 1, \ldots, T_p$,

$$x\left(\sum_{i=1}^{p-1}(T_i + 1) + t\right) = x^{(p)}(t) + (1 - d)^t \tilde{z}^{(p-1)}, \tag{9.55}$$

$$y\left(\sum_{i=1}^{p-1}(T_i + 1) + t - 1\right) = y^{(p)}(t - 1). \tag{9.56}$$

It is not difficult to see that for $k = 2$ our assumption holds. By (9.45),

$$(x^{(k)}(T_k), x^{(k+1)}(0) + z^{(k)}) \in \Omega, \ \xi^{(k)} \in \Lambda(x^{(k)}(T_k), x^{(k+1)}(0) + z^{(k)}), \tag{9.57}$$

$$\xi^{(k)} \leq x_{T_k}^{(k)}, \tag{9.58}$$

$$a(x^{(k+1)}(0) + z^{(k)} - (1 - d)x^{(k)}(T_k)) \leq 1, \tag{9.59}$$

$$a(x^{(k+1)}(0) + z^{(k)} - (1 - d)x^{(k)}(T_k)) + e\xi^{(k)} \leq 1. \tag{9.60}$$

In view of (9.55),

$$x\left(\sum_{i=1}^{k}(T_i + 1) - 1\right) = x^{(k)}(T_k) + (1 - d)^{T_k}\tilde{z}^{(k-1)}. \tag{9.61}$$

By (9.57) and (9.60),

$$a(x^{(k+1)}(0) + z^{(k)} - (1-d)^{T_k+1}\tilde{z}^{(k-1)}) - (1-d)x\left(\sum_{i=1}^{k}(T_i + 1) - 1\right) + e\xi^{(k)} \le 1.$$
(9.62)

It follows from (9.57)–(9.62) that

$$\left(x\left(\sum_{i=1}^{k}(T_i + 1) - 1\right), x^{(k+1)}(0) + z^{(k)} + (1-d)^{T_k+1}\tilde{z}^{(k-1)}\right) \in \Omega,$$
(9.63)

$$\xi^{(k)} \in \Lambda\left(x\left(\sum_{i=1}^{k}(T_i + 1) - 1\right), x^{(k+1)}(0) + z^{(k)} + (1-d)^{T_k+1}\tilde{z}^{(k-1)}\right).$$
(9.64)

Set

$$y\left(\sum_{i=1}^{k}(T_i + 1) - 1\right) = \xi^{(k)},$$
(9.65)

$$x\left(\sum_{i=1}^{k}(T_i + 1)\right) = x^{(k+1)}(0) + z^{(k)} + (1-d)^{T_k+1}\tilde{z}^{(k-1)},$$
(9.66)

for all $t = 1, \ldots, T_{k+1}$,

$$x\left(\sum_{i=1}^{k}(T_i + 1) + t\right) = x^{(k+1)}(t) + (1-d)^t z^{(k)} + (1-d)^{T_k+1+t}\tilde{z}^{(k-1)},$$
(9.67)

$$y\left(\sum_{i=1}^{k}(T_i + 1) + t - 1\right) = y^{(k+1)}(t - 1).$$
(9.68)

By (9.60) and (9.63)–(9.67),

$$x(t), \ t = 0, \ldots, \sum_{i=1}^{k+l}(T_i + 1) - 1, \ \ y(t), \ t = 0, \ldots, \sum_{i=1}^{k+1}(T_i + 1) - 2$$

is a program. Thus, the assumption made for k also holds for $k + 1$ with

$$\tilde{z}^{(k)} = z^{(k)} + (1-d)^{T_k+1}\tilde{z}^{(k-1)}.$$

Therefore by induction we constructed a program $\{x(t), y(t)\}_{t=0}^{\infty}$ such that (9.50) holds and (9.52)–(9.56) hold for all integers $p \geq 2$.

By (9.44), (9.50), and (9.58),

$$\sum_{t=0}^{T_1-1} w(by(t)) \geq T_1 w(b\widehat{y}) - 2^{-1}\epsilon \tag{9.69}$$

and for each integers $k \geq 2$,

$$\sum_{t=\sum_{i=1}^{k-1}(T_i+1)}^{\sum_{i=1}^{k}(T_i+1)-2} w(by(t)) = \sum_{t=0}^{T_k-1} w(by^{(k)}(t)) \geq T_k w(b\widehat{y}) - 2 \cdot 4^{-k}\epsilon. \tag{9.70}$$

In view of (9.49) and (9.53), for each integer $k \geq 1$,

$$w\left(by\left(\sum_{i=1}^{k}(T_i+1)-1\right)\right) = w(b\xi^{(k)}) \geq w(b\widehat{x}) - 4^{-k}\epsilon. \tag{9.71}$$

By (9.70) and (9.71),

$$\limsup_{T\to\infty} \sum_{t=0}^{T}(w(by(t)) - w(b\widehat{y})) > -\infty$$

and $\{x(t), y(t)\}_{t=0}^{\infty})$ is a good program. Therefore

$$\lim_{t\to\infty} \|x(t) - \widehat{x}\| = 0, \quad \lim_{t\to\infty} \|y(t) - \widehat{x}\| = 0.$$

This implies that there exists an integer $k_0 \geq 1$ such that for each integer $t \geq \sum_{i=1}^{k_0}(T_i+1)-1)$, we have

$$\|x(t) - \widehat{x}\| \leq \epsilon/8, \quad \|y(t) - \widehat{x}\| \leq \epsilon/8. \tag{9.72}$$

By (9.26), (9.27), and (9.46), for each integer $i \geq 1$,

$$\|z^{(i)}\| \leq \|x^{(i+1)}(0) + z^{(i)} - x^{(i+1)}(0)\|$$

$$\leq \|x^{(i+1)}(0) + z^{(i)} - \widehat{x}\| + \|\widehat{x} - x^{(i+1)}(0)\|$$

$$\leq \delta_{i+1} + 4^{-i}\epsilon \leq 2 \cdot 4^{-i}\epsilon$$

$$z^{(i)} \leq 2 \cdot 4^{-i}\epsilon e \text{ for all integers } i \geq 1. \tag{9.73}$$

In view of (9.52) and (9.73), for all integers $p \geq 1$,

$$\tilde{z}^{(p)} \leq \sum_{i=1}^{p} z^{(i)} \leq 2 \sum_{i=1}^{\infty} 4^{-i} \epsilon e = 2^{-1}(4/3)\epsilon e = (2/3)\epsilon e. \tag{9.74}$$

Let $k > k_0$ be an integer. It follows from (9.54), (9.55), and (9.74) that for all $t = 0, \ldots, T_k$,

$$0 \leq x \left(\sum_{i=1}^{k-1}(T_i + 1) + t \right) - x^{(k)}(t) \leq (2/3)\epsilon e. \tag{9.75}$$

By (9.56),

$$y^{(k)}(t) = y \left(\sum_{i=1}^{k}(T_i + 1) + t \right), \ t = 0, \ldots, T_k - 1. \tag{9.76}$$

In view of (9.72) and (9.76),

$$\|y^{(k)}(t) - \widehat{x}\| \leq \epsilon/8, \ t = 0, \ldots, T_k - 1. \tag{9.77}$$

Relations (9.72) and (9.75) imply that for all $t = 0, \ldots, T_k$,

$$\|x^{(k)}(t) - \widehat{x}\|_\infty \leq \|x^{(k)}(t) - x \left(\sum_{i=1}^{k}(T_i + 1) + t \right)\|_\infty$$

$$+ \|x \left(\sum_{i=1}^{k}(T_i + 1) + t \right) - \widehat{x}\|_\infty \leq (2/3)\epsilon + \epsilon/8 < \epsilon.$$

This contradicts (9.29). The contradiction we have reached proves Lemma 9.10.

Lemma 9.11 *Let*

$$M_0 > \max\{(a_i d)^{-1} : i = 1, \ldots, n\},$$

$M_1 > 0$, $\epsilon > 0$. *Then there is a natural number τ such that for each integer $T \geq \tau$, each program $(\{x(t)\}_{t=0}^{T}, \{y(t)\}_{t=0}^{T-1})$ satisfying*

$$x(0) \leq M_0 e,$$

$$\sum_{t=0}^{T-1} w(by(t)) \geq Tw(b\widehat{x}) - M_1$$

and each $S \in \{0, \ldots, T - \tau\}$ there is an integer $t \in [S, S + \tau - 1]$ such that

$$\|y(t) - \widehat{x}\|, \ \|x(t) - \widehat{x}\| \le \epsilon.$$

Proof Lemma 5.3 implies that

$$\text{if } (x, x') \in \Omega \text{ and } x \le M_0 e, \text{ then } x' \le M_0 e. \tag{9.78}$$

In view of Corollary 5.6, there exists $M_2 > 0$ such that the following property holds:

(a) for each integer $T \ge 1$, each program $\left(\{u(t)\}_{t=0}^{T}, \{v(t)\}_{t=0}^{T-1} \right)$ satisfying

$$u(0) \le M_0 e,$$

we have

$$\sum_{t=0}^{T-1} w(bv(t)) \le T\mu + M_2.$$

Lemma 9.9 implies that there exists a natural number τ such that for each program $\left(\{x(t)\}_{t=0}^{\tau}, \{y(t)\}_{t=0}^{\tau-1} \right)$ satisfying

$$x(0) \le M_0 e, \tag{9.79}$$

$$\sum_{t=0}^{\tau-1} w(by(t)) \ge \tau w(b\widehat{x}) - M_1 - 2M_2 \tag{9.80}$$

there is an integer $t \in [0, \tau - 1]$ such that

$$\|y(t) - \widehat{x}\|, \ \|x(t) - \widehat{x}\| \le \epsilon. \tag{9.81}$$

Assume that an integer $T \ge \tau$, $\left(\{x(t)\}_{t=0}^{T}, \{y(t)\}_{t=0}^{T-1} \right)$ is a program satisfying

$$x(0) \le M_0 e,$$

$$\sum_{t=0}^{T-1} w(by(t)) \ge T\mu(w) - M_1$$

and

$$S \in \{0, \dots, T - \tau\}.$$

By (9.78),

$$x(t) \le M_0 e, \ t = 0, \dots, T.$$

It follows from the relations above and property (a) that

$$\sum \{ w(by(t)) - w(b\widehat{y}) : \ t \in \{0, \dots, S\} \setminus \{S\} \} \le M_2,$$

$$\sum \{ w(by(t)) - w(b\widehat{y}) : \ t \in \{S + \tau, \dots, T\} \setminus \{T\} \} \le M_2,$$

$$\sum_{t=S}^{S+\tau-1} (w(by(t)) - w(b\widehat{y})) = \sum_{t=0}^{T-1} (w(by(t)) - w(b\widehat{y}))$$

$$- \sum \{ w(by(t)) - w(b\widehat{y}) : \ t \in \{0, \dots, S\} \setminus \{S\} \}$$

$$- \sum \{ w(by(t)) - w(b\widehat{y}) : \ t \in \{S + \tau, \dots, T\} \setminus \{T\} \}$$

$$\ge -M_1 - 2M_2.$$

By the relations above and the choice of τ (see (9.80) and (9.81)), there exists $t \in \{S, \dots, S + \tau - 1\}$ for which

$$\|x(t) - \widehat{x}\| \le \epsilon, \ \|y(t) - \widehat{x}\| \le \epsilon.$$

Lemma 9.11 is proved.

By the definition,

$$da\widehat{x} + e\widehat{x} \le 1.$$

This implies that $b\widehat{y} > 0$, $e\widehat{y} > 0$, and

$$a\widehat{x} < d^{-1}.$$

There exists $\Gamma_* \in (0, 1)$ such that

$$a\widehat{x} < \Gamma_* d^{-1}. \tag{9.82}$$

Proposition 9.12 *There is $m > 0$ such that for each $z \in R_+^n$ and each natural number T,*

$$U(z, T) \ge T w(b\widehat{x}) - m.$$

Proof By (9.82) and Proposition 4.5, there exists a natural number k_* such that the following property holds:

(b) for each $z_0 \in R_+^n$ there is a program $\left(\{x(t)\}_{t=0}^{k_*}, \{y(t)\}_{t=0}^{k_*-1}\right)$ such that $x(0) = z_0$, $x(k_*) \geq \widehat{x}$.

Set

$$m = k_*(w(b\widehat{y}) - w(0)). \tag{9.83}$$

Let $z \in R_+^n$. Property (b) implies that there exists a program

$$\left(\{x(t)\}_{t=0}^{k_*}, \{y(t)\}_{t=0}^{k_*-1}\right)$$

such that

$$x(0) = z, \quad x(k_*) \geq \widehat{x}.$$

For all integers $t \geq k_*$ define

$$y(t) = \widehat{y},$$
$$x(t+1) = (1-d)x(t) + d\widehat{x}.$$

It is not difficult to see that $\{x(t), y(t)\}_{t=0}^{\infty}$ is a program. For each natural number T, we have

$$U(z, T) \geq \sum_{t=0}^{T-1} w(by(t)).$$

By (9.83), for each natural number T,

$$U(z, T) - Tw(b\widehat{y}) \geq (w(0) - w(b\widehat{y}))k_* \geq -m.$$

Proposition 9.12 is proved.

Proposition 9.13 *Let $\Gamma \in (0, 1)$. Then there exists $m > 0$ such that for each $z_0 \in R_+^n$, each $z_1 \in R_+^n$ satisfying $az_1 \leq \Gamma d^{-1}$, and each natural number T,*

$$U(z_0, z_1, T) \geq Tw(b\widehat{x}) - m. \tag{9.84}$$

Proof By (9.82) and Proposition 4.5, there exists a natural number k_* such that the following property holds:

(c) for each $z_0 \in R_+^n$ there is a program $\left(\{x(t)\}_{t=0}^{k_*}, \{y(t)\}_{t=0}^{k_*-1}\right)$ such that $x(0) = z_0$, $x(k_*) \geq \widehat{x}$.

By Proposition 4.5, there exists a natural number k_1 such that the following property holds:

(d) for each $z_0 \in R_+^n$ and each $z_1 \in R_+^n$ satisfying $az_1 \leq \Gamma d^{-1}$, there is a program $\left(\{x(t)\}_{t=0}^{k_1}, \{y(t)\}_{t=0}^{k_1-1} \right)$ such that $x(0) = z_0$, $x(k_1) \geq z_1$.

Set

$$m = (k_* + k_1)(w(b\widehat{y}) - w(0)), \tag{9.85}$$

$$k_0 = k_1 + k_*. \tag{9.86}$$

Assume that $z_0 \in R_+^n$, $z_1 \in R_+^n$ satisfies $az_1 \leq \Gamma d^{-1}$ and that T is a natural number. We show that (9.84) holds. In view of (9.85) and (9.86), we may consider only the case

$$T > k_0.$$

Properties (c) and (d) imply that there exists a program $\left(\{x(t)\}_{t=0}^{T}, \{y(t)\}_{t=0}^{T-1} \right)$ such that

$$x(0) = z_0, \quad x(k_*) \geq \widehat{x},$$

for all integers $t = k_*, \ldots, T - k_1 - 1$,

$$y(t) = \widehat{y},$$
$$x(t+1) = (1-d)x(t) + d\widehat{x},$$
$$x(T) \geq z_1.$$

It is not difficult to see that

$$U(z_0, z_1, T) - Tw(b\widehat{y}) \geq (w(0) - w(b\widehat{y}))(k_* + k_1) \geq -m.$$

Proposition 9.13 is proved.

9.3   Proof of Theorem 9.3

By Lemma 9.10, there exists

$$\gamma \in (0, \min\{1, \epsilon/2\})$$

such that the following property holds:

(i) for each integer $S \geq 2$ and each program $\left(\{x(t)\}_{t=0}^{S}, \{y(t)\}_{t=0}^{S-1} \right)$ satisfying

$$\|x(0) - \widehat{x}\| \leq \gamma, \ \|x(S) - \widehat{x}\| \leq \gamma,$$

$$U(x(0), x(S), S) \leq \sum_{t=0}^{S-1} w(by(t)) + \gamma$$

the inequalities

$$\|x(t) - \widehat{x}\| \leq \epsilon, \ t = 0, \ldots, S,$$

$$\|y(t) - \widehat{x}\| \leq \epsilon, \ t = 0, \ldots, S-1$$

hold.

Proposition 9.13 implies that there exists $m_0 > 0$ such that the following property holds:

(ii) for each $z_0 \in R^n_+$, each $z_1 \in R^n_+$ satisfying $az_1 \leq \Gamma d^{-1}$, and each natural number T,

$$U(z_0, z_1, T) \geq Tw(b\widehat{x}) - m_0.$$

In view of Corollary 5.6, there exists $m_1 > 0$ such that the following property holds:

(iii) for each integer $T \geq 1$ and each program $\left(\{u(t)\}_{t=0}^{T}, \{v(t)\}_{t=0}^{T-1}\right)$ satisfying

$$x(0) \leq Me$$

we have

$$\sum_{t=0}^{T-1} w(by(t)) \leq m_1 + Tw(b\widehat{y}).$$

By Lemma 9.11, there is a natural number L such that the following property holds:

(iv) for each integer $\tau \geq L$, each program $\left(\{x(t)\}_{t=0}^{\tau}, \{y(t)\}_{t=0}^{\tau-1}\right)$ satisfying

$$x(0) \leq M_0 e,$$

$$\sum_{t=0}^{\tau-1} w(by(t)) \geq \tau w(b\widehat{x}) - m_0 - 2m_1 - 1$$

and each $S \in \{0, \ldots, \tau - L\}$, there is an integer $t \in [S, S + L - 1]$ such that

$$\|y(t) - \widehat{x}\|, \ \|x(t) - \widehat{x}\| \leq \gamma.$$

Assume that an integer $T > 2L + 1$, $z_0, z_1 \in R_+^n$,

$$z_0 \le Me, \quad az_1 \le \Gamma d^{-1} \tag{9.87}$$

and that a $\left(\{x(t)\}_{t=0}^T, \{y(t)\}_{t=0}^{T-1} \right)$ satisfies

$$x(0) = z_0, \ x(T) \ge z_1, \tag{9.88}$$

$$\sum_{t=0}^{T-1} w(by(t)) \ge U(z_0, z_1, T) - \gamma. \tag{9.89}$$

Lemma 5.3, (9.87), and (9.88) imply that

$$x(t) \le Me, \ t = 0, \dots, T. \tag{9.90}$$

By property (ii) and (9.87)–(9.89),

$$\sum_{t=0}^{T-1} w(by(t)) \ge Tw(b\widehat{y}) - m_0 - 1. \tag{9.91}$$

Let integers S_1, S_2 satisfy $0 \le S_1 < S_2 \le T$. Property (iii) and (9.91) imply that

$$
\sum_{t=S_1}^{S_2-1} (w(by(t)) - w(b\widehat{y}))
$$

$$
= \sum_{t=0}^{T-1} (w(by(t)) - w(b\widehat{y}))
$$

$$
- \sum \{w(by(t)) - w(b\widehat{y}) : t \in \{0, \dots, S_1\} \setminus \{S_1\}\}
$$

$$
- \sum \{w(by(t)) - w(b\widehat{y}) : t \in \{S_2, \dots, T\} \setminus \{T\}\}
$$

$$
\ge -m_0 - 1 - 2m_1. \tag{9.92}
$$

By (9.90), (9.92), and property (iv), there exist

$$\tau_1 \in \{0, \dots, L\}, \ \tau_2 \in \{T - L, \dots, T\}$$

such that

$$\|x(\tau_i) - \widehat{x}\| \le \gamma, \ i = 1, 2. \tag{9.93}$$

If $\|x(0) - \widehat{x}\| \leq \gamma$, we may assume that $\tau_1 = 0$, and if $\|x(T) - \widehat{x}\| \leq \gamma$, we may assume that $\tau_2 = T$. Property (i), (9.88), (9.89), and (9.93) imply that

$$\|x(t) - \widehat{x}\| \leq \epsilon, \ t = \tau_1, \dots, \tau_2,$$

$$\|y(t) - \widehat{x}\| \leq \epsilon, \ t = \tau_1, \dots, \tau_2 - 1$$

Theorem 9.3 is proved.

9.4   Proofs of Theorems 9.5 and 9.7

Proof of Theorem 9.7 By Lemma 9.10, there exists

$$\gamma \in (0, \min\{1, \epsilon/2\})$$

such that the following property holds:

(i) for each integer $S \geq 2$ and each program $\left(\{x(t)\}_{t=0}^{S}, \{y(t)\}_{t=0}^{S-1}\right)$ satisfying

$$\|x(0) - \widehat{x}\| \leq \gamma, \ \|x(S) - \widehat{x}\| \leq \gamma,$$

$$U(x(0), x(S), S) \leq \sum_{t=0}^{S-1} w(by(t)) + \gamma$$

the inequalities

$$\|x(t) - \widehat{x}\| \leq \epsilon, \ t = 0, \dots, S,$$

$$\|y(t) - \widehat{x}\| \leq \epsilon, \ t = 0, \dots, S - 1$$

hold.

   In view of Corollary 5.6, there exists $m_1 > 0$ such that the following property holds:

(ii) for each integer $T \geq 1$ and each program $\left(\{u(t)\}_{t=0}^{T}, \{v(t)\}_{t=0}^{T-1}\right)$ satisfying

$$u(0) \leq Me$$

we have

$$\sum_{t=0}^{T-1} w(by(t)) \leq m_1 + Tw(b\widehat{y}).$$

By Lemma 9.11, there is a natural number $L_0 \geq 4$ such that the following property holds:

(iii) for each integer $\tau \geq L_0$, each program $\left(\{x(t)\}_{t=0}^{\tau}, \{y(t)\}_{t=0}^{\tau-1} \right)$ satisfying

$$x(0) \leq M_0 e,$$

$$\sum_{t=0}^{\tau-1} w(by(t)) \geq \tau w(b\widehat{x}) - M_1 - 2m_1 - 1$$

and each $S \in \{0, \ldots, \tau - L_0\}$, there is an integer $t \in [S, \ldots, S + L_0 - 1]$ such that

$$\|y(t) - \widehat{x}\|, \ \|x(t) - \widehat{x}\| \leq \gamma.$$

Choose a natural number

$$L > 4(L_0 + 4) + (2L_0 + 2)(\gamma^{-1}(M_1 + m_1) + 2). \tag{9.94}$$

Assume that an integer $T \geq L$ and that a program $\left(\{x(t)\}_{t=0}^{T}, \{y(t)\}_{t=0}^{T-1} \right)$ satisfies

$$x(0) \leq M_0 e, \tag{9.95}$$

$$\sum_{t=0}^{T-1} w(by(t)) \geq T w(b\widehat{y}) - M_1. \tag{9.96}$$

Lemma 5.3 and (9.95) imply that

$$x(t) \leq M_0 e, \ t = 0, \ldots, T. \tag{9.97}$$

Set

$$t_0 = 0. \tag{9.98}$$

By induction, using property (iii) and (9.94)–(9.96), we can construct a finite sequence of natural numbers $\{t_i\}_{i=1}^{q}$ where $q \geq 2$ is an integer, such that

$$t_q = T$$

and that for each $i \in \{0, \ldots, q - 2\}$,

$$t_{i+1} - t_i \in [L_0, 2L_0], \tag{9.99}$$

$$t_q - t_{q-1} < 2L_0, \tag{9.100}$$

$$\|x(t_i) - \hat{x}\| \leq \gamma, \ i = 1, \ldots, q - 1. \tag{9.101}$$

It is clear that there exists a program $\left(\{\tilde{x}(t)\}_{t=0}^{t_1}, \{\tilde{y}(t)\}_{t=0}^{t_1-1}\right)$ such that

$$\tilde{x}(0) = x(0), \ \tilde{x}(t_1) \geq x(t_1), \tag{9.102}$$

$$\sum_{t=0}^{t_1-1} w(b\tilde{y}(t)) = U(x(0), x(t_1), t_1). \tag{9.103}$$

Assume that $k \in \{1, \ldots, q - 1\}$ and we defined a program

$$\left(\{\tilde{x}(t)\}_{t=0}^{t_k}, \{\tilde{y}(t)\}_{t=0}^{t_k-1}\right)$$

such that (9.102) and (9.103) hold, for all $i = 1, \ldots, k$,

$$\tilde{x}(t_i) \geq x(t_i), \tag{9.104}$$

and for all $i = 0 \ldots, k - 1$,

$$\sum_{t=t_i}^{t_{i+1}-1} w(b\tilde{y}(t)) = U(\tilde{x}(t_i), x(t_{i+1}), t_{i+1} - t_i) \geq \sum_{t=t_i}^{t_{i+1}-1} w(by(t)). \tag{9.105}$$

In view of (9.102) and (9.103), our assumption holds for $k = 1$. By (9.104),

$$\tilde{x}(t_k) \geq x(t_k).$$

It is not difficult to see that

$$\left(\{x(t) + (1-d)^{t-t_k}(\tilde{x}(t_k) - x(t_k))\}_{t=t_k}^{t_{k+1}}, \{y(t)\}_{t=t_k}^{t_{k+1}-1}\right)$$

is a program and there exists a program

$$\left(\{\tilde{x}(t)\}_{t=t_k}^{t_{k+1}}, \{\tilde{y}(t)\}_{t=t_k}^{t_{k+1}-1}\right)$$

such that

$$\tilde{x}(t_{k+1}) \geq x(t_{k+1}), \tag{9.106}$$

and

$$\sum_{t=t_k}^{t_{k+1}-1} w(b\tilde{y}(t)) = U(\tilde{x}(t_k), x(t_{k+1}), t_{k+1} - t_k) \geq \sum_{t=t_k}^{t_{k+1}-1} w(by(t)). \qquad (9.107)$$

In view of (9.106) and (9.107), the assumption made for k holds for $k+1$.

Thus by induction, we have constructed a program $\left(\{\tilde{x}(t)\}_{t=0}^{T}, \{\tilde{y}(t)\}_{t=0}^{T-1}\right)$ such that

$$\tilde{x}(0) = x(0), \tilde{x}(T) \geq x(T), \qquad (9.108)$$

$$\tilde{x}(t_i) \geq x(t_i), \quad i = 1, \ldots, q, \qquad (9.109)$$

and for all $i = 0, \ldots, q - 1$,

$$\sum_{t=t_i}^{t_{i+1}-1} w(b\tilde{y}(t)) = U(\tilde{x}(t_i), x(t_{i+1}), t_{i+1} - t_i) \geq \sum_{t=t_i}^{t_{i+1}-1} w(by(t)). \qquad (9.110)$$

Proposition 4.9 and (9.108)–(9.10) imply that for all $i = 0, \ldots, q - 1$,

$$\sum_{t=t_i}^{t_{i+1}-1} w(b\tilde{y}(t)) = U(\tilde{x}(t_i), x(t_{i+1}), t_{i+1} - t_i)$$

$$\geq U(x(t_i), x(t_{i+1}), t_{i+1} - t_i). \qquad (9.111)$$

By property (ii) and (9.95),

$$\sum_{t=0}^{T-1} w(b\tilde{y}(t)) \leq Tw(b\hat{y}) + m_1. \qquad (9.112)$$

In view of (9.96) and (9.112),

$$\sum_{t=0}^{T-1} (w(b\tilde{y}(t)) - w(b\tilde{y}(t)))$$

$$\leq Tw(\hat{y}) + m_1 - Tw(b\hat{y}) + M_1 = m_1 + M_1. \qquad (9.113)$$

By (9.111) and (9.113),

$$M_1 + m_1 \geq \sum_{t=0}^{T-1} (w(b\tilde{y}(t)) - w(by(t)))$$

$$= \sum_{i=0}^{q-1} U(\tilde{x}(t_i), x(t_{i+1}), t_{i+1} - t_i) - \sum_{t=0}^{T-1} w(by(t))$$

$$\geq \sum_{i=0}^{q-1} (U(x(t_i), x(t_{i+1}), t_{i+1} - t_i) - \sum_{t=t_i}^{t_{i+1}-1} w(by(t)))$$

$$\geq \gamma \text{Card}(\{i \in \{0, \ldots, q-1\}:$$

$$U(x(t_i), x(t_{i+1}), t_{i+1} - t_i) - \sum_{t=t_i}^{t_{i+1}-1} w(by(t)) \geq \gamma\}. \qquad (9.114)$$

Set

$$E = \{i \in \{0, \ldots, q-1\}: U(x(t_i), x(t_{i+1}), t_{i+1} - t_i) - \sum_{t=t_i}^{t_{i+1}-1} w(by(t)) \geq \gamma\}.$$

$$(9.115)$$

By (9.114) and (9.115),

$$\text{Carrd}(E) \leq \gamma^{-1}(M_1 + m_1). \qquad (9.116)$$

Property (iv), (9.101), (9.110), and (9.115) imply that if $i \in \{1, \ldots, q-1\} \setminus \{q-1\}$ and $i \notin E$, then

$$\|x(t) - \widehat{x}\| \leq \epsilon, \ t = t_i, \ldots, t_{i+1},$$

$$\|y(t) - \widehat{x}\| \leq \epsilon, \ t = t_i, \ldots, t_{i+1} - 1.$$

This implies that

$$\{t \in \{0, \ldots, T-1\}: \max\{\|x(t) - \widehat{x}\|, \ \|y(t) - \widehat{x}\|\} > \epsilon\}$$

$$\subset \{0, \ldots, t_1\} \cup \{t_{q-1}, \ldots, T\} \cup \{\{t_i, \ldots, t_{i+1}\}: \ i \in E\}.$$

Together with (9.94), (9.99), (9.100), and (9.116), this implies that

$$\text{Card}\{t \in \{0, \ldots, T-1\}: \max\{\|x(t) - \widehat{x}\|, \ \|y(t) - \widehat{x}\|\} > \epsilon\}$$

$$\leq 2(2L_0 + 1) + \text{Card}(E)(2L_0 + 1) + \text{Card}(E)$$

$$\leq \text{Card}(E)(2L_0 + 2) + 2(2L_0 + 1)$$

$$\leq (\gamma^{-1}(M_1 + m_1) + 2)(2L_0 + 2) < L.$$

Theorem 9.7 is proved.

9.5   Generalizations of the Turnpike Results

Assume that for each good program $\{x(t), y(t)\}_{t=0}^{\infty}$,

$$\lim_{t \to \infty} (x(t), y(t)) = (\widehat{x}, \widehat{x}). \tag{9.117}$$

Theorem 9.14 *Let*

$$M > \max\{(a_i d)^{-1} : i = 1, \ldots, n\},$$

ϵ be a positive number and $\Gamma \in (0, 1)$. Then there exist a natural number L and a positive number γ such that for each integer $T > 2L$, each $z_0, z_1 \in R_+^n$ satisfying $z_0 \leq Me$ and $az_1 \leq \Gamma d^{-1}$, and each program $\left(\{x(t)\}_{t=0}^T, \{y(t)\}_{t=0}^{T-1}\right)$ which satisfies

$$x(0) = z_0, \ x(T) \geq z_1,$$

$$\sum_{t=\tau}^{\tau+L-1} w(by(t)) \geq U(x(\tau), x(\tau + L), L) - \gamma \text{ for all } \tau \in \{0, \ldots, T - L\}$$

and

$$\sum_{t=T-L}^{T-1} w(by(t)) \geq U(x(T - L), z_1, L) - \gamma$$

there are integers τ_1, τ_2 such that

$$\tau_1 \in [0, L], \ \tau_2 \in [T - L, T],$$

$$\|x(t) - \widehat{x}\| \leq \epsilon \text{ for all } t = \tau_1, \ldots, \tau_2,$$

$$\|y(t) - \widehat{x}\| \leq \epsilon \text{ for all } t = \tau_1, \ldots, \tau_2 - 1.$$

Moreover if $\|x(0) - \widehat{x}\| \leq \gamma$, then $\tau_1 = 0$, and if $\|x(T) - \widehat{x}\| \leq \gamma$, then $\tau_2 = T$.

Theorem 9.15 *Let*

$$M > \max\{(a_i d)^{-1} : i = 1, \ldots, n\}$$

and ϵ be a positive number. Then there exist a natural number L and a positive number γ such that for each integer $T > 2L$, each $z_0 \in R_+^n$ satisfying $z_0 \leq Me$, and each program $(\{x(t)\}_{t=0}^T, \{y(t)\}_{t=0}^{T-1})$ which satisfies

$$x(0) = z_0,$$

$$\sum_{t=\tau}^{\tau+L-1} w(by(t)) \geq U(x(\tau), x(\tau+L), L) - \gamma \text{ for all } \tau \in \{0, \dots, T-L\}$$

and

$$\sum_{t=T-L}^{T-1} w(by(t)) \geq U(x(T-L), L) - \gamma,$$

there are integers τ_1, τ_2 *such that*

$$\tau_1 \in [0, L], \quad \tau_2 \in [T-L, T],$$

$$\|x(t) - \widehat{x}\| \leq \epsilon \text{ for all } t = \tau_1, \dots, \tau_2,$$

$$\|y(t) - \widehat{x}\| \leq \epsilon \text{ for all } t = \tau_1, \dots, \tau_2 - 1.$$

Moreover if $\|x(0) - \widehat{x}\| \leq \gamma$, *then* $\tau_1 = 0$, *and if* $\|x(T) - \widehat{x}\| \leq \gamma$, *then* $\tau_2 = T$.

Theorem 9.15 follows from Theorem 9.14 applied with $z_2 = 0$.

9.6 Proof of Theorem 9.14

Recall (see (9.82)) that $\Gamma_* \in (0, 1)$ satisfies

$$a\widehat{x} < \Gamma_* d^{-1}. \tag{9.118}$$

We may assume that

$$\Gamma_* < \Gamma. \tag{9.119}$$

By Lemma 9.10, (9.118), and (9.119), there exists

$$\gamma \in (0, \min\{1, \epsilon/2\})$$

such that

$$a(\widehat{x} + \gamma e) < \Gamma d^{-1} \tag{9.120}$$

and the following property holds:

(i) for each integer $S \geq 2$ and each program $\left(\{x(t)\}_{t=0}^{S}, \{y(t)\}_{t=0}^{S-1}\right)$ satisfying

$$\|x(0) - \widehat{x}\| \leq \gamma, \ \|x(S) - \widehat{x}\| \leq \gamma,$$

$$U(x(0), x(S), S) \leq \sum_{t=0}^{S-1} w(by(t)) + \gamma$$

the inequalities

$$\|x(t) - \widehat{x}\| \leq \epsilon, \ t = 0, \ldots, S,$$

$$\|y(t) - \widehat{x}\| \leq \epsilon, \ t = 0, \ldots, S-1$$

are valid.

By Proposition 9.13, there exists $m_0 > 0$ such that the following property holds:

(ii) for each $z_0 \in R_+^n$, each $z_1 \in R_+^n$ satisfying $az_1 \leq \Gamma d^{-1}$, and each natural number T,

$$U(z_0, z_1, T) \geq Tw(b\widehat{x}) - m_0.$$

By Lemma 9.11, there is a natural number L_0 such that the following property holds:

(iii) for each integer $\tau \geq L_0$, each program $\left(\{x(t)\}_{t=0}^{\tau}, \{y(t)\}_{t=0}^{\tau-1} \right)$ satisfying

$$x(0) \leq M_0 e,$$

$$\sum_{t=0}^{\tau-1} w(by(t)) \geq \tau w(b\widehat{x}) - m_0 - 1$$

and each $S \in \{0, \ldots, \tau - L_0\}$, there is an integer $t \in [S, \ldots, S + L_0 - 1]$ such that

$$\|y(t) - \widehat{x}\|, \ \|x(t) - \widehat{x}\| \leq \gamma.$$

Choose a natural number

$$L > 8(L_0 + 1). \tag{9.121}$$

Assume that an integer $T > 2L$, $z_0, z_1 \in R_+^n$ satisfies

$$z_0 \leq Me, \ az_1 \leq \Gamma d^{-1} \tag{9.122}$$

and that a program $\left(\{x(t)\}_{t=0}^{T}, \{y(t)\}_{t=0}^{T-1} \right)$ satisfies

$$x(0) = z_0, \ x(T) \geq z_1, \tag{9.123}$$

$$\sum_{t=\tau}^{\tau+L-1} w(by(t)) \geq U(x(\tau), x(\tau + L), L) - \gamma \text{ for all } \tau \in \{0, \ldots, T-L\}$$

$$\tag{9.124}$$

for all $\tau \in \{0, \ldots, T-L\}$ and

$$\sum_{t=T-L}^{T-1} w(by(t)) \geq U(x(T-L), z_1, L) - \gamma. \tag{9.125}$$

Lemma 5.3, (9.122), and (9.123) imply that

$$x(t) \leq M_0 e, \ t = 0, \ldots, T. \tag{9.126}$$

Consider the program $\left(\{x(t)\}_{t=T-L}^{T}, \{y(t)\}_{t=T-L}^{T-1} \right)$. Property (ii), (9.122), (9.123), and (9.125) imply that

$$\sum_{t=T-L}^{T-1} w(by(t)) \geq U(x(T-L), z_1, L) - 1 \geq Lw(b\widehat{y}) - m_0 - 1. \tag{9.127}$$

Property (iii), (9.121), (9.126), and (9.127) imply that

$$t_0 \in \{T - L_0, \ldots, T\}, \ t_1 \in \{T - 4L_0, \ldots, T - 3L_0\} \tag{9.128}$$

such that

$$\|x(t_i) - \widehat{x}\|, \ \|y(t_i) - \widehat{x}\| \leq \gamma, \ i = 0, 1. \tag{9.129}$$

If $\|x(T) - \widehat{x}\| \leq \gamma$, then we may assume that $t_0 = T$.

Assume that $k \geq 1$ is an integer and that we defined a strictly decreasing sequence of natural numbers $t_i, \ i = 0, \ldots, k$ such that

$$t_{i+1} < t_i, \ i = 0, \ldots, k-1,$$

for all $i = 0, \ldots, k-1$,

$$L_0 \leq t_i - t_{i+1} \leq 4L_0, \tag{9.130}$$

$$\|x(t_i) - \widehat{x}\| \leq \gamma, \ i = 0, \ldots, k, \tag{9.131}$$

$$t_k \geq 4L_0 + 4. \tag{9.132}$$

(Clearly, for $k = 1$ our assumption holds.) It follows from (9.120) and (9.131) that for $i = 0, \ldots, k$,

$$ax(t_i) \leq a(\widehat{x} + \gamma e) < \Gamma d^{-1}. \tag{9.133}$$

Property (iii), (9.132) and (9.133) imply that

$$U(x(t_k - 4L_0 - 4), x(t_k), 4L_0 + 4) > (4L_0 + 4)w(b\widehat{y}) - m_0. \tag{9.134}$$

Relations (9.121) and (9.124) imply that

$$\sum_{t=t_k-4L_0-4}^{t_k-1} w(by(t)) \geq (4L_0 + 4)w(b\widehat{y}) - m_0 - 1. \tag{9.135}$$

There are two cases:

$$t_k > 6L_0 + 6; \tag{9.136}$$

$$t_k \leq 6L_0 + 6, \ \|x(0) - \widehat{x}\| > \gamma; \tag{9.137}$$

$$t_k \leq 6L_0 + 6, \ \|x(0) - \widehat{x}\| \leq \gamma. \tag{9.138}$$

If (9.137) holds, then the construction is completed. If (9.138) is true, then we set $t_{k+1} = 0$, and the construction is completed too.

Assume that (9.136) holds. Property (iii), (9.135), and (9.136) imply that

$$t_{k+1} \in \{t_k - 2L_0, \ldots, t_k - L_0\} \tag{9.139}$$

such that

$$\|x(t_{k+1}) - \widehat{x}\| \leq \gamma. \tag{9.140}$$

In view of (9.136) and (9.139),

$$t_{k+1} > 6L_0 + 6 - 2L_0 > 4L_0 + 4.$$

Thus the assumption we made for k also holds for $k + 1$.

Therefore by induction (see (9.128)–(9.132)), we constructed the finite strictly decreasing sequence of nonnegative integers t_i, $i = 0, \ldots, q$ such that

$$T \geq t_0 \geq T - L_0 \text{ and if } \|x(T) - \widehat{x}\| \leq \gamma, \text{ then } t_0 = T,$$

for all $i = 0, \ldots, q$, we have

$$\|x(t_i) - \widehat{x}\| \leq \gamma, \tag{9.141}$$

for all $i = 0, \ldots, q - 1$,

$$L_0 \le t_i - t_{i+1} \le 6L_0 + 6, \tag{9.142}$$

$$t_q \le 6L_0 + 6, \tag{9.143}$$

if $\|x(0) - \widehat{x}\| \le \gamma$, then $t_q = 0$.

Let $i \in \{1, \ldots, q\}$. By (9.121), (9.124), (9.141), and (9.142),

$$\sum_{t=t_i}^{t_{i-1}-1} w(by(t)) \ge U(x(t_i), x(t_{i-1}), t_{i-1} - t_i) + \gamma.$$

Property (i), (9.141), (9.142), and the inequality above imply that

$$\|x(t) - \widehat{x}\| \le \epsilon, \ t = t_i, \ldots, t_{i-1},$$

$$\|y(t) - \widehat{x}\| \le \epsilon, \ t = t_i, \ldots, t_{i-1} - 1.$$

This implies that

$$\|x(t) - \widehat{x}\| \le \epsilon, \ t \in [t_q, t_0],$$

$$\|y(t) - \widehat{x}\| \le \epsilon, \ t \in [t_q, t_0 - 1].$$

Theorem 9.14 is proved.

9.7 Stability Results

For every positive number M and every function $\phi : R_+^n \to R^1$, define

$$\|\phi\|_M = \sup\{|\phi(z)| : z \in R_+^n \text{ such that and } 0 \le z \le Me\}. \tag{9.144}$$

Let integers T_1, T_2 satisfy $0 \le T_1 < T_2$ and $w_i : R_+^n \to R^1$, $i = T_1, \ldots, T_2 - 1$ be bounded on bounded subsets of R_+^n functions. For every pair of points $z_0, z_1 \in R_+^n$, define

$$U\left(\{w_t\}_{t=T_1}^{T_2-1}, z_0, z_1\right) = \sup\left\{\sum_{t=T_1}^{T_2-1} w_t(y(t)) : \right.$$

$$\left(\{x(t)\}_{t=T_1}^{T_2}, \{y(t)\}_{t=T_1}^{T_2-1}\right) \text{ is a program such that } x(T_1) = z_0, \ x(T_2) \ge z_1 \right\},$$

$$\tag{9.145}$$

$$U\left(\{w_t\}_{t=T_1}^{T_2-1}, z_0\right) = \sup \left\{ \sum_{t=T_1}^{T_2-1} w_t(y(t)) : \right.$$

$$\left. \left(\{x(t)\}_{t=T_1}^{T_2}, \{y(t)\}_{t=T_1}^{T_2-1}\right) \text{ is a program such that } x(T_1) = z_0 \right\}. \tag{9.146}$$

If $M_0 > 0$ and $w_t = w$, $t = T_1, \ldots, T_2 - 1$, where $w : R_+^n \to R^1$ is a bounded on bounded subsets of R_+^n function, set

$$U(w, z_0, z_1, T_1, T_2) = U\left(\{w_t\}_{t=T_1}^{T_2-1}, z_0, z_1\right),$$

$$U(w, z_0, T_1, T_2) = U\left(\{w_t\}_{t=T_1}^{T_2-1}, z_0\right),$$

$$U(w, M_0, T) = \sup \left\{ \sum_{t=0}^{T-1} w(y(t)) : \right.$$

$$\left. \left(\{x(t)\}_{t=0}^{T}, \{y(t)\}_{t=0}^{T-1}\right) \text{ is a program such that } x(0) \le M_0 e \right\} \tag{9.147}$$

for every natural number T. (Here we assume that supremum over empty set is $-\infty$.) It is not difficult to see that the following result holds.

Lemma 9.16 *Let integers T_1, T_2 satisfy $0 \le T_1 < T_2$ and $w_i : R_+^n \to R^1$, $i = T_1, \ldots, T_2 - 1$ be bounded on bounded subsets of R_+^n upper semicontinuous functions. Then the following assertions hold.*

1. *For every point $z_0 \in R_+^n$, there exists a program $\left(\{x(t)\}_{t=T_1}^{T_2}, \{y(t)\}_{T_1}^{T_2-1}\right)$ such that*

$$x(T_1) = z_0, \quad \sum_{t=T_1}^{T_2-1} w_t(y(t)) = U(\{w_t\}_{t=T_1}^{T_2-1}, z_0).$$

2. *For every pair of points $z_0, z_1 \in R_+^n$ such that $U(\{w_t\}_{t=T_1}^{T_2-1}, z_0, z_1)$ is finite, there exists a program $(\{x(t)\}_{t=T_1}^{T_2}, \{y(t)\}_{t=T_1}^{T_2-1})$ such that $x(0) = z_0$, $x(T_2) \ge z_1$ and*

$$\sum_{t=T_1}^{T_2-1} w_t(y(t)) = U(\{w_t\}_{t=T_1}^{T_2-1}, z_0, z_1).$$

The following stability results hold.

Theorem 9.17 *Let $M > \max\{(a_i d)^{-1} : i = 1, \ldots, n\}$, $\epsilon > 0$ and $\Gamma \in (0, 1)$. Then there exist a natural number L and a positive number $\tilde{\gamma}$ such that for each integer $T > 2L$, each $z_0, z_1 \in R_+^n$ satisfying $z_0 \leq Me$ and $az_1 \leq \Gamma d^{-1}$, each finite sequence of functions $w_i : R_+^n \to R^1$, $i = 0, \ldots, T - 1$ which are bounded on bounded subsets of R_+^n and such that*

$$\|w_i - w(b(\cdot))\|_M \leq \tilde{\gamma}$$

for every integer $i \in \{0, \ldots, T - 1\}$ and every program $\left(\{x(t)\}_{t=0}^T, \{y(t)\}_{t=0}^{T-1}\right)$ such that

$$x(0) = z_0, \ x(T) \geq z_1,$$

$$\sum_{t=\tau}^{\tau+L-1} w_t(y(t)) \geq U(\{w_t\}_{t=\tau}^{\tau+L-1}, x(\tau), x(\tau + L)) - \tilde{\gamma}$$

for every $\tau \in \{0, \ldots, T - L\}$ and

$$\sum_{t=T-L}^{T-1} w_t(y(t)) \geq U(\{w_t\}_{t=T-L}^{T-1}, x(T - L), z_1) - \tilde{\gamma},$$

there exist integers τ_1, τ_2 such that

$$\tau_1 \in [0, L], \ \tau_2 \in [T - L, T],$$

$$\|x(t) - \hat{x}\|, \ \|y(t) - \hat{x}\| \leq \epsilon \text{ for all } t = \tau_1, \ldots, \tau_2 - 1.$$

Moreover if $|x(0) - \hat{x}\| \leq \tilde{\gamma}$, then $\tau_1 = 0$, and if $\|x(T) - \hat{x}\| \leq \tilde{\gamma}$, then $\tau_2 = T$.

Proof Theorem 9.17 follows easily form Theorem 9.14. Namely, let a natural number L and $\gamma > 0$ be as guaranteed by Theorem 9.14. Set

$$\tilde{\gamma} = \gamma(4^{-1}(L + 1))^{-1}.$$

Now it easy to see that the assertion of Theorem 9.17 holds.

Theorem 9.17 applied with $z_1 = 0$ implies the following result.

Theorem 9.18 *Let $M > \max\{(a_i d)^{-1} : i = 1, \ldots, n\}$ and $\epsilon > 0$. Then there exist a natural number L and a positive number $\tilde{\gamma}$ such that for each integer $T > 2L$, each $z_0 \in R_+^n$ satisfying $z_0 \leq Me$, each finite sequence of functions $w_i : R_+^n \to R^1$, $i = 0, \ldots, T - 1$ which are bounded on bounded subsets of R_+^n and such that*

$$\|w_i - w(b(\cdot))\|_M \leq \tilde{\gamma}$$

for each $i \in \{0, \dots, T - 1\}$ and each program $\left(\{x(t)\}_{t=0}^{T}, \{y(t)\}_{t=0}^{T-1}\right)$ which satisfies

$$x(0) = z_0,$$

$$\sum_{t=\tau}^{\tau+L-1} w_t(y(t)) \geq U(\{w_t\}_{t=\tau}^{\tau+L-1}, x(\tau), x(\tau + L)) - \tilde{\gamma},$$

for each integer $\tau \in \{0, \dots, T - L\}$ and

$$\sum_{t=T-L}^{T-1} w_t(y(t)) \geq U(\{w_t\}_{t=T-L}^{T-1}, x(T - L)) - \tilde{\gamma}$$

there are integers τ_1, τ_2 such that

$$\tau_1 \in [0, L], \quad \tau_2 \in [T - L, T],$$

$$\|x(t) - \widehat{x}\|, \quad \|y(t) - \widehat{x}\| \leq \epsilon \text{ for all } t = \tau_1, \dots, \tau_2 - 1.$$

Moreover if $\|x(0) - \widehat{x}\| \leq \gamma$, then $\tau_1 = 0$, and if $\|x(T) - \widehat{x}\| \leq \gamma$, then $\tau_2 = T$.

The following result is proved in Section 9.8.

Theorem 9.19 Let $M_0 > \max\{(a_i d)^{-1} : i = 1, \dots, n\}$, $M_1 > 0$, $\epsilon > 0$ and $\Gamma \in (0, 1)$. Then there exist a natural number L and a positive number γ such that for each integer $T > L$, each $z_0, z_1 \in R_+^n$ satisfying $z_0 \leq Me$ and $az_1 \leq \Gamma d^{-1}$, each finite sequence of functions $w_i : R_+^n \to R^1$, $i = 0, \dots, T - 1$ which are bounded on bounded subsets of R_+^n and such that

$$\|w_i - w(b(\cdot))\|_{M_0} \leq \gamma$$

for each $i \in \{0, \dots, T - 1\}$ and each program $(\{x(t)\}_{t=0}^{T}, \{y(t)\}_{t=0}^{T-1})$ such that

$$x(0) = z_0, \quad x(T) \geq z_1,$$

$$\sum_{t=0}^{T-1} w_t(y(t)) \geq U(\{w_t\}_{t=0}^{T-1}, z_0, z_1) - M_1$$

the following inequality holds:

$$\text{Card}(\{t \in \{0, \dots, T - 1\} : \max\{\|x(t) - \widehat{x}\|, \|y(t) - \widehat{x}\|\} > \epsilon\}) \leq L.$$

Theorem 9.19, applied with $z_1 = 0$, implies the following result.

Theorem 9.20 *Let $M_0 > \max\{(a_i d)^{-1} : i = 1, \ldots, n\}$, $M_1 > 0$ and $\epsilon > 0$. Then there exist a natural number L and a positive number γ such that for each integer $T > L$, each $z_0 \in R_+^n$ satisfying $z_0 \leq M\epsilon$, each finite sequence of functions $w_i : R_+^n \to R^1$, $i = 0, \ldots, T - 1$ which are bounded on bounded subsets of R_+^n and such that*

$$\|w_i - w(b(\cdot))\|_{M_0} \leq \gamma$$

for each $i \in \{0, \ldots, T - 1\}$ and each program $\left(\{x(t)\}_{t=0}^{T}, \{y(t)\}_{t=0}^{T-1}\right)$ which satisfies

$$x(0) = z_0,$$

$$\sum_{t=0}^{T-1} w_t(y(t)) \geq U(\{w_t\}_{t=0}^{T-1}, z_0) - M_1$$

the following inequality holds:

$$Card(\{t \in \{0, \ldots, T - 1\} : \max\{\|x(t) - \widehat{x}\|, \|y(t) - \widehat{x}\|\} > \epsilon\}) \leq L.$$

9.8   Proof of Theorem 9.19

Recall (see (9.82)) that $\Gamma_* \in (0, 1)$ satisfies

$$a\widehat{x} < \Gamma_* d^{-1}. \tag{9.148}$$

We may assume that

$$\Gamma_* < \Gamma. \tag{9.149}$$

By Lemma 9.10, (9.148), and (9.149), there exists

$$\gamma_0 \in (0, \min\{1, \epsilon/2\})$$

such that

$$a(\widehat{x} + \gamma e) < \Gamma d^{-1} \tag{9.150}$$

and that the following property holds:

(i) for each integer $S \geq 2$ and each program $\left(\{x(t)\}_{t=0}^{S}, \{y(t)\}_{t=0}^{S-1}\right)$ satisfying

$$\|x(0) - \widehat{x}\| \le \gamma_0, \ \|x(S) - \widehat{x}\| \le \gamma_0,$$

$$U(x(0), x(S), S) \le \sum_{t=0}^{S-1} w(by(t)) + 8\gamma_0$$

the inequalities

$$\|x(t) - \widehat{x}\| \le \epsilon, \ t = 0, \dots, S,$$

$$\|y(t) - \widehat{x}\| \le \epsilon, \ t = 0, \dots, S - 1$$

are valid.

By Proposition 9.13, there exists $m_0 > 0$ such that the following property holds:

(ii) for each $z_0 \in R_+^n$, each $z_1 \in R_+^n$ satisfying $az_1 \le \Gamma d^{-1}$, and each natural number T,

$$U(z_0, z_1, T) \ge Tw(b\widehat{x}) - m_0.$$

By Lemma 9.11, there is a natural number $L_0 > 4$ such that the following property holds:

(iii) for each integer $\tau \ge L_0$, each program $\left(\{x(t)\}_{t=0}^{\tau}, \{y(t)\}_{t=0}^{\tau-1} \right)$ satisfying

$$x(0) \le M_0 e,$$

$$\sum_{t=0}^{\tau-1} w(by(t)) \ge \tau w(b\widehat{x}) - 2M_1 - m_0 - 1$$

and each $S \in \{0, \dots, \tau - L_0\}$, there is an integer $t \in [S, \dots, S + L_0 - 1]$ such that

$$\|y(t) - \widehat{x}\|, \ \|x(t) - \widehat{x}\| \le \gamma_0.$$

Choose a natural number

$$L > 8(L_0 + 1) + 2L_0 M_1 \gamma_0^{-1} \tag{9.151}$$

and $\gamma \in (0, \gamma_0/2)$ such that

$$\gamma \le (2L + 2)^{-1} \gamma_0. \tag{9.152}$$

Assume that an integer $T > L$, $z_0, z_1 \in R_+^n$ satisfies

$$z_0 \le Me, \ az_1 \le \Gamma d^{-1}, \tag{9.153}$$

$w_i : R_+^n \to R^1, i = 0, \ldots, T-1$ are bounded on bounded subsets of R_+^n functions such that

$$\|w_i - w(b(\cdot))\|_{M_0} \le \gamma, \ i = 0, \ldots, T-1 \tag{9.154}$$

and that a program $\left(\{x(t)\}_{t=0}^T, \{y(t)\}_{t=0}^{T-1}\right)$ satisfies

$$x(0) = z_0, \ x(T) \ge z_1, \tag{9.155}$$

$$\sum_{t=0}^{T-1} w_t(y(t)) \ge U(\{w_t\}_{t=0}^{T-1}, z_0, z_1) - M_1. \tag{9.156}$$

Lemma 5.3, (9.153), and (9.155) imply that

$$x(t) \le M_0 e, \ t = 0, \ldots, T. \tag{9.157}$$

By (9.150), (9.152), (9.154), and (9.157),

$$\left\| \sum_{t=T-L}^{T-1} w(by(t)) - \sum_{t=T-L}^{T-1} w_t(y(t)) \right\| \le L\gamma \le 2^{-1}, \tag{9.158}$$

$$\left| U\left(\{w_t\}_{t=T-L}^{T-1}, z_0, z_1\right) - U(w(b(\cdot)), z_0, z_1, T-L, T) \right| \le L\gamma \le 2^{-1}. \tag{9.159}$$

Property (iii), (9.153), (9.156), and (9.158) imply that

$$\sum_{t=T-L}^{T-1} w(by(t)) \ge U(w(b(\cdot)), z_0, z_1, T-L, T) - M_1 - 1$$

$$\ge Lw(b\widehat{y}) - m_0 - M_1 - 1. \tag{9.160}$$

Property (iii), (9.151), (9.157), and (9.160) imply that there exist

$$t_0 \in \{T-L_0, \ldots, T\}, \ t_1 \in \{T-4L_0, \ldots, T-3L_0\} \tag{9.161}$$

such that

$$\|x(t_i) - \widehat{x}\| \le \gamma_0, \ i = 0, 1. \tag{9.162}$$

Assume that $k \ge 1$ is an integer and that we defined a strictly decreasing sequence of natural numbers $t_i, i = 0, \ldots, k$ such that

$$t_{i+1} < t_i, \ i = 0, \ldots, k-1,$$

for all $i = 0, \ldots, k - 1$,

$$L_0 \leq t_i - t_{i+1} \leq 4L_0, \qquad (9.163)$$

$$\|x(t_i) - \widehat{x}\| \leq \gamma_0, \; i = 0, \ldots, k, \qquad (9.164)$$

$$t_k \geq 4L_0 + 4. \qquad (9.165)$$

(Clearly, for $k = 1$ our assumption holds.) It follows from (9.150) and (9.164) that

$$ax(t_k) \leq a(\widehat{x} + \gamma_0 e) < \Gamma d^{-1}. \qquad (9.166)$$

Property (ii), (9.165), and (9.166) imply that

$$U(w(b(\cdot)), x(t_k - 4L_0 - 4), x(t_k), t_k - 4L_0 - 4, t_k) \geq (4L_0 + 4)w(b\widehat{y}) - m_0. \qquad (9.167)$$

Relations (9.151), (9.154), (9.156), (9.157), (9.165), and (9.167) imply that

$$\sum_{t=t_k-4L_0-4}^{t_k-1} w(by(t)) \geq \sum_{t=t_k-4L_0-4}^{t_k-1} w_t(y(t)) - L\gamma$$

$$\geq U(\{w_t\}_{t=t_k-4L_0-4}^{t_k-1}, x(t_k - 4L_0 - 4), x(t_k)) - M_1 - L\gamma$$

$$\geq U(w(b(\cdot)), x(t_k - 4L_0 - 4), x(t_k), t_k - 4L_0 - 4, t_k) - 2\gamma L - M_1$$

$$\geq U(w(b(\cdot)), x(t_k - 4L_0 - 4), x(t_k), t_k - 4L_0 - 4, t_k) - 1 - M_1$$

$$\geq (4L_0 + 4)w(b\widehat{y}) - m_0 - M_1 - 1. \qquad (9.168)$$

Property (iii), (9.157), and (9.168) imply that there exist

$$t_{k+1} \in \{t_k - 2L_0, \ldots, t_k - L_0\} \qquad (9.169)$$

such that

$$\|x(t_{k+1}) - \widehat{x}\| \leq \gamma_0. \qquad (9.170)$$

By (9.165) and (9.169),

$$t_{k+1} \geq 2L_0 + 4. \qquad (9.171)$$

If

$$t_{k+1} < 4L_0 + 4,$$

then the construction is completed. Otherwise the assumption made for k also holds for $k + 1$.

Therefore by induction we constructed the finite strictly decreasing sequence of natural numbers t_i, $i = 0, \ldots, q$ such that

$$T \geq t_0 \geq T - L_0,$$

$$t_{i+1} < t_i, \quad i = 0, \ldots, q - 1,$$

$$L_0 \leq t_i - t_{i+1} \leq 4L_0, \quad i = 0, \ldots, q - 1, \tag{9.172}$$

$$\|x(t_i) - \hat{x}\| \leq \gamma_0, \quad i = 0, \ldots, q, \tag{9.173}$$

$$4L_0 + 4 > t_q \geq 2L_0 + 4.$$

Set

$$E = \{i \in \{1, \ldots, q\} :$$

$$U\left(\{w_t\}_{t=t_i}^{t_{i-1}-1}, x(t_i), x(t_{i-1})\right) - \sum_{t=t_i}^{t_{i-1}-1} w_t(y(t)) \geq 4\gamma_0\}. \tag{9.174}$$

By induction we can construct a program $(\{\tilde{x}(t)\}_{t=0}^{T}, \{\tilde{y}(t)\}_{t=0}^{T-1})$ such that

$$\tilde{x}(t) = x(t), \ t = 0, \ldots, t_q, \ \tilde{y}(t) = y(t), \ t = 0, \ldots, t_q - 1, \tag{9.175}$$

for $i = q, \ldots, 1,$

$$\tilde{x}(t_i) \geq x(t_i), \tag{9.176}$$

if $i \in \{1, \ldots, q\} \setminus E$, then

$$\tilde{x}(t_i) \geq x(t_i), \ t = t_i, \ldots, t_{i-1}, \ \tilde{y}(t_i) = y(t_i), \ t = t_i, \ldots, t_{i-1} - 1, \tag{9.177}$$

if $i \in E$, then

$$\sum_{t=t_i}^{t_{i-1}-1} w_t(\tilde{y}(t)) \geq U(\{w_t\}_{t=t_i}^{t_{i-1}-1}, \tilde{x}(t_i), x(t_{i-1})) - \gamma_0, \tag{9.178}$$

$$\tilde{y}(t) = y(t), \ t \in \{t_0, \ldots, T\} \setminus \{T\}, \tag{9.179}$$

$$\tilde{x}(T) \geq x(T). \tag{9.180}$$

Proposition 4.9, (9.156), and (9.174)–(9.180) imply that

$$M_1 \geq U(\{w_t\}_{t=0}^{T-1}, z_0, z_1) - \sum_{t=0}^{T-1} w_t(y(t))$$

$$\geq \sum_{t=0}^{T-1} w_t(\tilde{y}(t)) - \sum_{t=0}^{T-1} w_t(y(t))$$

$$\geq \sum \left\{ \sum_{t=t_i}^{t_{i-1}-1} w_t(\tilde{y}(t)) - \sum_{t=t_i}^{t_{i-1}-1} w_t(y(t)) : i \in E \right\}$$

$$\geq \sum \left\{ U\left(\{w_t\}_{t=t_i}^{t_{i-1}-1}, \tilde{x}(t_i), x(t_{i-1})\right) - \gamma_0 - \sum_{t=t_i}^{t_{i-1}-1} w_t(y(t)) : i \in E \right\}$$

$$\geq \sum \left\{ U\left(\{w_t\}_{t=t_i}^{t_{i-1}-1}, x(t_i), x(t_{i-1})\right) - \gamma_0 - \sum_{t=t_i}^{t_{i-1}-1} w_t(y(t)) : i \in E \right\}$$

$$\geq 3\gamma_0 \mathrm{Card}(E)$$

and

$$\mathrm{Card}(E) \leq 3^{-1} \gamma_0^{-1} M_1. \tag{9.181}$$

Let

$$i \in \{1, \ldots, q\} \setminus E. \tag{9.182}$$

In view of (9.173),

$$\|x(t_i) - \widehat{x}\| \leq \gamma_0, \quad \|x(t_{i-1}) - \widehat{x}\| \leq \gamma_0. \tag{9.183}$$

Property (i), (9.152), (9.154), (9.155), (9.172), (9.174), and (9.182) imply that

$$\sum_{t=t_i}^{t_{i-1}-1} w_t(y(t)) \geq U(\{w_t\}_{t=t_i}^{t_{i-1}-1}, x(t_i), x(t_{i-1})) - 4\gamma_0$$

$$\geq U(w(b(\cdot)), x(t_i), x(t_{i-1}), t_i, t_{i-1}) - 4\gamma_0 - 4\gamma L_0$$

$$\geq U(w(b(\cdot)), x(t_i), x(t_{i-1}), t_i, t_{i-1}) - 6\gamma_0. \tag{9.184}$$

By (9.183) and (9.184),

$$\|x(t) - \widehat{x}\| \leq \epsilon, \quad t = t_i, \ldots, t_{i-1},$$

$$\|y(t) - \widehat{x}\| \leq \epsilon, \quad t = t_i, \ldots, t_{i-1} - 1.$$

By the relation above,

$$\{t \in \{0, \dots, T-1\} : \max\{\|x(t) - \widehat{x}\|, \|y(t) - \widehat{x}\|\} > \epsilon\}$$

$$\subset \{0, \dots, t_q\} \cup \{t_0, \dots, T\} \cup \{\{t_i, \dots, t_{i-1}\} : i \in E\}.$$

Together with (9.157), (9.172), (9.174), and (9.181), this implies that

$$\text{Card}(\{t \in \{0, \dots, T-1\} : \max\{\|x(t) - \widehat{x}\|, \|y(t) - \widehat{x}\|\} > \epsilon\})$$

$$\leq 4L_0 + 4 + L_0 + 4L_0\text{Card}(E)$$

$$\leq 4L_0\text{Card}(E) + 5L_0 + 4$$

$$\leq 2L_0 M_1 \gamma_0^{-1} + 5L_0 + 4 < L.$$

This completes the proof of Theorem 9.19.

9.9 An Auxiliary Result

Proposition 9.21 *Let* $M_0 > 0$,

$$d + a_\sigma^{-1} \neq 2, \tag{9.185}$$

$(\{u(t), v(t)\}_{t=-\infty}^{\infty} \subset R_+^n \times R_+^n$ *and for all integers* t.

$$u(t) \leq M_0 e, \tag{9.186}$$

$$(u(t), u(t+1)) \in \Omega, \ v(t) \in \Lambda(u(t), u(t+1)), \tag{9.187}$$

$$bv(t) = b_\sigma(1 + da_\sigma)^{-1}. \tag{9.188}$$

Then for all integers t,

$$u(t) = v(t) = (1 + da_\sigma)^{-1} e_\sigma = \widehat{x}.$$

Proof Consider the RSS model with the function

$$w(t) = t, \ t \in [0, \infty).$$

By Theorem 3.1, for all (w)-good programs $\{x(t), y(t)\}_{t=0}^{\infty}$,

$$\lim_{t \to \infty} x(t) = \lim_{t \to \infty} y(t) = (1 + da_\sigma)^{-1} e_\sigma = \widehat{x} \tag{9.189}$$

and the results of Chapters 3 and 4 can be applied to the RSS model with the function

$$w(t) = t, \ t \in [0, \infty).$$

For $i = 1, \ldots, n$ set

$$\widehat{p}_i = \widehat{q}_i = a_i b_i / (1 + d a_i).$$

For any $(x, x') \in \Omega$ and any $y \in \Lambda(x, x')$, set (see (3.12))

$$\delta(x, y, x') = \widehat{p}(x - x') - (w(by) - w(b\widehat{x})). \tag{9.190}$$

Lemma 2.2 implies that

$$\delta(x, y, x') \geq 0 \text{ for every } (x, x') \in \Omega \text{ and every } y \in \Lambda(x, x'). \tag{9.191}$$

By Lemma 3.8 and (9.185)–(9.187), in order to prove the proposition, it is sufficient to show that

$$\delta(u(t), v(t), u(t + 1)) = 0$$

for all integers t.

In view of (9.188), for each pair of integers $T_2 > T_1$,

$$\sum_{t=T_1}^{T_2-1} (bv(t) - b\widehat{x}) = 0.$$

Combined with Lemma 4.11, this implies that the following property holds:

(a) for each $\epsilon > 0$ there exist sequences of integers $\{t_k\}_{k=1}^{\infty}$, $\{s_k\}_{k=1}^{\infty}$ such that

$$t_k > 0, \ s_k < 0 \text{ for all integers } k \geq 1,$$

$$\lim_{k \to \infty} t_k = \infty, \ \lim_{k \to \infty} s_k = -\infty, \tag{9.192}$$

$$\|u(t_k) - \widehat{x}\|, \ \|u(s_k) - \widehat{x}\| \leq \epsilon, \ k = 1, 2, \ldots. \tag{9.193}$$

Let $\epsilon > 0$ and sequences of integers $\{t_k\}_{k=1}^{\infty}$, $\{s_k\}_{k=1}^{\infty}$ be as guaranteed by property (a).

Let $k \geq 1$ be an integer. By (9.188), (9.190), and (9.193),

$$\sum_{t=s_k}^{t_k-1} \delta(u(t), v(t), u(t + 1))$$

$$= \sum_{t=s_k}^{t_k-1} \widehat{p}(u(t) - u(t + 1)) = \widehat{p}((u(s_k) - u(t_k)) \leq 2\|\widehat{p}\|\epsilon.$$

Since ϵ is an arbitrary positive number and k is an arbitrary natural number, we have

$$\sum_{t=-\infty}^{\infty} \delta(u(t), v(t), u(t+1)) = 0.$$

Proposition 9.21 is proved.

9.10 Perturbations

Let $w : [0, \infty) \to [0, \infty)$ be a continuous strictly increasing function satisfying

$$w(0) > 0. \tag{9.194}$$

Assume that

$$d + a_\sigma^{-1} \neq 2 \tag{9.195}$$

and that

$$\sup\{w(by) : \ y \in \Lambda(x, x), \ x \in R_+^n, \ (x, x) \in \Omega\} = \mu(w). \tag{9.196}$$

Theorem 9.2 implies that

$$\{(x, y) \in R_+^n \times R_+^n : \ (x, x) \in \Omega, \ y \in \Lambda(x, x), \ w(by) = \mu(w)\}$$
$$= \left\{ (1 + da_\sigma)^{-1} e(\sigma), (1 + da_\sigma)^{-1} e(\sigma) \right\}. \tag{9.197}$$

Let $\lambda \in (0, 1)$ satisfy

$$w(0) > b_\sigma (1 + da_\sigma)^{-1}(1 - \lambda). \tag{9.198}$$

Define a function $w_\lambda : [0, \infty) \to [0, \infty)$ as follows:

$$w_\lambda(t) = w(t) + (t - b_\sigma(1 + da_\sigma)^{-1})(1 - \lambda) \tag{9.199}$$

for all

$$t \in [0, b_\sigma(1 + da_\sigma)^{-1}]$$

and

$$w_\lambda(t) = w(b_\sigma(1 + da_\sigma)^{-1}) + \lambda(w(t) - w(b_\sigma(1 + da_\sigma)^{-1}))$$
$$= \lambda w(t) + (1 - \lambda)w(b_\sigma(1 + da_\sigma)^{-1}) \tag{9.200}$$

for all

$$t > b_\sigma(1 + da_\sigma)^{-1}.$$

It follows from (9.198)–(9.200) that w_λ is a continuous strictly increasing function,

$$w_\lambda(0) > 0, \ w_\lambda(t) \leq w(t), \ t \in [0, \infty), \tag{9.201}$$

$$\mu(w_\lambda) \leq \mu(w). \tag{9.202}$$

$$\sup\{w_\lambda(by) : \ y \in \Lambda(x, x), \ x \in R_+^n, \ (x, x) \in \Omega\}$$
$$\leq \sup\{w(by) : \ y \in \Lambda(x, x), \ x \in R_+^n, \ (x, x) \in \Omega\}. \tag{9.203}$$

In view of (9.199),

$$w_\lambda(b_\sigma(1 + da_\sigma)^{-1}) = w(b_\sigma(1 + da_\sigma)^{-1}). \tag{9.204}$$

By (9.196), (9.197), and (9.202)–(9.204),

$$\mu(w_\lambda) = \mu(w) = w(b_\sigma(1 + da_\sigma)^{-1}). \tag{9.205}$$

Theorem 9.2 and (9.205) imply that

$$\{(x, y) \in R_+^n \times R_+^n : \ (x, x) \in \Omega, \ y \in \Lambda(x, x), \ w_\lambda(by) = \mu(w_\lambda)\}$$
$$= \{(1 + da_\sigma)^{-1}e(\sigma), (1 + da_\sigma)^{-1}e(\sigma)\}. \tag{9.206}$$

Theorem 9.22 *Let $\{x(t), y(t)\}_{t=0}^\infty$ be a (w_λ)-good program. Then*

$$\lim_{t \to \infty} (x(t), y(t)) = ((1 + da_\sigma)^{-1}e(\sigma), (1 + da_\sigma)^{-1}e(\sigma)) = (\widehat{x}, \widehat{x}).$$

Proof Lemma 5.3 implies that there exists $M_0 > 0$ such that

$$x(t) \leq M_0 \text{ for all integers } t \geq 0. \tag{9.207}$$

Assume that $\{t_k\}_{k=1}^\infty$ is an increasing sequence of natural numbers such that there exists

$$(\widehat{x}_0, \widehat{y}_0, \widehat{x}_1) = \lim_{k \to \infty} (x(t_k), y(t_k), x(t_k + 1)). \tag{9.208}$$

In order to complete the proof, it is sufficient to show that

$$\widehat{x}_0 = \widehat{y}_0 = \widehat{x}_1 = \widehat{x} = (1 + da_\sigma)^{-1} e(\sigma). \tag{9.209}$$

First, we show that

$$b\widehat{y}_0 = b_\sigma (1 + da_\sigma)^{-1}. \tag{9.210}$$

Since the program $\{x(t), y(t)\}_{t=0}^\infty$ is (w_λ)-good, we have that the sequence

$$\left\{ \sum_{t=0}^{T-1} (w_\lambda(by(t)) - \mu(w_\lambda)) \right\}_{T=1}^\infty \quad \text{is bounded.} \tag{9.211}$$

If

$$\lim_{T \to \infty} \sum_{t=0}^{T-1} (w(by(t)) - \mu(w)) = -\infty,$$

then in view of (9.202) and (9.205),

$$\lim_{T \to \infty} \sum_{t=0}^{T-1} (w_\lambda(by(t)) - \mu(w_\lambda)) = -\infty.$$

This contradicts (9.211). The contradiction we have reached proves that the sequence

$$\left\{ \sum_{t=0}^{T-1} (w(by(t)) - \mu(w)) \right\}_{T=1}^\infty$$

is bounded too. Together with (9.205) and (9.211), this implies that there exists $M_1 > 0$ such that for each natural number T,

$$\left| \sum_{t=0}^{T-1} (w_\lambda(by(t)) - \mu(w_\lambda)) \right| \le M_1, \tag{9.212}$$

$$\left| \sum_{t=0}^{T-1} (w(by(t)) - \mu(w)) \right| \le M_1. \tag{9.213}$$

Assume that

$$b\widehat{y}_0 > b_\sigma (1 + da_\sigma)^{-1}. \tag{9.214}$$

Fix

$$\Delta \in (b_\sigma (1 + da_\sigma)^{-1}, b\widehat{y_0}).$$ (9.215)

In view of (9.208) and (9.215), we may assume without loss of generality that for all natural numbers k,

$$by(t_k) > \Delta.$$ (9.216)

By (9.200), (9.215), and (9.216), there exists $\delta > 0$ such that for each natural number k,

$$w(by(t_k)) \geq w_\lambda (by(t_k)) + \delta.$$ (9.217)

It follows from (9.213) and (9.217) that for each natural number k,

$$M_1 \geq \sum_{t=0}^{t_k} (w(by(t)) - \mu(w))$$

$$= \sum_{i=1}^{k} (w(by(t_i)) - \mu(w))$$

$$+ \sum \{w(by(t)) - \mu(w) : t \in \{0, \ldots, t_k\} \setminus \{t_1, \ldots, t_k\}\}$$

$$\geq \sum_{i=1}^{k} (w(by(t_i)) - w_\lambda(by(t_i)))$$

$$+ \sum_{i=1}^{k} (w_\lambda(by(t_i)) - \mu(w_\lambda))$$

$$+ \sum \{w_\lambda(by(t)) - \mu(w_\lambda) : t \in \{0, \ldots, t_k\} \setminus \{t_1, \ldots, t_k\}\}$$

$$\geq \delta k + \sum_{t=0}^{t_k} (w_\lambda(by(t)) - \mu(w_\lambda)) \geq \delta k - M_1$$

and

$$2M_1 \geq \delta k \to \infty$$

as $k \to \infty$. The contradiction we have reached proves that

$$b\widehat{y_0} \leq b_\sigma (1 + da_\sigma)^{-1}.$$ (9.218)

Assume that

$$b\widehat{y_0} < b_\sigma (1 + da_\sigma)^{-1}. \tag{9.219}$$

Fix

$$\Delta \in (b\widehat{y_0}, b_\sigma (1 + da_\sigma)^{-1}). \tag{9.220}$$

By (9.208) and (9.220), we may assume without loss of generality that for each natural number k,

$$by(t_k) < \Delta. \tag{9.221}$$

By (9.199), (9.220), and (9.221), there exists $\delta > 0$ such that for each natural number k,

$$w(by(t_k)) \geq w_\lambda (by(t_k)) + \delta. \tag{9.222}$$

It follows from (9.201), (9.212), (9.213), and (9.222) that for each natural number k,

$$M_1 \geq \sum_{t=0}^{t_k}(w(by(t)) - \mu(w))$$

$$= \sum_{i=1}^{k}(w(by(t_i)) - \mu(w))$$

$$+ \sum \{w(by(t)) - \mu(w) : t \in \{0, \ldots, t_k\} \setminus \{t_1, \ldots, t_k\}\}$$

$$\geq \sum_{i=1}^{k}(w(by(t_i)) - w_\lambda(by(t_i)))$$

$$+ \sum_{i=1}^{k}(w_\lambda(by(t_i)) - \mu(w_\lambda))$$

$$+ \sum \{w_\lambda(by(t)) - (\mu(w_\lambda)) : t \in \{0, \ldots, t_k\} \setminus \{t_1, \ldots, t_k\}\}$$

$$\geq \delta k + \sum_{t=0}^{t_k}(w_\lambda(by(t)) - \mu(w_\lambda)) \geq \delta k - M_1$$

and

$$2M_1 \geq \delta k \to \infty$$

as $k \to \infty$. The contradiction we have reached proves that

$$b\widehat{y}_0 \geq b_\sigma (1 + da_\sigma)^{-1}.$$

Together with (9.218) this implies that

$$b\widehat{y}_0 = b_\sigma (1 + da_\sigma)^{-1}.$$

This implies that

$$\lim_{t \to \infty} by(t) = b_\sigma (1 + da_\sigma)^{-1}. \tag{9.223}$$

In view of (9.213), for each pair of integers $T_2 > T_1 \geq 0$,

$$\left| \sum_{t=T_1}^{T_2-1} (w_\lambda(by(t)) - \mu(w_\lambda)) \right| \leq 2M_1. \tag{9.224}$$

For each integer $k \geq 1$, define $\{x^{(k)}(t),\, y^{(k)}(t)\}_{t=-t_k}^{\infty}$ by

$$x^{(k)}(t) = x(t + t_k),\ \ y^{(k)}(t) = y(t + t_k). \tag{9.225}$$

Extracting subsequences and using (9.207) and the diagonalization process, we obtain that there exist a strictly increasing sequence of natural numbers $\{k_p\}_{p=1}^{\infty}$ such that for each integer t, there exist

$$u_t = \lim_{p \to \infty} x^{(k_p)}(t),\ \ v_t = \lim_{p \to \infty} y^{(k_p)}(t). \tag{9.226}$$

By (9.208), (9.225), and (9.226),

$$u_0 = \widehat{x}_0,\ \ u_1 = \widehat{x}_1,\ \ v_0 = \widehat{y}_0, \tag{9.227}$$

for each integer t,

$$(u_t, u_{t+1}) \in \Omega,\ \ v_t \in \Lambda(u_t, u_{t+1}). \tag{9.228}$$

It follows from (9.223), (9.225), and (9.226) that for every integer t,

$$bv_t = b_\sigma (1 + da_\sigma)^{-1}. \tag{9.229}$$

Proposition 9.21, (9.228), and (9.229) imply that for all integers t,

$$u_t = v_t = (1 + da_\sigma)^{-1} e(\sigma) = \widehat{x}.$$

Together with (9.227) this implies that

$$\widehat{x}_0 = \widehat{x}_1 = \widehat{y}_0 = (1 + da_\sigma)^{-1} e(\sigma).$$

Theorem 9.22 is proved.

9.11 Generic Results

Denote by $\mathcal{A}$ the set of all continuous increasing functions $w : [0, \infty) \to [0, \infty)$. For every pair of positive numbers ϵ, M, set

$$\mathcal{U}(\epsilon, M) = \{(w_1, w_2) \in \mathcal{A} : |w_1(z) - w_2(z)| \le \epsilon \text{ for all } z \in [0, M]\}. \tag{9.230}$$

We equip the set $\mathcal{A}$ with the uniformity determined by the base $\mathcal{U}(\epsilon, M)$, where $\epsilon, M > 0$. It is not difficult to see that the uniform space $\mathcal{A}$ is metrizable (by a metric $\rho_\mathcal{A}$) and complete.

Let $w \in \mathcal{A}$ and $x \in R^n_+$. Set

$$\mu(w, x) = \sup \left\{ \limsup_{T \to \infty} T^{-1} \sum_{t=0}^{T-1} w(by(t)) : \right.$$

$$\left. \{x(t), y(t)\}_{t=0}^\infty \text{ is a program such that } x(0) = x \right\} \tag{9.231}$$

and

$$\mu(w) = \sup\{\mu(w, x) : x \in R^n_+\}.$$

Let M_0 be a positive number and let $T \ge 1$ be an integer. Set

$$U(w, M_0, T) = \sup \left\{ \sum_{t=0}^{T-1} w(by(t)) : \right.$$

$$\left. \left(\{x(t)\}_{t=0}^T, \{y(t)\}_{t=0}^{T-1}\right) \text{ is a program such that } x(0) \le M_0 e \right\}. \tag{9.232}$$

Let

$$M_* > \max\{(a_i d)^{-1} : i = 1, \dots, n\}. \tag{9.233}$$

Lemma 5.3 and (9.233) imply that

$$\text{if } (x, x') \in \Omega \text{ and } x \le M_* e, \text{ then } x' \le M_* e. \tag{9.234}$$

Let $x \in R_+^n$. Set

$$a(x) = \{x' \in R_+^n : (x, x') \in \Omega\}$$
$$= \{x \in R_+^n : x' \ge (1-d)x, \ a(x' - (1-d)x) \le 1\}$$
$$= (1-d)x + \{u \in R_+^n : au \le 1\}. \tag{9.235}$$

For each pair of nonempty sets $A, B \subset R_+^n$, its Hausdorff distance is defined by

$$H(A, B) = \max \left\{ \sup_{x \in A} \inf_{y \in B} \|x - y\|, \ \sup_{y \in B} \inf_{x \in A} \|x - y\| \right\}. \tag{9.236}$$

In view of (9.235) and (9.236), for all $x, y \in R_+^n$,

$$H(a(x), a(y)) \le (1-d)\|x - y\|. \tag{9.237}$$

Relations (9.234) and (9.237) imply the following result.

Proposition 9.23 *Let $\{x(t), y(t)\}_{t=0}^{\infty}$ be a program. Then for all sufficiently large natural numbers t,*

$$x(t) \le (M_* + 1)e.$$

Proposition 9.23 implies the following result.

Proposition 9.24 *For every $w \in \mathcal{A}$,*

$$\mu(w) = \sup\{\mu(w, x) : x \in R_+^n \text{ and } x \le (M_* + 1)e\}.$$

Suppose that

$$d + a_\sigma^{-1} \ne 2. \tag{9.238}$$

Denote by $\mathcal{A}_0$ the set of all $w \in \mathcal{A}$ such that

$$\mu(w) = w(b_\sigma(1 + da_\sigma)^{-1}). \tag{9.239}$$

Proposition 9.25 *$\mathcal{A}_0$ is a closed set in $(\mathcal{A}, \rho_\mathcal{A})$.*

Proof Let $\{w_k\}_{k=1}^{\infty} \subset \mathcal{A}_0$, $w \in \mathcal{A}$ and

$$\lim_{k \to \infty} \rho_A(w_k, w) = 0. \tag{9.240}$$

Proposition 9.24 and (9.240) imply that

$$w(b_\sigma(1 + da_\sigma)^{-1}) = \lim_{k \to \infty} w_k(b_\sigma(1 + da_\sigma)^{-1}). \tag{9.241}$$

It follows from (9.230), (9.231), and (9.240) that for every natural number k,

$$|\mu(w_k) - \mu(w)|$$
$$\times |\sup\{\mu(w_k, x) : x \in R_+^n \text{ and } x \le (M_* + 1)e\}$$
$$- \sup\{\mu(w, x) : x \in R_+^n \text{ and } x \le (M_* + 1)e\}|$$
$$\le \sup\{|\mu(w, x) - \mu(w_k, x)| : x \in R_+^n \text{ and } x \le (M_* + 1)e\}$$
$$\le \sup\{|w(bz) - w_k(bz)| : z \in R_+^n \text{ and } z \le (M_* + 1)e\} \to 0$$

as $k \to \infty$. This implies that

$$\lim_{k \to \infty} \mu(w_k) = \mu(w).$$

Together with (9.239) and (9.241), this implies that

$$\mu(w) = w(b_\sigma(1 + da_\sigma)^{-1}).$$

Thus $w \in \mathcal{A}_0$. Proposition 9.25 is proved.

We consider the complete metric space $(\mathcal{A}_0, \rho_A)$. Clearly, the function

$$w_*(t) = t, \ t \in [0, \infty) \text{ belongs to } \mathcal{A}_0. \tag{9.242}$$

It is not difficult to see that the following two propositions hold.

Proposition 9.26 *Let $w \in \mathcal{A}_0$ and $\alpha \in (0, 1]$. Then the function*

$$\alpha w(z) + (1 - \alpha), \ z \in [0, \infty)$$

belongs to $\mathcal{A}_0$.

Proposition 9.27 *Let $w \in \mathcal{A}_0$ and $\alpha \in (0, 1)$. Then the function*

$$\alpha w(z) + (1 - \alpha)w(b\widehat{x})z(b\widehat{x})^{-1}, \ z \in [0, \infty)$$

belongs to $\mathcal{A}_0$.

Proposition 9.27 implies the following result.

Proposition 9.28 *There exists a set $\mathcal{B} \subset \mathcal{A}_0$ which is a countable intersection of open everywhere dense subsets of $(\mathcal{A}_0, \rho_\mathcal{A})$ such that each $w \in \mathcal{B}$ is strictly increasing.*

Let $\mathcal{B} \subset \mathcal{A}_0$ be as guaranteed by Proposition 9.28.
Proposition 9.28 and Theorem 9.22 imply the following result.

Proposition 9.29 *There exists a set $\mathcal{B}_0 \subset \mathcal{B}$ which is everywhere dense in $(\mathcal{A}_0, \rho_\mathcal{A})$ such that for each $w \in \mathcal{B}_0$ and each (w)-good program $\{x(t), y(t)\}_{t=0}^{\infty}$,*

$$\lim_{t \to \infty} x(t) = \lim_{t \to \infty} y(t) = (1 + da_\sigma)^{-1} e(\sigma).$$

Theorem 9.30 *There exists a set $\mathcal{G} \subset \mathcal{A}_0$ which is a countable intersection of open everywhere dense sets in $(\mathcal{A}_0, \rho_\mathcal{A})$ such that $\mathcal{G} \subset \mathcal{B}$ and for each $w \in \mathcal{G}$ and each (w)-good program $\{x(t), y(t)\}_{t=0}^{\infty}$,*

$$\lim_{t \to \infty} x(t) = \lim_{t \to \infty} y(t) = (1 + da_\sigma)^{-1} e(\sigma).$$

Proof Let the set $\mathcal{B}_0$ be as guaranteed by Proposition 9.29. Assume that $u \in \mathcal{B}_0$ and $q_0, q_1 \geq 1$ be integers. By Theorem 9.19, there exist a natural number $L(u, q_0, q_1)$ and an open neighborhood $\mathcal{U}((u, q_0, q_1)$ of u in $\mathcal{A}_0$ such that the following property holds:

(a) for each integer $T > L(u, q_0, q_1)$, each $w \in \mathcal{U}((u, q_0, q_1)$, and each program $\left(\{x(t)\}_{t=0}^{T}, \{y(t)\}_{t=0}^{T-1} \right)$ which satisfies

$$x(0) \leq q_0 e,$$

$$\sum_{t=0}^{T-1} w(by(t)) \geq U(w, q_0, T) - q_1$$

the following inequality holds:

Card($\{t \in \{0, \dots, T-1\}$:

$$\max\{\|x(t) - (1 + da_\sigma)^{-1} e(\sigma)\|, \ \|y(t) - (1 + da_\sigma)^{-1} e(\sigma)\|\} > q_1^{-1} n\})$$
$$\leq L(u, q_0, q_1).$$

Set

$$\mathcal{G} = \mathcal{B} \cap (\cap_{q_0=1}^{\infty} \cap_{q_1=1}^{\infty} \cup \{\mathcal{U}((u, q_0, q_1) : \ u \in \mathcal{B}_0\}). \tag{9.243}$$

Clearly, $\mathcal{G}$ is a countable intersection of open everywhere dense sets in $\mathcal{A}_0$.
Let $w \in \mathcal{G}$. The inclusion $\mathcal{G} \subset \mathcal{B}$ implies that w is strictly increasing.

Let a program $\{x(t), y(t)\}_{t=0}^{\infty}$ be (w)-good, and let an integer

$$q_0 > \max\{(da_i)^{-1} : i = 1, \dots, n\} \tag{9.244}$$

be such that

$$x(0) \le q_0 e. \tag{9.245}$$

Lemma 5.3, (9.244), and (9.245) imply that

$$x(t) \le q_0 e, \ t = 0, 1, \dots. \tag{9.246}$$

By Theorem 5.5, there exists an integer $q_1 > 0$ such that for each integer $p \ge 1$,

$$U(w, q_0 p) \le \sum_{t=0}^{p-1} w(by(t)) + q_1. \tag{9.247}$$

Let $\epsilon > 0$. Choose a natural number q_2 such that

$$4q_2^{-1} < \epsilon, \ q_2 > 4q_1. \tag{9.248}$$

In view of (9.243), there exists

$$u \in \mathcal{B}_0 \tag{9.249}$$

such that

$$w \in \mathcal{U}(u, q_0, q_2). \tag{9.250}$$

Property (a), (9.246), and (9.248)–(9.250) imply that for all integers $T > L(u, q_0, q_2)$,

Card($\{t \in \{0, \dots, T-1\}$:

$\max\{\|x(t) - (1 + da_\sigma)^{-1} e(\sigma)\|, \ \|y(t) - (1 + da_\sigma)^{-1} e(\sigma)\|\} > \epsilon\}$)

$\le L(u, q_0, q_2).$

Therefore

Card($\{t \in \{0, 1, \dots\}$:

$\max\{\|x(t) - (1 + da_\sigma)^{-1} e(\sigma)\|, \ \|y(t) - (1 + da_\sigma)^{-1} e(\sigma)\|\} > \epsilon\}$)

$\le L(u, q_0, q_2).$

This implies that

$$\lim_{t \to \infty} x(t) = \lim_{t \to \infty} y(t) = \widehat{x}.$$

Theorem 9.30 is proved.

Chapter 10
An Autonomous One-Dimensional Model

The one-dimensional Robinson–Solow–Srinivasan model was studied by T. Mitra and M. Ali Khan in [39–45]. In this chapter we discuss the results obtained in [39] which was the starting point of their research. In [39] the value-loss approach of Radner–Gale–McKenzie was used in order to show a multiplicity of optimal programs under certain conditions, and a theory of undiscounted dynamic programming was used to derive properties of the optimal policy correspondence.

10.1 The Model

Let $a > 0, d \in (0, 1)$ and

$$\Omega = \{(x, x') \in R_+^2 : x' \geq (1 - d)x \text{ and } a(x' - (1 - d)x) \leq 1\}.$$

For every $(x, x') \in \Omega$ define

$$\Lambda(x, x') = \{y \in R_+^1 : y \leq x \text{ and } y \leq 1 - a(x' - (1 - d)x)\}$$

and

$$u(x, x') = \sup\{y \in \Lambda(x, x')\}.$$

Let $x_0 \in R_+^1$. A sequence $\{x(t), y(t)\}_{t=0}^{\infty}$ is called a feasible program from x_0 if

$$x(0) = x_0$$

and for all nonnegative integers t,

© The Author(s), under exclusive license to Springer Nature Switzerland AG 2020   341
A. J. Zaslavski, *Turnpike Theory for the Robinson–Solow–Srinivasan Model*,
Springer Optimization and Its Applications 166,
https://doi.org/10.1007/978-3-030-60307-6_10

$$(x(t), x(t+1)) \in \Omega \text{ and } y(t) \in \Lambda(x(t), x(t+1)).$$

A sequence $\{x(t), y(t)\}_{t=0}^{\infty}$ is called a program from x_0 if

$$x(0) = x_0$$

and for all nonnegative integers t,

$$(x(t), x(t+1)) \in \Omega$$

and

$$y(t) = u(x(t), x(t+1)).$$

A sequence $\{x(t), y(t)\}_{t=0}^{\infty}$ is called a program if it is a program from $x(0)$. Let $\{x(t), y(t)\}_{t=0}^{\infty}$ be a program. We associate with it a gross investment sequence $\{z(t+1)\}_{t=0}^{\infty}$ and a consumption sequence $\{c(t+1)\}_{t=0}^{\infty}$ defined by

$$z(t+1) = x(t+1) - (1-d)x(t) \text{ and } c(t+1) = y(t)$$

for all integers $t \geq 0$. It is not difficult to see that every program $\{x(t), y(t)\}_{t=0}^{\infty}$ is bounded by

$$M(x(0)) := \max\{x(0), \bar{x}\},$$

where

$$\bar{x} = (ad)^{-1}$$

is the maximum sustainable capital stock. A program $\{x(t), y(t)\}_{t=0}^{\infty}$ is called stationary if

$$(x(t), x(t+1)) = (x(t+1), x(t+2))$$

for all nonnegative integers t. For a stationary program $\{x(t), y(t)\}_{t=0}^{\infty}$, we have $x(t) \leq \bar{x}$ for all nonnegative integers t.

A program $\{x^*(t), y^*(t)\}_{t=0}^{\infty}$ is called weakly optimal if

$$\liminf_{T \to \infty} \sum_{t=0}^{T} [u(x(t), x(t+1)) - u(x^*(t), x^*(t+1))] \leq 0$$

for every program $\{x(t), y(t)\}_{t=0}^{\infty}$ such that $x(0) = x^*(0))$. A stationary weakly optimal program is a program which is stationary and weakly optimal.

A program $\{x(t), y(t)\}_{t=0}^{\infty}$ is full-employment if for all integers $t \geq 0$,

$$y(t) + a(x(t+1) - (1-d)x(t)) = 1.$$

It is not difficult to see that the following result holds (see Proposition 1 of [39].

Proposition 10.1

(1) *For every program* $\{x(t), y(t)\}_{t=0}^{\infty}$, *there exists a full-employment program* $\{x'(t), y'(t)\}_{t=0}^{\infty}$ *such that* $x(0) = x'(0))$ *and* $y'(t) = y(t)$ *for all integers* $t \geq 0$.
(2) *If* $\{x(t), y(t)\}_{t=0}^{\infty}$ *is a program from* x *and* $x' > x$, *then there exists a full-employment program* $\{x'(t), y'(t)\}_{t=0}^{\infty}$ *from* x' *such that* $y'(t) \geq y(t)$ *for all integers* $t \geq 0$ *and* $y'(t) > y(t)$ *for some integer* $t \geq 0$.
(3) *If* $\{x(t), y(t)\}_{t=0}^{\infty}$ *is a program from* x *which is not a full-employment program, then there is a full-employment program* $\{x't), y'(t)\}_{t=0}^{\infty}$ *from* x *such that* $y'(t) \geq y(t)$ *for all integers* $t \geq 0$ *and* $y'(t) > y(t)$ *for some integer* $t \geq 0$.

10.2 A Golden Rule

We define a stock $\widehat{x} \in R_+^1$ as a golden rule if $(\widehat{x}, \widehat{x}) \in \Omega$ and

$$u(\widehat{x}, \widehat{x}) \geq u(x, x')$$

for all $(x, x') \in \Omega$ such that $x' \geq x$.

The following result is proved in [39]. Its proof is given in Section 10.6.

Proposition 10.2

(1) *The pair*

$$(\widehat{x}, \widehat{p}) = ((1 + ad)^{-1}, a(1 + ad)^{-1})$$

satisfies

$$(\widehat{x}, \widehat{x}) \in \Omega$$

and

$$u(\widehat{x}, \widehat{x}) \geq u(x, x') + \widehat{p}x' - \widehat{p}x \text{ for all } (x, x') \in \Omega. \tag{10.1}$$

(2) $\widehat{x}$ *is a unique golden-rule stock.*

We refer to $\widehat{p}$, given by Proposition 10.2, as the golden price and to the pair $(\widehat{x}, \widehat{p})$ as the golden rule. The value loss (relative to the golden rule) from operating at $(x, x') \in \Omega$ and $y = u(x, x')$ is defined by

$$\delta(x, x', y) = u(\widehat{x}, \widehat{x}) - [u(x, x') + \widehat{p}x' - \widehat{p}x] \geq 0. \tag{10.2}$$

We split the value loss in (10.2) as follows:

$$\alpha(x, x', y) = \widehat{p}d(x - y) \geq 0,$$

$$\beta(x, x', y) = (\widehat{p}/a)(1 - y - a(x' - (1 - d)x)) \geq 0,$$

$$\delta(x, x', y) = \alpha(x, x', y) + \beta(x, x', y). \tag{10.3}$$

Let $\{x(t), y(t)\}_{t=0}^{\infty}$ be a program. Then for all integers $t \geq 0$,

$$(x(t), x(t + 1)) \in \Omega \text{ and } y(t) = u(x(t), x(t + 1)).$$

It follows from (10.2) and (10.3) that for all integers $t \geq 0$,

$$u(\widehat{x}, \widehat{x}) = u(x(t), x(t + 1)) + \widehat{p}x(t + 1) - \widehat{p}x(t) + \alpha(t) + \beta(t)$$

$$= u(x(t), x(t + 1)) + \widehat{p}x(t + 1) - \widehat{p}x(t) + \beta(t) \tag{10.4}$$

where

$$\alpha(t) = \alpha(x(t), x(t + 1), y(t)),$$

$$\beta(t) = \beta(x(t), x(t + 1), y(t)),$$

$$\delta(t) = \delta(x(t), x(t + 1), y(t))$$

for every integer $t \geq 0$. This yields a useful identity, relating the sum of value losses to the sum of utility differences from the golden-rule utility level, along any program $\{x(t), y(t)\}_{t=0}^{\infty}$ and any integer $T \geq 0$:

$$\sum_{t=0}^{T} [u(x(t), x(t + 1)) - u(\widehat{x}, \widehat{x})]$$

$$= \widehat{p}x(0) - \widehat{p}x(T + 1) - \sum_{t=0}^{T} \delta(t). \tag{SVL}$$

We can eliminate the variable $y(t) = u(x(t), x(t+1))$ from (10.4), by using (10.3), to obtain for all integers $t \geq 0$,

$$x(t + 1) = a^{-1} - \xi x(t) + Ax(t) - B(t)$$

where

$$\xi = a^{-1} - (1 - d)$$

and

$$A(t) = (1/ad\widehat{p})\alpha(t), \quad B(t) = (1/\widehat{p})\beta(t)$$

for all integers $t \geq 0$. On measuring capital stocks relative to the golden-rule stock, $X(t) = x(t) - \widehat{x}$, and on noting that

$$\widehat{x} = a^{-1} - \xi\widehat{x},$$

we obtain that

$$X(t+1) - \xi X(t) + A(t) - B(t) \text{ for all integers } t \geq 0. \tag{10.5}$$

In this chapter we assume that

$$\xi > 0.$$

Recall that a program $\{x(t), y(t)\}_{t=0}^{\infty}$ is good if there is a number G such that for all integers $T \geq 1$,

$$\sum_{t=0}^{T}[u(x(t), x(t+1)) - u(\widehat{x}, \widehat{x})] \geq G.$$

The following result is proved in [39]. Its proof is given in Section 10.6.

Theorem 10.3

(1) *For every $x_0 \in R_+^n$ there exists a good program $\{x(t), y(t)\}_{t=0}^{\infty}$ satisfying $x(0) = x_0$.*
(2) *If $\{x(t), y(t)\}_{t=0}^{\infty}$ is a good program, then $\sum_{t=0}^{\infty} \delta(t) < \infty$.*
(3) *If $\xi \neq 1$ and $\{x(t), y(t)\}_{t=0}^{\infty}$ is a good program, then*

$$(x(t), y(t)) \to (\widehat{x}, \widehat{x}) \text{ as } t \to \infty$$

and

$$\sum_{t=0}^{T}[u(x(t), x(t+1)) - u(\widehat{x}, \widehat{x})]$$

$$= \widehat{p}x(0) - \widehat{p}\widehat{x} - \sum_{t=0}^{\infty} \delta(t). \tag{US}$$

(4) *Suppose that* $\xi = 1$, $\{x(t), y(t)\}_{t=0}^{\infty}$ *is a good program and that* $\tilde{x}$ *is a limit point of the sequence* $\{x(t)\}_{t=0}^{\infty}$. *Then either* $x(t)$ *converges to* $\tilde{x}$ *for all odd periods and to* $2\hat{x} - \tilde{x}$ *for all even periods or* $x(t)$ *converges to* $\tilde{x}$ *for all even periods and to* $2\hat{x} - \tilde{x}$ *for all odd periods.*

Note that Assertions 1–3 of Theorem 10.3 were proved in Chapters 2 and 3 for the multidimensional model (see Propositions 2.8 and 2.9 and Theorem 3.2) while Assertion 4 does not have a multidimensional analog.

10.3   Optimality and Value-Loss Minimization

The following result plays an important role in the study in [39]. It follows from Propositions 2.10 and 2.11.

Proposition 10.4

(1) *If* $\{x(t), y(t)\}_{t=0}^{\infty}$ *is a program such that*

$$\sum_{t=0}^{\infty} \delta(t) \leq \sum_{t=0}^{\infty} \delta'(t) \qquad\qquad (VLM)$$

for every program $\{x'(t), y'(t)\}_{t=0}^{\infty}$ *satisfying* $x(0) = x'(0)$, *then* $\{x(t), y(t)\}_{t=0}^{\infty}$ *is weakly optimal.*
(2) *For every* $x \in R_+^1$ *there exists a program* $\{x(t), y(t)\}_{t=0}^{\infty}$ *from x such that*

$$\sum_{t=0}^{\infty} \delta(t) \leq \sum_{t=0}^{\infty} \delta'(t)$$

for every program $\{x'(t), y'(t)\}_{t=0}^{\infty}$ *from x.*
(3) *For every* $x \in R_+^1$ *there exists a weakly optimal program from x.*

The next auxiliary result obtained [39] is proved in Section 10.6.

Lemma 10.5

(1) *For every full-employment program* $\{x(t), y(t)\}_{t=0}^{\infty}$ *and for every integer* $T > 1$,

$$\sum_{t=0}^{T} [\alpha(t)(-\xi)^{-t}] = ad\xi[\hat{p}X(0) - \hat{p}(X(T+1)(-\xi)^{-T-1})]. \qquad (10.6)$$

(2) *Every weakly optimal program is a full-employment program and satisfies (10.6).*

10.4 Optimality Does Not Imply Value-Loss Minimization

The following two results obtained [39] are proved in Section 10.6.

Proposition 10.6 *Let* $\{x(t), y(t)\}_{t=0}^{\infty}$ *be a weakly optimal program. Then it is good. If* $\xi \neq 1$, *then*

$$\lim_{t \to \infty} \widehat{p}(x(t) - \widehat{x}) = 0 \tag{10.7}$$

and

$$\sum_{t=0}^{\infty} \delta \leq \sum_{t=0}^{\infty} \delta'(t) \tag{VLM}$$

for every program $\{x'(t), y'(t)\}_{t=0}^{\infty}$ *satisfying* $x'(0) = x(0)$. *Moreover,*

$$\sum_{t=0}^{T} [u(x(t), x(t+1)) - u(\widehat{x}, \widehat{x})]$$

$$= \widehat{p}x(0) - \widehat{p}\widehat{x} - \sum_{t=0}^{\infty} \delta(t). \tag{US}$$

If $\xi \neq 1$, (US) of Proposition 10.6 can be rewritten as

$$\sum_{t=0}^{T} [u(x(t), x(t+1)) - u(\widehat{x}, \widehat{x})] = \widehat{p}x(0) - \widehat{p}\widehat{x} - \delta(x) \tag{10.8}$$

where

$$\delta(x) = \sum_{t=0}^{\infty} \delta(t).$$

By Proposition 10.6, $\delta(x)$ is well-defined and depends only on x.

Theorem 10.7 *Assume that* $\xi = 1$ *and* $x_0 \in (\widehat{x}, 1)$.

(1) The full-employment program $\{x(t), y(t)\}_{t=0}^{\infty}$ *from* x_0 *satisfying* $x(t) = \widehat{x}$ *for all integers* $t \geq 1$ *is weakly optimal from* x_0.
(2) The full-employment program $\{x'(t), y'(t)\}_{t=0}^{\infty}$ *from* x_0 *satisfying* $x'(t) = x_0$ *for all even* t *and* $x'(t) = 2\widehat{x} - x_0$ *for all odd* t *is weakly optimal from* x_0.
(3) There is a continuum of weakly optimal programs from x_0.

10.5   Optimal Policy Function

The optimal policy correspondence is a correspondence h from R_+^1 to subsets of R_+^1 such that, given any $x \in R_+^1$, there is a weakly optimal program $\{x(t), y(t)\}_{t=0}^{\infty}$ from x such that $x(1) \in h(x)$.

Define a function $V : R_+^1 \to R_+^1$ such that for every $x \in R_+^1$,

$$V(x) = \sum_{t=0}^{T} [u(x(t), x(t+1)) - u(\widehat{x}, \widehat{x})] \tag{10.9}$$

where $\{x(t), y(t)\}_{t=0}^{\infty}$ is any weakly optimal program from x. Clearly, $V(x)$ is independent of the choice of the weakly optimal program (see (10.8)).

It follows from (US) of Theorem 10.3(3), (10.8), and (10.9) that for every good program $\{x'(t), y'(t)\}_{t=0}^{\infty}$ from x,

$$V(x) \geq \sum_{t=0}^{T} [u(x'(t), x'(t+1)) - u(\widehat{x}, \widehat{x})]. \tag{10.10}$$

For every $x \in R_+^1$ define

$$\Omega(x) = \{x' \in R_+^1 : (x, x') \in \Omega\}.$$

The next result obtained in [39] is proved in Section 10.6.

Proposition 10.8

(1) The value function V is a concave and strictly increasing function on R_+^1 and continuous on $(0, \infty)$ satisfying $V(\widehat{x}) = 0$.

(2) V satisfies the equation

$$V(x) = \max\{u(x, x') - u(\widehat{x}, \widehat{x}) + V(x') : x' \in \Omega(x)\}$$

for all $x \in R_+^1$.

(3) $\{x(t), y(t)\}_{t=0}^{\infty}$ is a weakly optimal program if and only if for every integer $t \geq 0$,

$$V(x(t)) = u(x(t), x(t+1)) - u(\widehat{x}, \widehat{x}) + V(x(t+1)).$$

Since any program starting from a point in $X = [0, 1/ad]$ remains in X, we assume that X is our space of states and confirm our solution of the optimal policy correspondence to this domain.

When $\xi \neq 1$ we can use Proposition 10.8 in order to describe the optimal policy correspondence for certain subsets of X. More precisely, set

$$k = \widehat{x}(1-d)^{-1}, \quad A = [0, \widehat{x}], \quad B = (\widehat{x}, k), \quad C = (k, 1/ad]. \tag{10.11}$$

Proposition 10.8 implies the following results.

Corollary 10.9 *Assume that* $\xi \neq 1$. *Then the optimal policy correspondence* h : $X \to R_+^1$ *satisfies*

$$h(x) = a^{-1} - \xi x, \quad x \in A,$$
$$h(x) = (1-d)x, \quad x \in C. \tag{10.12}$$

Corollary 10.10

(1) Assume that $\xi > 1$. *Then the optimal policy correspondence* h : $X \to R_+^1$ *satisfies*

$$h(x) = \widehat{x}, \quad x \in (\widehat{x}, k). \tag{10.13}$$

(2) Assume that $\xi < 1$. *Then the optimal policy correspondence* h : $X \to R_+^1$ *satisfies*

$$h(x) = a^{-1} - \xi x, \quad x \in (\widehat{x}, 1),$$
$$h(x) = (1-d)x, \quad x \in [1, k). \tag{10.14}$$

Corollary 10.11 *Assume that* $\xi = 1$. *Then the optimal policy correspondence* h : $X \to R_+^1$ *satisfies*

$$a^{-1} - \xi x \in h(x), \quad x \in A$$
$$[a^{-1} - \xi x, \widehat{x}] \subset h(x), \quad x \in (\widehat{x}, 1)$$
$$[(1-d)x, \widehat{x}] \subset h(x), \quad x \in [1, k),$$
$$(1-d)x \in h(x), \quad x \in C. \tag{10.15}$$

10.6 Proofs

Proof of Proposition 10.2

(1) It is not difficult to see that

$$(\widehat{x}, \widehat{x}) \in \Omega \text{ and } u(\widehat{x}, \widehat{x}) = \widehat{x}.$$

For every $(x, x') \in \Omega$ and every $y \in \Lambda(x, x')$, set

$$\alpha(x, x', y) = d\widehat{p}(x - y)$$

and

$$\beta(x, x', y) = (\widehat{p}/a)(1 - y - a(x' - (1 - d)x)).$$

Clearly,

$$\alpha(x, x', y) \geq 0 \text{ and } \beta(x, x', y) \geq 0.$$

We have

$$
\begin{aligned}
y + \widehat{p}x' - \widehat{p}x &= y + \widehat{p}(x' - (1 - d)x) - d\widehat{p}x \\
&= (1 + ad)^{-1}y + \widehat{p}(x' - (1 - d)x) - \alpha(x, x', y) \\
&= (1 + ad)^{-1}(y + a(x' - (1 - d)x)) - \alpha(x, x', y) \\
&= \widehat{x}(y + a(x - (1 - d)x)) - \alpha(x, x', y) \\
&= \widehat{x} - \widehat{x}(1 - y - a(x' - (1 - d)x)) - \alpha(x, x', y) \\
&= \widehat{x} - \beta(x, x', y) - \alpha(x, x', y). \qquad (10.16)
\end{aligned}
$$

Equation (10.16) implies the validity of Proposition 10.2(1).
(2) Assume that x is a golden-rule stock. Then

$$(x, x) \in \Omega \text{ and } \widehat{x} \in \Lambda(x, x).$$

In view of (10.16),

$$\beta(x, x, y) = \alpha(x, x, y) = 0$$

and $x = y = \widehat{x}$. Proposition 10.2(2) is proved.

Proof of Theorem 10.3

(1) Define

$$y(0) = 0$$

and

$$y(t + 1) = (1 - d)y(t) + d\widehat{x}$$

for all integers $t \geq 0$. Clearly, $\{y(t)\}_{t=0}^{\infty}$ is a monotonically non-decreasing sequence which converges to $\widehat{x}$ as $t \to \infty$. Define

$$z(t+1) = d\widehat{x}$$

for all integers $t \geq 0$.

Let $x \in R_+^1$ be given. Define $x(0) = x$ and

$$x(t+1) = (1-d)x(t) + z(t+1)$$

for all integers $t \geq 0$. It is easy to see that $\{x(t), y(t)\}_{t=0}^{\infty}$ is a program and that for all integers $t \geq 0$,

$$y(t) - \widehat{x} = (1-d)^t(y(0) - \widehat{x}).$$

This implies that the program $\{x(t), y(t)\}_{t=0}^{\infty}$ is good.

(2) In view of (SVL), for every integer $T \geq 0$, we have

$$\sum_{t=0}^{T}[u(x(t), x(t+1)) - u(\widehat{x}, \widehat{x})]$$

$$= \widehat{p}x(0) - \widehat{p}\widehat{x} - \sum_{t=0}^{\infty}\delta(t).$$

Since the program $\{x(t), y(t)\}_{t=0}^{\infty}$ is good, there exists a number G such that for all integers $T \geq 0$,

$$\sum_{t=0}^{T}[u(x(t), x(t+1)) - u(\widehat{x}, \widehat{x})] \geq G.$$

Hence

$$\sum_{t=0}^{T}\delta(t) \leq \widehat{p}x(0) - G$$

for all integers $T \geq 0$. This implies that $\sum_{t=0}^{\infty}\delta(t) < \infty$.

(3) Since the program $\{x(t), y(t)\}_{t=0}^{\infty}$ is good, it follows from assertion 2 and (10.3) that

$$\sum_{t=0}^{\infty}A(t) < \infty, \quad \sum_{t=0}^{\infty}B(t) < \infty.$$

Let $\epsilon > 0$ be given. There exists a natural number τ such that

$$\sum_{t=\tau}^{\infty} A(t) + \sum_{t=\tau}^{\infty} B(t) < \epsilon/3. \tag{10.17}$$

Set

$$D = 2M(x).$$

Since $x(t) \le M(x)$ for all integers $t \ge 0$ and $\widehat{x} \le \bar{x} \le M(x)$, we conclude that

$$|X(t)| \le D \text{ for all integers } t \ge 0.$$

The equation

$$0 < \xi = a^{-1} - (1 - d)$$

implies that we have two possible cases: (a) $0 < \xi < 1$; (b) $\xi > 1$. Hence, there exists a natural number ν such that

$$\min\{D\xi^{\nu}, D\xi^{-\nu}\} < \epsilon/3. \tag{10.18}$$

Let a natural number $T \ge \tau$ be given. Then for all natural numbers $n \ge \nu$, it follows from (10.5) that

$$X(T + n) = (-\xi)^n X(T)$$

$$+ \sum_{s=0}^{n-1} (-\xi)^{n-1-s} A(T + s) - \sum_{s=0}^{n-1} (-\xi)^{n-1-s} B(T + s). \tag{10.19}$$

In case (a), by (10.17)–(10.19), we have $|X(T + n)| < \epsilon$. Consider case (b). We divide (10.19) by $(-\xi)^n$ and obtain

$$(-\xi)^{-n} X(T + n) = X(T)$$

$$+ \sum_{s=0}^{n-1} (-\xi)^{-1-s} A(T + s) - \sum_{s=0}^{n-1} (-\xi)^{-1-s} B(T + s). \tag{10.20}$$

In view of (10.17), (10.18), and (10.20), $|X(T)| < \epsilon$. Since $T \ge \tau$ and $n \ge \nu$ are arbitrary, we conclude that in both cases

$$|x(t) - \widehat{x}| = |X(t)| \to 0 \text{ as } t \to \infty.$$

Proposition 10.1 implies that

$$y(t) = 1 - a(x(t + 1) - (1 - d)x(t))$$

for all integers $t \geq 0$ and

$$y(t) \to 1 - a(\widehat{x} - (1-d)\widehat{x}) = \widehat{x} \text{ as } t \to \infty.$$

It follows from assertion 2 and the relation

$$x(t) \to \widehat{x} \text{ as } t \to \infty$$

that the right-hand side of (SVL) has a limit as $T \to \infty$ and the left-hand side of (SVL) also has a limit as $T \to \infty$. Taking limits in (SVL), by letting $T \to \infty$, we obtain (US).

(4) Let $\epsilon > 0$ be given. As in the proof of (3), choose a natural number τ such that (10.17) holds. For every integer $T \geq \tau$, we sum (10.5) from T to $T + s$, except that for s odd, use the equation in (10.5) with a negative sign and conclude that

$$-\epsilon/3 \leq -\sum_{t=T}^{\infty} A(t) - \sum_{t=T}^{\infty} B(t)$$

$$\leq X(T) + (-1)^s X(T+s+1)$$

$$\leq \sum_{t=T}^{\infty} A(t) + \sum_{t=T}^{\infty} B(t) \leq \epsilon/3. \tag{10.21}$$

Let $\tilde{x}$ be an arbitrary limit point of the sequence $\{x(t)\}_{t=0}^{\infty}$. Set

$$\tilde{X} = \tilde{x} - \widehat{x}.$$

There exists a subsequence $\{t_r\}_{r=1}^{\infty}$ such that $\{X(t_r)\}_{r=1}^{\infty}$ converges to $\tilde{X}$. There is μ such that for all integers $r \geq \mu$, we have

$$t_r \geq \tau,$$

$$\tilde{X} - \epsilon/3 \leq X(t_r) \leq \tilde{X} + \epsilon/3.$$

By (10.21) for integers $r \geq \mu$ and integers $s \geq 0$,

$$\tilde{X} - 2\epsilon/3 \leq -X(t_r) - \epsilon/3 \leq (-1)^s X(t_r + s + 1)$$

$$\leq -X(t_r) + \epsilon/3 \leq -\tilde{X} + 2\epsilon/3.$$

Thus, for even positive integers s, we obtain that

$$-\tilde{X} - 2\epsilon/3 \leq X(t_\mu + s + 1) \leq -\tilde{X} + 2\epsilon/3,$$

so that $X(t_\mu + s)$ converges to $-\tilde{X}$ for all odd integers s. For odd natural numbers s, we obtain that

$$-\tilde{X} - 2\epsilon/3 \le -X(t_r + s + 1) \le -\tilde{X} + 2\epsilon/3.$$

This implies that

$$-\tilde{X} - 2\epsilon/3 \le X(t_r + s + 1) \le \tilde{X} + 2\epsilon/3$$

and that $X(t_\mu + s)$ converges to $\tilde{X}$ for all odd integers s.

If t_μ is odd, then $X(t)$ converges to $-\tilde{X}$ for even integers and $X(t)$ converging to $\tilde{X}$ for odd integers. If t_μ is even, then $X(t)$ converges to $-\tilde{X}$ for odd integers and $X(t)$ converging to $\tilde{X}$ for even integers. This completes the proof of Theorem 10.3.

Proof of Lemma 10.5 Let a sequence $\{x(t), y(t)\}_{t=0}^\infty$ be a full-employment program. Then $B(t) = 0$ for all integers $t \ge 0$. In view of (10.5),

$$A(t)\xi^{-t} = X(t+1)\xi^{-t} + \xi X(t)\xi^{-t} \text{ for all integers } t \ge 0. \tag{10.22}$$

Let a natural number T be given. We sum (10.22) from $t = 0$ to $t = T$, except that for t odd, use (10.22) and obtain

$$\sum_{t=0}^{T} A(t)(-\xi)^{-t} = \xi(X(0) - X(T+1)(-\xi)^{-T-1}). \tag{10.23}$$

Equation (10.6) follows from (10.23) and the equality

$$A(t)\alpha(t)/\widehat{p}\bar{a}d, \ t = 0, 1, 2, \ldots.$$

(2) Let $\{x(t), y(t)\}_{t=0}^\infty$ be a weakly optimal program. By Assertion (3) of Proposition 10.1, it is a full-employment program. In view of Assertion (1), it satisfies (10.6). Lemma 10.5 is proved.

Proof of Proposition 10.6 Let $\{x(t), y(t)\}_{t=0}^\infty$ be a weakly optimal program from x and $\{x'(t), y'(t)\}_{t=0}^\infty$ be a program from x. By (10.4), for every natural number T,

$$\sum_{t=0}^{T-1} [u(x'(t), x'(t+1)) - u(\widehat{x}, \widehat{x})]$$

$$= \widehat{p}x'(0) - \widehat{p}x'(T) - \sum_{t=0}^{T-1} \delta'(t) \tag{10.24}$$

and

$$\sum_{t=0}^{T-1} [u(x'(t), x'(t+1)) - u(x(t), x(t+1))]$$

$$= \widehat{p}x(T) - \widehat{p}x'(T) - \sum_{t=0}^{T-1} \delta'(t) + \sum_{t=0}^{T-1} \delta(t). \tag{10.25}$$

Assertion 1 of Theorem 10.3 implies that there exists a good program $\{\bar{x}(t), \bar{y}(t)\}_{t=0}^{\infty}$ satisfying $\sum_{t=0}^{\infty} \bar{\delta}(t) < \infty$. Using this program in place of $\{x'(t), y'(t)\}_{t=0}^{\infty}$ in (10.25) and the weak optimality of $\{x(t), y(t)\}_{t=0}^{\infty}$, we conclude that $\sum_{t=0}^{\infty} \delta(t) < \infty$. Using $\{x(t), y(t)\}_{t=0}^{\infty}$ in place of $\{x'(t), y'(t)\}_{t=0}^{\infty}$ in (10.24), we obtain that $\{x(t), y(t)\}_{t=0}^{\infty}$ is good. Now Assertion (3) of Theorem 10.3 implies (10.7).

Since the program $\{x(t), y(t)\}_{t=0}^{\infty}$ is good, we have

$$\sum_{t=0}^{\infty} \delta(t) < \infty.$$

If $\{x'(t), y'(t)\}_{t=0}^{\infty}$ is not good, then $\sum_{t=0}^{\infty} \delta'(t) = \infty$ and (VLM) holds. If $\{x'(t), y'(t)\}_{t=0}^{\infty}$ is good, then $\sum_{t=0}^{\infty} \delta'(t) < \infty$, by Theorem 10.3,

$$\lim_{t \to \infty} \widehat{p}x(t) = \widehat{px}$$

and

$$\lim_{t \to \infty} \widehat{p}x'(t) = \widehat{px}.$$

Thus the right-hand side of (10.25) has a limit as $T \to \infty$. Therefore the left-hand side of (10.25) has a limit as $T \to \infty$ which is equal to

$$\sum_{t=0}^{\infty} \delta(t) - \sum_{t=0}^{\infty} \delta'(t).$$

Now the weak optimality of $\{x(t), y(t)\}_{t=0}^{\infty}$ implies (VLM). Theorem 10.3 implies (US). Proposition 10.6 is proved.

Proof of Theorem 10.7 The equality $\xi = 1$ implies that

$$\widehat{x} = 1/2a, \quad \widehat{p} = 1/2, \quad 1/a = 1 + (1-d) > 1,$$

$$a \in (0, 1), \quad ad \in (0, 1).$$

(1) Since the program $\{x(t), y(t)\}_{t=0}^{\infty}$ is full-employment, we conclude that for all integers $t \geq 0$,

$$\beta = 0, \ \delta(t) = \alpha(t).$$

It is easy to see that for all integers $t \geq 1$,

$$X(t) = 0, \ \delta(t) = 0.$$

Lemma 10.5 and equality $\xi = 1$ imply that for every natural number T,

$$\sum_{t=0}^{T} \delta(t) = \delta(0) = \alpha(0) = \sum_{t=0}^{T} \alpha(t)(-\xi)^{-t} = ad\widehat{p}X(0). \tag{10.26}$$

If the program $\{x(t), y(t)\}_{t=0}^{\infty}$ is not weakly optimal, then there exist a positive number θ, an integer N, and a program $\{x'(t), y'(t)\}_{t=0}^{\infty}$ from x_0 such that for every integer $T > N$, we have

$$\theta \leq \sum_{t=0}^{T-1} [u(x'(t), x'(t+1)) - u(x(t), x(t+1))]. \tag{10.27}$$

In view of Proposition 10.1, we may assume without loss of generality that $\{x'(t), y'(t)\}_{t=0}^{\infty}$ is a full-employment program from x_0. Therefore for all integers $t \geq 0$,

$$\beta'(t) = 0, \ \delta'(t) = \alpha'(t).$$

Lemma 10.5 and the equality $\xi = 1$ imply for every natural number T,

$$\sum_{t=0}^{T} \delta'(t) \geq \sum_{t=0}^{T} \alpha'(t)(-1)^{t}$$

$$= ad\widehat{p}X(0) - ad\widehat{p}X'(T+1)(-1)^{T+1}. \tag{10.28}$$

By (10.26) arfd (10.28),

$$\sum_{t=0}^{T} \delta(t) - \sum_{t=0}^{T} \delta'(t) \leq ad\widehat{p}X'(T)(-1)^{T}. \tag{10.29}$$

It follows from (10.25), (10.27), and (10.29) that for every integer $T > N$,

$$-\widehat{p}X'(T) + ad\widehat{p}X'(T)(-1)^{T} \geq \theta. \tag{10.30}$$

In view of (10.27), the program $\{x'(t), y'(t)\}_{t=0}^{\infty}$ is good. By (10.30), for odd integer $T > N$,

$$-X'(T) \geq \theta/\widehat{p}(1+ad) = \theta/a.$$

Theorem 10.3 implies that for even $T > N$, $X'(T)$ converges to a positive number as $T \to \infty$. On the other hand, for even $T > N$, by (10.30),

$$\widehat{p}X'(T)(ad-1) \geq \theta,$$

$$X'(T) \leq -\theta/\widehat{p}(1-ad) < 0,$$

a contradiction. Thus $\{x(t), y(t)\}_{t=0}^{\infty}$ is a weakly optimal program.

(2) It is easy to see that for the program $\{x'(t), y'(t)\}_{t=0}^{\infty}$ for every integer $t \geq 0$,

$$\alpha'(t) = \beta'(t) = 0$$

and (VLM) holds. Proposition 10.4 implies that it is weakly optimal.

(3) Since the function h is convex, there is a continuum of weakly optimal programs from x_0. This completes the proof of Theorem 10.7. $\qquad \blacksquare$

Proof of Proposition 10.8 The golden-rule program defined by

$$(\widehat{x}(t), \widehat{y}(t)) = (\widehat{x}, \widehat{x}), \quad t = 0, 1, \ldots$$

has zero value loss in each period, and it is weakly optimal in view of Proposition 10.4. It follows from (10.10) that

$$V(\widehat{x}) = 0.$$

If $\{x(t), y(t)\}_{t=0}^{\infty}$ is a weakly optimal program from x and $x' > x$, then we set

$$x'(0) = x',$$

$$x'(t+1) = x'(t) + z(t+1) \text{ for every integer } t \geq 0$$

and

$$y'(t) = y(t) \text{ for every integer } t \geq 0,$$

where $\{z(t+1)\}_{t=0}^{\infty}$ is the investment sequence associated with the program $\{x(t), y(t)\}_{t=0}^{\infty}$. It is not difficult to see that $x'(t) > x(t)$ for every integer $t \geq 0$ and $\{x'(t), y'(t)\}_{t=0}^{\infty}$ is a program from x'. Therefore, V is non-decreasing on R_+^1.

Let us show that V is strictly increasing on R_+^1. If $z(t+1) = 0$ for all integers $t \geq 0$, then $x(t) \to 0$ as $t \to \infty$, a contradiction. Therefore $z(t+1) > 0$ for some integer $t \geq 0$. Let T be the first period for which $z(T+1) > 0$. If $x' > x$, we define

$$x'(0) = x',$$

$$x'(t+1) = x'(t) + z(t+1) \text{ if } t \neq T,$$

$$x'(T+1) = x'(T) + z'(T+1),$$

where

$$z'(T+1) = z(T+1) - \epsilon,$$

$$0 < \epsilon < z(T+1)$$

and ϵ is sufficiently close to zero so that

$$x'(T+1) > x(T+1).$$

Set

$$y'(t) = y(t) \text{ if } t \neq T,$$

$$y'(T) = y(T) + \epsilon.$$

It is not difficult to see that $x'(t) > x(t)$ for all integers $t \geq 0$ and that $\{x'(t), y'(t)\}_{t=0}^{\infty}$ is a program from x'. Hence, the function V is strictly increasing on R_+^1.

Since the set Ω is convex and the function u is concave, we conclude that the value function is concave on R_+^1 and continuous on $(0, \infty)$.

(2) Assume that $\{x(t), y(t)\}_{t=0}^{\infty}$ is a weakly optimal program from x. By definition,

$$V(x) = [u(x, x(1)) - u(\widehat{x}, \widehat{x})] + \sum_{t=1}^{\infty}[u(x(t), x(t+1)) - u(\widehat{x}, \widehat{x})].$$

For all integers $t \geq 0$, define

$$(x'(t), y'(t)) = (x(t+1), y(t+1)).$$

It is easy to see that $\{x'(t), y'(t)\}_{t=0}^{\infty}$ is a good program and that

$$\sum_{t=1}^{\infty}[u(x(t), x(t+1)) - u(\widehat{x}, \widehat{x})]$$

$$= \sum_{t=0}^{\infty}[u(x'(t), x'(t+1)) - u(\widehat{x}, \widehat{x})] \leq V(x(1))$$

and

$$V(x) \leq [u(x, x(1)) - u(\widehat{x}, \widehat{x})] + V(x(1)). \tag{10.31}$$

Assume that $(x, x') \in \Omega$ and that $\{x'(t), y'(t)\}_{t=0}^{\infty}$ is a weakly optimal program from x'. Define

$$(x(0), y(0)) = (x, \max \Lambda(x, x')),$$

$$(x(t), y(t)) = (x'(t-1), y'(t-1)) \text{ for all integers } t \geq 1.$$

It is not difficult to see that $\{x(t), y(t)\}_{t=0}^{\infty}$ is a good program from x. By (10.9) and (10.10), we have

$$V(x) \geq [u(x, x') - u(\widehat{x}, \widehat{x})] + \sum_{t=1}^{\infty} [u(x(t), x(t+1)) - u(\widehat{x}, \widehat{x})]$$

$$= [u(x, x') - u(\widehat{x}, \widehat{x})] + \sum_{t=1}^{\infty} [u(x'(t-1), x'(t)) - u(\widehat{x}, \widehat{x})]$$

$$= [u(x, x') - u(\widehat{x}, \widehat{x})] + \sum_{t=0}^{\infty} [u(x'(t), x'(t+1)) - u(\widehat{x}, \widehat{x})]$$

$$= [u(x, x') - u(\widehat{x}, \widehat{x})] + V(x'). \tag{10.32}$$

Clearly, if $\{x(t), y(t)\}_{t=0}^{\infty}$ is a weakly optimal program, then equality holds in (10.31). This implies that

$$V(x(1)) = \sum_{t=1}^{\infty} [u(x(t), x(t+1)) - u(\widehat{x}, \widehat{x})],$$

for all integers $t \geq 0$,

$$V(x(t)) = [u(x(t), x(t+1)) - u(\widehat{x}, \widehat{x})] + V(x(t+1)) \tag{10.33}$$

and that

$$V(x) = \max\{[u(x, x') - u(\widehat{x}, \widehat{x})] + V(x') : (x, x') \in \Omega\}.$$

(3) We already proved one half of (iii) in (10.33). Let us now prove the other half. Assume that a program $\{x(t), y(t)\}_{t=0}^{\infty}$ satisfies for every integer $t \geq 0$,

$$V(x(t)) = [u(x(t), x(t+1)) - u(\widehat{x}, \widehat{x})] + V(x(t+1)).$$

Then, for every integer $T \geq 0$,

$$V(x(0)) = \sum_{t=0}^{T}[u(x(t), x(t+1)) - u(\widehat{x}, \widehat{x})] + V(x(T+1)). \qquad (10.34)$$

By monotonicity of V on X and the inequality $x(t) \leq M(x(0))$, $t = 0, 1, \ldots$, we define

$$m = V(M(x(0)))$$

and obtain that

$$V(x(t)) \leq V(M(x(0))) = m$$

for all integers $t \geq 0$. Together with (10.34) this implies that

$$\sum_{t=0}^{T}[u(x(t), x(t+1)) - u(\widehat{x}, \widehat{x})] = V(x(0)) - V(x(T+1)) \geq V(x(0)) - m.$$

Therefore the program $\{x(t), y(t)\}_{t=0}^{\infty}$ is good and

$$x(t) \to \widehat{x} \text{ as } t \to \infty.$$

It follows from the continuity of V that

$$V(x(t)) \to V(\widehat{x}) = 0 \text{ as } t \to \infty.$$

Together with (10.34) this implies that

$$\lim_{T \to \infty} \sum_{t=0}^{T}[u(x(t), x(t+1)) - u(\widehat{x}, \widehat{x})]$$

exists, is finite, and satisfies

$$V(x) = \sum_{t=0}^{T}[u(x(t), x(t+1)) - u(\widehat{x}, \widehat{x})].$$

By (10.8), (10.9), (US), and Theorem 10.3,

$$\delta(x(0)) = \sum_{t=0}^{\infty}\delta(t)$$

and $\{x(t), y(t)\}_{t=0}^{\infty}$ is a weakly optimal program. Proposition 10.8 is proved.

Proof of Corollary 10.9 In view of (10.9) and concavity of V,

$$V(x) - V(\widehat{x}) \le \widehat{p}(x - \widehat{x})$$

for all $x > \widehat{x}$,

$$V'_+(\widehat{x}) \le \widehat{p}$$

and

$$V'_+(x) \le V'_+(\widehat{x}) \le \widehat{p} \text{ for all } x > \widehat{x}.$$

Consider the case with $x \in C$. Then for $(x, x') \in \Omega$ we have

$$\bar{x} := (1 - d)x \le x'.$$

Assume, contrary to the claim of the corollary, that there exists $x' > \bar{x}$ such that

$$x' \in h(x).$$

Then

$$x' > \widehat{x},$$
$$V(x') - V(\bar{x}) \le V'_+(\bar{x})(x' - \bar{x}) \le \widehat{p}(x' - \bar{x}) < a(x' - \bar{x}).$$

Together with Proposition 10.8(3), this implies that

$$u(x, x') + V(x') = 1 - a(x' - (1 - d)x) + (V(x') - V(\bar{x})) + V(\bar{x})$$
$$< 1 - a(x' - \bar{x}) + a(x' - \bar{x}) + V(\bar{x})$$
$$= u(x, \bar{x}) + V(\bar{x}) \le V(x) + u(\widehat{x}, \widehat{x}). \tag{10.35}$$

Clearly, (10.35) contradicts Proposition 10.8(2).
    Consider the case $x \in A$. Set

$$\bar{x} = a^{-1} - \xi x.$$

We have

$$\bar{x} \ge \widehat{x}.$$

Assume, contrary to the corollary, that there exists

$$x' \in h(x)$$

satisfying

$$x' \neq \bar{x}.$$

If $x' < \bar{x}$, then for $y \in \Lambda(x, x')$,

$$a(x' - (1 - d)x) + y$$
$$< a(\bar{x} - (1 - d)x) + x = 1,$$

so labor is not fully employed, a contradiction to Lemma 10.5(2).
Assume that $x' > \bar{x}$. Then

$$x' > \bar{x} \geq \widehat{x},$$
$$V(x') - V(\bar{x}) \leq V'_+(\bar{x})(x' - \bar{x}) \leq \widehat{p}(x' - \bar{x}).$$

Together with Proposition 10.8(3), this implies that

$$u(x, x') + V(x') = 1 - a(x' - (1 - d)x) + (V(x') - V(\bar{x})) + V(\bar{x})$$
$$\leq 1 - a(\bar{x} - (1 - d)x) - a(x' - \bar{x}) + \widehat{p}(x' - \bar{x}) + V(\bar{x})$$
$$< x + V(\bar{x})$$
$$= u(x, \bar{x}) + V(\bar{x}) \leq V(x) + u(\widehat{x}, \widehat{x}). \tag{10.36}$$

Clearly, (10.36) contradicts Proposition 10.8(2). Corollary 10.9 is proved.

Proof of Corollary 10.10

(1) Set

$$\bar{x}(0) = x,$$
$$\bar{x}(t) = \widehat{x} \text{ for all integers } t \geq 0.$$

Clearly,

$$(\bar{x}(t), \bar{x}(t+1)) \in \Omega \text{ for all integers } t \geq 0.$$

Set

$$\bar{y}(t) = u(\bar{x}(t), \bar{x}(t+1)) \text{ for all integers } t \geq 0.$$

It is easy to see that $\{\bar{x}(t), \bar{y}(t)\}_{t=0}^{\infty}$ is a full-employment program, for all
integers $t \geq 0$,

$$\bar{\beta}(t) = 0, \quad \bar{\delta}(t) = \bar{\alpha}(t)$$

and for all integers $t \geq 1$,

$$\bar{X}(t) = 0, \quad \bar{\delta}(t) = 0.$$

Lemma 10.5 implies that

$$\sum_{t=0}^{\infty} \bar{\delta}(t) = \bar{\delta}(0) = \bar{\alpha}(0)$$

$$= \sum_{t=0}^{\infty} \bar{\alpha}(t)(-\xi)^{-t} = ad\xi \, \hat{p} \bar{X}(0). \tag{10.37}$$

Assume that $\{x'(t), y'(t)\}_{t=0}^{\infty}$ is a weakly optimal program satisfying $x(0) = x$. Then $\{x'(t), y'(t)\}_{t=0}^{\infty}$ is a full-employment program. Lemma 10.5 and the inequality $\xi > 1$ imply that

$$\sum_{t=0}^{\infty} \delta'(t) \geq \sum_{t=0}^{\infty} \alpha'(t) \geq \sum_{t=0}^{\infty} \alpha'(t)\xi^{-t}$$

$$\geq \sum_{t=0}^{\infty} \alpha'(t)(-\xi)^{-t} = ad\xi \, \hat{p} X(0). \tag{10.38}$$

Proposition 10.6 implies that

$$\sum_{t=0}^{\infty} \delta'(t) \leq \sum_{t=0}^{\infty} \delta(t) \tag{10.39}$$

for every program $\{x(t), y(t)\}$ satisfying $x(0) = x$. In view of (10.37) and (10.38),

$$\sum_{t=0}^{\infty} \delta'(t) \geq \sum_{t=0}^{\infty} \bar{\delta}(t).$$

Thus

$$\sum_{t=0}^{\infty} \delta'(t) = \sum_{t=0}^{\infty} \bar{\delta}(t).$$

Together with (10.39) this implies that

$$\sum_{t=0}^{\infty} \bar{\delta}(t) \le \sum_{t=0}^{\infty} \delta(t)$$

for every program $\{x(t), y(t)\}$ satisfying $x(0) = x$. Proposition 10.4 implies that the program $\{\bar{x}(t), \bar{y}(t)\}$ is weakly optimal. Since

$$\sum_{t=0}^{\infty} \delta'(t) = \sum_{t=0}^{\infty} \bar{\delta}(t)$$

we have equality in all the inequalities of (10.38). Together with (10.5) this implies that for all integers $t \ge 1$,

$$\alpha'(t) = 0, \ \ X'(t+1) = (-\xi)X'(t).$$

Since $x'(t) \le M(x)$ for all integers $t \ge 0$ and $\xi > 1$, we conclude that $X'(1) = 0$, $x'(1) = \hat{x}$ and that $\hat{x} = h(x)$.

(2) Consider the case $x \in (\hat{x}, 1)$. Set

$$x(0) = x,$$

$$x(t+1) = -\xi(x(t) - \hat{x}) + \hat{x} \text{ for all integers } t \ge 0.$$

It is easy to see that $(x(t), x(t+1)) \in \Omega$ for all integers $t \ge 0$. Set

$$y(t) = u(x(t), x(t+1)) \text{ for all integers } t \ge 0.$$

Then $\{x(t), y(t)\}_{t=0}^{\infty}$ is a full-employment program with $\delta(t) = 0$ for all integers $t \ge 0$ in view of (10.5). Proposition 10.4 implies that the program $\{x(t), y(t)\}_{t=0}^{\infty}$ is weakly optimal and

$$x(1) = a^{-1} - \xi x \in h(x).$$

Proposition 10.6 implies that if $\{x'(t), y'(t)\}_{t=0}^{\infty}$ is a weakly optimal program satisfying $x'(0) = x$, then

$$\delta'(t) = 0 \text{ for all integers } t \ge 0$$

and in view of (10.5),

$$x'(1) = -\xi(x - \hat{x}) + \hat{x} = a^{-1} - \xi x.$$

Thus

$$h(x) = a^{-1} - \xi x.$$

Consider the case $x \in [1, k)$. Set

$$x(0) = x,$$
$$x(1) = (1 - d)x,$$
$$x(t + 1) = -\xi(x(t) - \hat{x}) + \hat{x} \text{ for all integers } t \geq 1.$$

It is easy to see that $(x(t), x(t + 1)) \in \Omega$ for all integers $t \geq 0$. Set

$$y(t) = u(x(t), x(t + 1)) \text{ for all integers } t \geq 0.$$

Then $\{x(t), y(t)\}_{t=0}^{\infty}$ is a full-employment program with $y(0) = 1$, $\delta(t) = 0$ for all integers $t \geq 1$ in view of (10.5). If $\{x'(t), y'(t)\}_{t=0}^{\infty}$ is a program satisfying $x'(0) = x$, then

$$x'(1) \geq (1 - d)x,$$
$$y'(1) \leq 1 = y(t),$$
$$\delta'(0) \geq \alpha'(0) \geq \alpha(0) = \delta(0).$$

Proposition 10.4 implies that the program $\{x(t), y(t)\}_{t=0}^{\infty}$ is weakly optimal and

$$x(1) = (1 - d)x \in h(x).$$

Proposition 10.6 implies that if $\{x'(t), y'(t)\}_{t=0}^{\infty}$ is a weakly optimal program satisfying $x'(0) = x$, then

$$\delta'(0) = \delta(0),$$
$$\delta'(t) = 0 \text{ for all integers } t \geq 1.$$

Thus

$$x'(1) = x(1) = (1 - d)x$$

and

$$h(x) = (1 - d)x.$$

Corollary 10.10 is proved.

Proof of Corollary 10.11 Consider the case $x \in (\hat{x}, 1)$. Following the proof of Theorem 10.7, we obtain that

$$[a^{-1} - \xi x, \hat{x}] \subset h(x).$$

Consider the case $x \in [1, k)$. Following the proof of Corollary 10.10(2), we obtain that

$$(1 - d)x \in h(x).$$

Following the proof of Theorem 10.7(1), we obtain that

$$\widehat{x} \in h(x).$$

Since the function h is convex, we conclude that

$$[(1 - d)x, \widehat{x}] \subset h(x).$$

Consider the case $x \in C$. Let $T \geq 0$ be the smallest integer such that

$$\bar{x} = (1 - d)^T x < 1.$$

Then $T \geq 1$ and

$$(1 - d)^T x \geq (1 - d).$$

Set

$$x(t) = (1 - d)^t x, \ t = 0, \ldots, T,$$
$$x(t + 1) = 2\widehat{x} - x(t) \text{ for all integers } t \geq T.$$

It is easy to see that $(x(t), x(t + 1)) \in \Omega$ for all integers $t \geq 0$. Set

$$y(t) = u(x(t), x(t + 1)) \text{ for all integers } t \geq 0.$$

Then $\{x(t), y(t)\}_{t=0}^{\infty}$ is a full-employment program with $\delta(t) = 0$ for all integers $t \geq T$. If $\{x'(t), y'(t)\}_{t=0}^{\infty}$ is a program satisfying $x'(0) = x$, then for all integers $t = 0, \ldots, T - 1$

$$x'(t) \geq x(t), \ y'(t) \leq 1 = y(t),$$
$$\delta'(t) \geq \alpha'(t) \geq \alpha(t) = \delta(t).$$

Proposition 10.4 implies that the program $\{x(t), y(t)\}_{t=0}^{\infty}$ is weakly optimal and

$$x(1) = (1 - d)x \in h(x).$$

Consider the case $x \in A$. Set

$$\bar{x} = a^{-1} - \xi x.$$

Then $\bar{x} \geq \hat{x}$. If $\bar{x} \in (\hat{x}, 1)$, define the program $\{x(t), y(t)\}_{t=0}^{\infty}$ as in the proof of Corollary 10.10(2), dealing with $x \in (\hat{x}, 1)$. If $\bar{x} \in [1, k)$, define the program $\{x(t), y(t)\}_{t=0}^{\infty}$ as in the proof of Corollary 10.10(2), dealing with $x \in ([1, k)$. If $\bar{x} \in C$, define the program $\{x(t), y(t)\}_{t=0}^{\infty}$ as in the above paragraph. Then, in each case, consider the program $\{x'(t), y'(t)\}_{t=0}^{\infty}$ defined by

$$(x'(0), y'(0)) = (x, x)$$

and

$$(x'(t), y'(t)) = (x(t-1), y(t-1))$$

for all integers $t \geq 1$. It is not difficult to see that in each case this defines a weakly optimal program. Thus

$$a^{-1} - \xi x \in h(x)$$

in each case.

Chapter 11
The Continuous-Time Robinson–Solow–Srinivasan Model

In this chapter we study the continuous-time Robinson–Solow–Srinivasan model. We establish a convergence of good programs to the golden-rule stock, show the existence of overtaking optimal programs and analyze their convergence to the golden-rule stock, and consider some properties of good programs. We are also interested in turnpike properties of the approximate solutions which are independent of the length of the interval, for all sufficiently large intervals.

11.1 Infinite Horizon Problems

Let R^1 (R^1_+) be the set of real (nonnegative) numbers, and let R^n be a finite-dimensional Euclidean space with nonnegative orthant $R^n_+ = \{x \in R^n : x_i \geq 0, \ i = 1, \ldots, n\}$. For any $x, y \in R^n$, let the inner product $xy = \sum_{i=1}^n x_i y_i$, and $x \gg y, x > y, x \geq y$ have their usual meaning. Let $e(i), i = 1, \ldots, n$ be the ith unit vector in R^n and e be an element of R^n_+ all of whose coordinates are unity. For any $x \in R^n$, let $||x||$ denote the Euclidean norm of x.

Denote by $\text{mes}(E)$ the Lebesgue measure of a Lebesgue measurable set $E \subset R^1$.

Let $a = (a_1, \ldots, a_n) \gg 0, b = (b_1, \ldots, b_n) \gg 0, b_1 \geq b_2 \cdots \geq b_n, d \in (0, 1)$, $c_i = b_i/(1 + da_i), \ i = 1, \ldots, n$. We suppose:

There exists $\sigma \in \{1, \ldots, n\}$ such that for all

$$i \in \{1, \ldots, n\} \setminus \{\sigma\}, \ c_\sigma > c_i. \tag{11.1}$$

We now give a formal description of our technological structure.
Set

$$\Omega = \{(x, z) \in R^n_+ \times R^n : z + dx \geq 0 \text{ and } a(z + dx) \leq 1\}. \tag{11.2}$$

For every point $(x, z) \in \Omega$, set

$$\Lambda(x, z) = \left\{ y \in R_+^n : \ y \leq x \ \text{and} \ ey \leq 1 - a(z + dx) \right\}. \tag{11.3}$$

Let I be either $[0, \infty)$ or $[0, T]$ with a positive number T. A pair of functions $(x(\cdot), y(\cdot))$ is called a program if $x : I \rightarrow R^n$ is an absolutely continuous (a.c.) function on any finite subinterval of I, $y : I \rightarrow R^n$ is a Lebesgue measurable function and if

$$(x(t), x'(t)) \in \Omega \ \text{for almost every} \ t \in I, \tag{11.4}$$

$$y(t) \in \Lambda(x(t), x'(t)) \ \text{for almost every} \ t \in I. \tag{11.5}$$

In the sequel if $I = [0, T]$, then the program $(x(\cdot), y(\cdot))$ is denoted by $(x(t), y(t))_{t=0}^T$, and if $I = [0, \infty)$, then the program $(x(\cdot), y(\cdot))$ is denoted by $(x(t), y(t))_{t=0}^\infty$.

Let $w : [0, \infty) \rightarrow [0, \infty)$ be a continuous strictly increasing concave and differentiable function which represents the preferences of the planner.

For every point $(x, z) \in \Omega$, set

$$u(x, z) = \max\{w(by) : \ y \in \Lambda(x, z)\}.$$

A golden-rule stock is a vector $\widehat{x} \in R_+^n$ such that a point $(\widehat{x}, 0)$ is a solution to the problem:

maximize $u(x, z)$ subject to

(i) $z \geq 0$; (ii) $(x, z) \in \Omega$.

By Theorem 2.3, there exists a unique golden-rule stock

$$\widehat{x} = (1/(1 + da_\sigma))e(\sigma). \tag{11.6}$$

It is easy see that $\widehat{x}$ is a solution of the problem

$$w(by) \rightarrow \max, \ y \in \Lambda(\widehat{x}, 0).$$

Put

$$\widehat{y} = \widehat{x}. \tag{11.7}$$

For all integers $i = 1, \ldots, n$, put

$$\widehat{q}_i = a_i b_i (1 + da_i)^{-1}, \ \widehat{p}_i = w'(b\widehat{x})\widehat{q}_i. \tag{11.8}$$

Set

$$\xi_\sigma = 1 - d - 1/a_\sigma. \tag{11.9}$$

By Lemma 2.2,

$$w(b\widehat{x}) \geq w(by) + \widehat{p}z \qquad (11.10)$$

for every $(x, z) \in \Omega$ and for every $y \in \Lambda(x, z)$.

We will prove the following three propositions obtained in [101].

Proposition 11.1 *Let m_0 be a positive number. Then there exists a positive number m_1 such that for every positive number T and every program $(x(t), y(t))_{t=0}^{T}$ which satisfies $x(0) \leq m_0 e$, the inequality $x(t) \leq m_1 e$ is valid for every number $t \in [0, T]$.*

We use the following notion of good programs.

A program $(x(t), y(t))_{t=0}^{\infty}$ is called good if there exists $M \in R^1$ such that

$$\int_0^T (w(by(t)) - w(b\widehat{y}))dt \geq M \text{ for all } T \geq 0.$$

A program is called bad if

$$\lim_{T \to \infty} \int_0^T (w(by(t)) - w(b\widehat{y}))dt = -\infty.$$

Proposition 11.2 *Any program $(x(t), y(t))_{t=0}^{\infty}$ that is not good is bad.*

Proposition 11.3 *For every point $x_0 \in R_+^n$, there exists a good program $(x(t), y(t))_{t=0}^{\infty}$ satisfying $x(0) = x_0$.*

In the sequel we use a notion of an overtaking optimal program.

A program $(\tilde{x}(t), \tilde{y}(t))_{t=0}^{\infty}$ is overtaking optimal if for every program $(x(t), y(t))_{t=0}^{\infty}$ satisfying $x(0) = \tilde{x}(0)$, the inequality

$$\limsup_{T \to \infty} \left[\int_0^T w(by(t))dt - \int_0^T w(b\tilde{y}(t))dt \right] \leq 0$$

holds.

We prove the following two theorems obtained in [101].

Theorem 11.4 *Assume that a program $(x(t), y(t))_{t=0}^{\infty}$ is good. Then*

$$(i) \lim_{t \to \infty} x(t) = \widehat{x}.$$

(ii) Let $\epsilon \in (0, 1)$ and $L > 1$. Then there exists a positive number T_0 such that for every number $T \geq T_0$,

$$mes([T, T + L] : \|y(t) - \widehat{x}\| > \epsilon) \leq \epsilon.$$

Theorem 11.5 *For every* $x_0 \in R_+^n$ *there exists an overtaking optimal program* $(x(t), y(t))_{t=0}^\infty$ *such that* $x(0) = x$.

11.2 Proofs of Propositions 11.1–11.3

Proof of Proposition 11.1 Fix a number $m_1 > 0$ such that

$$m_1 > 4m_0,$$

$$\min\{a_i : i = 1, \ldots, n\}m_1 > 8m_0 \sum_{i=1}^n a_i, \tag{11.11}$$

$$d \min\{a_i : i = 1, \ldots, n\}m_1 > 16. \tag{11.12}$$

Assume that T is a positive number and that $(x(t), y(t))_{t=0}^T$ is a program which satisfies

$$x(0) \le m_0 e. \tag{11.13}$$

We claim that the inequality $x(t) \le m_1 e$ is valid for all numbers $t \in [0, T]$. Assume the contrary. Then there exists a number

$$t_0 \in (0, T]$$

such that

$$m_1 e - x(t_0) \notin R_+^n. \tag{11.14}$$

In view of (11.14),

$$a(x(t_0)) \ge \min\{a_i : i = 1, \ldots, n\}m_1. \tag{11.15}$$

Equations (11.11), (11.13), and (11.15) imply that there exists a number t_1 for which

$$t_1 \in (0, t_0), \ a(x(t_1)) = 4^{-1} \min\{a_i : i = 1, \ldots, n\}m_1, \tag{11.16}$$

$$a(x(t)) \ge 4^{-1} \min\{a_i : i = 1, \ldots, n\}m_1 \text{ for all } t \in [t_1, t_0]. \tag{11.17}$$

Since $(x(t), y(t))_{t=0}^T$ is a program for almost every $t \in [0, T]$, we have

$$x'(t) + dx(t) \ge 0, \ a(x'(t) + dx(t)) \le 1. \tag{11.18}$$

In view of (11.12), (11.17), and (11.18), for almost every $t \in [t_1, t_0]$, we have

$$ax'(t) \le 1 - adx(t) \le 1 - 4^{-1}d\min\{a_i : i = 1, \ldots, n\}m_1 < -1.$$

Therefore the function $ax(\cdot)$ is decreasing on $[t_1, t_0]$. On the other hand, it follows from (11.15) and (11.16) that $ax(t_0) > ax(t_1)$. The contradiction we have reached completes the Proof of Proposition 11.1.

For every point $(x, z) \in \Omega$ and every $y \in \Lambda(x, z)$, define

$$\delta(x, y, z) = w(b\widehat{y}) - w(by) - \widehat{p}z. \tag{11.19}$$

It follows from (11.10) and (11.19) that

$$\delta(x, y, z) \ge 0 \text{ for every } (x, z) \in \Omega \text{ and every } y \in \Lambda(x, z). \tag{11.20}$$

It is not difficult to prove the following result.

Lemma 11.6 *Let T be a positive number and let $(x(t), y(t))_{t=0}^{T}$ be a program. Then*

$$\int_0^T (w(by(t)) - w(b\widehat{y}))dt = -\int_0^T \delta(x(t), y(t), x'(t))dt - \widehat{p}(x(T) - x(0)).$$

Proposition 11.2 now easily follows from Proposition 11.1 and Lemma 11.6.

Proof of Proposition 11.3 Let $x_0 \in R_+^n$ be given. Fix a number h satisfying

$$h + (a_\sigma d)^{-1} = (x_0)_\sigma. \tag{11.21}$$

There exists a number $T_0 > 4$ for which

$$he^{-dT_0} + (a_\sigma d)^{-1} > (a_\sigma d + 1)^{-1}. \tag{11.22}$$

Set

$$x_i(t) = (x_0)_i e^{-dt}, \ t \in [0, T_0], \ i \in \{1, \ldots, n\} \setminus \{\sigma\}, \tag{11.23}$$

$$x_\sigma(t) = he^{-dt} + (a_\sigma d)^{-1}, \ t \in [0, T_0], \tag{11.24}$$

$$y(t) = 0, \ t \in [0, T_0]. \tag{11.25}$$

In view of (11.4), (11.5), (11.21), and (11.23)–(11.25), $(x(t), y(t))_{t=0}^{T_0}$ is a program and

$$x(0) = x_0. \tag{11.26}$$

By (11.22) and (11.24), we have

$$x_\sigma(T_0) > (a_\sigma d + 1)^{-1} = \widehat{x}_\sigma. \tag{11.27}$$

Define $x(t)$, $y(t)$ for all numbers $t > T_0$ as follows:

$$y_i(t) = 0, \ x_i(t) = (x_0)_i e^{-dt} \text{ for all } t \in (T_0, \infty) \text{ and all } i \in \{1, \ldots, n\} \setminus \{\sigma\},$$

$$y_\sigma(t) = \widehat{x}_\sigma, \ x_\sigma(t) = e^{-d(t-T_0)}(x_\sigma(T_0) - (1 + da_\sigma)^{-1})$$

$$+ (1 + da_\sigma)^{-1}, \ t \in (T_0, \infty).$$

, It is not difficult to see that $(x(t), y(t))_{t=0}^\infty$ is a good program. This completes the Proof of Proposition 11.3.

Lemma 11.6 and Proposition 11.1 imply the following result.

Proposition 11.7 *A program* $(x(t), y(t))_{t=0}^\infty$ *is good if and only if*

$$\int_0^\infty \delta(x(t), y(t), x'(t))dt := \lim_{T \to \infty} \int_0^T \delta(x(t), y(t), x'(t))dt$$

is finite.

11.3   Auxiliary Results

Proposition 11.8 *Let* $T > 0$, $m_0 > 0$, *and let* $\{(x^{(i)}(t), y^{(i)}(t))_{t=0}^T\}_{i=1}^\infty$ *be a sequence of programs satisfying*

$$x^{(i)}(0) \le m_0 e \text{ for all integers } i \ge 0. \tag{11.28}$$

Then there exist a program $(x(t), y(t))_{t=0}^T$ *and a strictly increasing sequence of natural numbers* $\{i_k\}_{k=1}^\infty$ *such that*

$$x^{(i_k)}(t) \to x(t) \text{ as } k \to \infty \text{ uniformly on } [0, T], \tag{11.29}$$

$$\left(x^{(i_k)}\right)' \to x' \text{ as } k \to \infty \text{ weakly in } L^2([0, T]; R^n), \tag{11.30}$$

$$y^{(i_k)} \to y \text{ as } k \to \infty \text{ weakly in } L^2([0, T]; R^n). \tag{11.31}$$

Proof In view of Proposition 11.1 and (11.28), there exists a positive number m_1 such that

$$x^{(i)}(t) \le m_1 e \text{ for all } t \in [0, T] \text{ and for all integers } i \ge 1. \tag{11.32}$$

By (11.3), (11.5), and (11.32),

$$y^{(i)}(t) \le m_1 e \text{ for all } t \in [0, T] \text{ and for all integers } i \ge 1. \tag{11.33}$$

By (11.2), (11.4), and (11.32), for all integers $i \ge 1$ and for almost every $t \in [0, T]$, we have

$$\left(x^{(i)}\right)'(t) \ge -dx^{(i)}(t) \ge -dm_1 e, \tag{11.34}$$

$$a\left(x^{(i)}\right)'(t) \le 1 - adx^{(i)}(t) \le 1. \tag{11.35}$$

In view of (11.35), there exists a positive number m_2 such that for all integers $i \ge 1$ and almost every $t \in [0, T]$, we have

$$\left\|\left(x^{(i)}\right)'(t)\right\| \le m_2. \tag{11.36}$$

Note that the space $L^2([0, T]); R^n)$ is Hilbert. Equations (11.33) and (11.36) imply that $\{(x^{(i)})'\}_{t=1}^{\infty}, \{y^{(i)}\}_{i=1}^{\infty}$ are bounded sequences in $L^2([0, T]; R^n)$. Therefore there exist a strictly increasing sequence of natural numbers $\{i_k\}_{k=1}^{\infty}$,

$$y \in L^2([0, T]; R^n) \text{ and } u \in L^2([0, T]; R^n)$$

such that

$$\left(x^{(i_k)}\right)' \to u \text{ as } k \to \infty \text{ in } L^2([0, T]; R^n), \tag{11.37}$$

$$y^{(i_k)} \to y \text{ as } k \to \infty \text{ in } L^2([0, T]; R^n) \tag{11.38}$$

in the weak topology, and there exists

$$\lim_{k\to\infty} x^{(i_k)}(0). \tag{11.39}$$

For every number $\tau \in [0, T]$, define

$$x(\tau) = \lim_{k\to\infty} x^{(i_k)}(0) + \int_0^\tau u(t)dt. \tag{11.40}$$

It follows from (11.37) and (11.40) that relation (11.29) holds. In order to complete the proof of the proposition, it is sufficient to show that $(x(t), y(t))_{t=0}^T$ is a program.

Let $\epsilon \in (0, 1)$ be given. Fix number $\epsilon_0 > 0$ such that

$$2\epsilon_0 \left(2n + 2 + \sum_{i=1}^n a_i\right) < \epsilon. \tag{11.41}$$

In view of (11.29), there exists an integer $k_0 \ge 1$ such that for every natural number $k \ge k_0$, we have

$$\|x^{(i_k)}(t) - x(t)\| \le \epsilon_0, \ t \in [0, T]. \tag{11.42}$$

By (11.37), (11.38), and (11.40), (x', y) is a limit point of the convex hull of the set

$$\left\{ \left((x^{(i_k)})', \ y^{(i_k)} \right) : \ k \text{ is an integer and } k \ge k_0 \right\}$$

in the norm topology of the space $L^2([0, T]; R^n) \times L^2([0, T]; R^n)$. Therefore there exist an integer $k_1 > k_0$ and nonnegative numbers $\alpha_k, k = k_0, \ldots, k_1$ such that

$$\sum_{k=k_0}^{k_1} \alpha_k = 1, \tag{11.43}$$

$$\int_0^T \|\sum_{k=k_0}^{k_1} \alpha_k \left(x^{(i_k)} \right)'(t) - x'(t)\|^2 dt < \epsilon_0^4, \tag{11.44}$$

$$\int_0^T \|\sum_{k=k_0}^{k_1} \alpha_k y^{(i_k)}(t) - y(t)\|^2 dt < \epsilon_0^4. \tag{11.45}$$

Define

$$E_1 = \left\{ t \in [0, T] : \ \|\sum_{k=k_0}^{k_1} \alpha_k (x^{(i_k)})'(t) - x'(t)\| \le \epsilon_0 \right\}, \tag{11.46}$$

$$E_2 = \left\{ t \in [0, T] : \ \|\sum_{k=k_0}^{k_1} \alpha_k y^{(i_k)}(t) - y(t)\| \le \epsilon_0 \right\}. \tag{11.47}$$

It follows from (11.41) and (11.44)–(11.47) that

$$\epsilon_0^2 \text{mes}([0, T] \setminus E_1) \le \int_0^T \|\sum_{k=k_0}^{k_1} \alpha_k (x^{(i_k)})'(t) - x'(t)\|^2 dt < \epsilon_0^4, \tag{11.48}$$

$$\epsilon_0^2 \ \text{mes}([0, T] \setminus E_2) \le \int_0^T \|\sum_{k=k_0}^{k_1} \alpha_k y^{(i_k)}(t) - y(t)\|^2 dt < \epsilon_0^4 \tag{11.49}$$

and

$$\text{mes}([0, T] \setminus E_1), \ \text{mes}([0, T] \setminus E_2) < \epsilon_0^2 < \epsilon_0,$$
$$\text{mes}([0, T] \setminus (E_1 \cap E_2)) < 2\epsilon_0 < \epsilon. \tag{11.50}$$

Equations (11.2)–(11.5) imply that for almost every $t \in E_1 \cap E_2$ and all integers $k = k_0, \ldots, k_1$, we have

$$\left(x^{i_k}\right)'(t) + dx^{(i_k)}(t) \geq 0, \ a\left(\left(x^{(i_k)}\right)'(t) + dx^{(i_k)}(t)\right) \leq 1, \tag{11.51}$$

$$0 \leq y^{(i_k)}(t) \leq x^{(i_k)}(t), \tag{11.52}$$

$$ey^{(i_k)}(t) \leq 1 - a\left(\left(x^{(i_k)}\right)'(t) + dx^{(i_k)}(t)\right). \tag{11.53}$$

In view of (11.42), (11.43), (11.46), and (11.51), for almost every $t \in E_1 \cap E_2$, we have

$$x'(t) + dx(t) \geq \sum_{k=k_0}^{k_1} \alpha_k \left(x^{(i_k)}\right)'(t) - \epsilon_0 e$$

$$+ d\left(\sum_{k=k_0}^{k_1} \alpha_k(x^{i_k})(t) - \epsilon_0 e\right) \geq -(1+d)\epsilon_0 e. \tag{11.54}$$

It follows from (11.42), (11.44), and (11.51) that for almost every $t \in E_1 \cap E_2$, we have

$$a(x'(t) + dx(t)) \leq a\left[\sum_{k=k_0}^{k_1} \alpha_k \left(x^{(i_k)}\right)'(t) + \epsilon_0 e + d\left(\sum_{k=k_0}^{k_1} \alpha_k x^{(i_k)}(t) + \epsilon_0 e\right)\right]$$

$$= \sum_{k=k_0}^{k_1} \alpha_k \left[a\left(x^{(i_k)}\right)'(t) + dx^{(l_k)}(t)\right] + a(\epsilon_0 e + d\epsilon_0 e)$$

$$\leq 1 + \epsilon_0(1+d) \sum_{i=1}^{n} a_i. \tag{11.55}$$

By (11.42), (11.43), (11.46), (11.47), and (11.52), for almost every $t \in E_1 \cap E_2$, we have

$$-\epsilon_0 e \leq -\epsilon_0 e + \sum_{k=k_0}^{k_1} \alpha_k y^{(i_k)}(t) \leq y(t) \leq \epsilon_0 e + \sum_{k=k_0}^{k_1} \alpha_k y^{(i_k)}(t)$$

$$\leq \epsilon_0 e + \sum_{k=k_0}^{k_1} \alpha_k x^{(i_k)}(t) \leq \epsilon_0 e + x(t) + \epsilon_0 e = x(t) + 2\epsilon_0 e. \tag{11.56}$$

In view of (11.2), (11.4), (11.42), (11.43), (11.46), (11.47), and (11.53), for almost every $t \in E_1 \cap E_2$,

$$ey(t) \leq e\left(\sum_{k=k_0}^{k_1} \alpha_k y^{(i_k)}(t) + \epsilon_0 e\right) = \epsilon_0 n + \sum_{k=k_0}^{k_1} \alpha_k e y^{(i_k)}(t)$$

$$\leq \epsilon_0 n + \sum_{k=k_0}^{k_1} \alpha_k\left[1 - a\big((x^{(i_k)})'(t) + dx^{(i_k)}(t)\big)\right]$$

$$= \epsilon_0 n + 1 - a\sum_{k=k_0}^{k_1} \alpha_k\big(x^{(i_k)}\big)'(t) - da\sum_{k=k_0}^{k_1} \alpha_k x^{(i_k)}(t) \leq \epsilon_0 n + 1$$

$$- a(x'(t) - \epsilon_0 e) - da(x(t) - \epsilon_0 e)$$

$$\leq 1 + \epsilon_0 n + \epsilon_0 ae + d\epsilon_0 ae - a(x'(t) + dx(t)). \tag{11.57}$$

It follows from (11.41) and (11.54)–(11.57) that for almost every $t \in E_1 \cap E_2$,

$$x'(t) + dx(t) \geq -(1+d)\epsilon_0 e \geq -\epsilon e, \tag{11.58}$$

$$a\big(x'(t) + dx(t)\big) \leq 1 + \epsilon_0(1+d)\sum_{i=1}^{n} a_i \leq 1 + \epsilon, \tag{11.59}$$

$$-\epsilon e < -\epsilon_0 e \leq y(t) \leq x(t) + 2\epsilon_0 e \leq x(t) + \epsilon e,$$

$$ey(t) \leq 1 - a\big(x'(t) + dx(t)\big) + \epsilon_0(n + ae(1+d)) \tag{11.60}$$

$$\leq 1 - a(x'(t) + dx(t)) + \epsilon.$$

Since ϵ is an arbitrary number from the open interval $(0, 1)$, it follows from (11.50) that for every integer $k \geq 1$, there exists a Lebesgue measurable set $F_k \subset [0, T]$ such that

$$\mathrm{mes}([0, T] \setminus F_k) \leq 2^{-k} \tag{11.61}$$

and that for almost every $t \in F_k$, the following inequalities hold:

$$x'(t) + dx(t) \geq -2^{-k}e, \tag{11.62}$$

$$a(x'(t) + dx(t)) \leq 1 + 2^{-k}, \tag{11.63}$$

$$-2^{-k}e \leq y(t) \leq x(t) + 2^{-k}e, \tag{11.64}$$

$$ey(t) \leq 1 - a(x'(t) + dx(t)) + 2^{-k}e. \tag{11.65}$$

Define

$$F = \cup_{j=1}^{\infty} \cap_{k=j}^{\infty} F_k. \tag{11.66}$$

By (11.61) and (11.66), mes($[0, T] \setminus F$) = 0. In view of (11.62)–(11.66), for almost every $t \in F$,

$$x'(t) + dx(t) \geq 0, \quad a(x'(t) + dx(t)) \leq 1,$$

$$0 \leq y(t) \leq x(t),$$

$$ey(t) \leq 1 - a(x'(t) + dx(t)).$$

These inequalities imply that $(x(t), y(t))_{t=0}^{T}$ is a program. Proposition 11.8 is proved.

Proposition 11.9 *Let $T > 0$, $m_0 > 0$, $\{(x^{(i)}(t), y^{(i)}(t))_{t=0}^{T}\}_{i=1}^{\infty}$ be a sequence of programs satisfying*

$$x^{(i)}(0) \leq m_0 e \text{ for all integers } i \geq 0 \tag{11.67}$$

and let $(x(t), y(t))_{t=0}^{T}$ be a program such that

$$x^{(i)}(t) \to x(t) \text{ as } i \to \infty \text{ uniformly on } [0, T], \tag{11.68}$$

$$\left(x^{(i)}\right)' \to x' \text{ as } i \to \infty \text{ weakly in } L^2([0, T]; R^n), \tag{11.69}$$

$$y^{(i)} \to y \text{ as } i \to \infty \text{ weakly in } L^2([0, T]; R^n). \tag{11.70}$$

Then

$$\int_0^T w(by(t))dt \geq \limsup_{i \to \infty} \int_0^T w\left(by^{(i)}(t)\right)dt.$$

Proof Proposition 11.1 and (11.67) imply that there exists a positive number m_1 such that

$$x^{(i)}(t) \leq m_1 e \text{ for all } t \in [0, T] \text{ and all natural numbers } i. \tag{11.71}$$

By (11.3), (11.5), and (11.71),

$$y^{(i)}(t) \leq m_1 e \text{ for almost all } t \in [0, T] \text{ and all natural numbers } i. \tag{11.72}$$

In view of (11.72), we have

$$\int_0^T w\left(by^{(i)}(t)\right)dt \leq Tw(m_1 be) \text{ for all natural numbers } i.$$

We may assume without loss of generality that there exists

$$\lim_{i \to \infty} \int_0^T w\big(by^{(i)}(t)\big)dt.$$

Let $\epsilon \in (0, 1)$ be given. Fix a number $\epsilon_0 > 0$ for which

$$\epsilon_0 < \epsilon, \quad 2\epsilon_0(w(m_1be) + 4) < \epsilon/8, \tag{11.73}$$

$$|w(by) - w(by')| \le \epsilon(8(T + 1))^{-1} \tag{11.74}$$

for each y, y' satisfying $0 \le y$, $y' \le m_1e$ and $\|y - y'\| \le 2\epsilon_0$.

There exists an integer $k_0 \ge 1$ such that for every natural number $k \ge k_0$, we have

$$\left| \int_0^T w\big(by^{(k)}(t)\big)dt - \lim_{i \to \infty} \int_0^T w\big(by^{(i)}(t)\big)dt \right| \le \epsilon_0. \tag{11.75}$$

Equations (11.69) and (11.70) imply that (x', y) is a limit point of the convex hull of the set

$$\left\{ \big((x^{(i)})', y^{(i)}\big) : \ i \text{ is an integer and } i \ge k_0 \right\}$$

in the norm topology of the space $L^2([0, T]; R^n) \times L^2([0, T]; R^n)$. This implies that there exist a natural number $k_1 > k_0$ and nonnegative numbers α_k, $k = k_0, \ldots, k_1$ such that

$$\sum_{k=k_0}^{k_1} \alpha_k = 1, \tag{11.76}$$

$$\int_0^T \Big\| \sum_{k=k_0}^{k_1} \alpha_k y^{(k)}(t) - y(t) \Big\|^2 dt < \epsilon_0^4. \tag{11.77}$$

Define

$$E = \left\{ t \in [0, T] : \ \Big\| \sum_{k=k_0}^{k_1} \alpha_k y^{(k)}(t) - y(t) \Big\| \le \epsilon_0 \right\}. \tag{11.78}$$

In view of (11.77) and (11.78), we have

$$\epsilon_0^2 \operatorname{mes}([0, T] \setminus E) \le \int_0^T \Big\| \sum_{k=k_0}^{k_1} \alpha_k y^{(k)}(t) - y(t) \Big\|^2 dt < \epsilon_0^4$$

and

$$\text{mes}([0, T] \setminus E) < \epsilon_0^2. \tag{11.79}$$

By (11.72) and (11.79), we have for all integers $k = k_0, \ldots, k_1$

$$\left| \int_0^T w\big(by^{(k)}(t)\big)dt - \int_E w\big(by^{(k)}(t)\big)dt \right|$$

$$\leq \text{mes}([0, T] \setminus E)w(m_1be) < \epsilon_0^2 w(m_1be). \tag{11.80}$$

In view of (11.3), (11.5), (11.62), and (11.71), we have

$$0 \leq y(t) \leq x(t) \leq m_1e \text{ for a.e. } t \in [0, T]. \tag{11.81}$$

It follows from (11.79) and (11.81) that

$$\left| \int_0^T w(by(t))dt - \int_F w(by(t))dt \right|$$

$$\leq \text{mes}([0, T] \setminus E)w(m_1be) < \epsilon_0^2 w(m_1be). \tag{11.82}$$

By (11.74), (11.78), and (11.81), for almost every $t \in E$, we have

$$w(by(t)) \geq w\left(b\left(\sum_{k=k_0}^{k_1} \alpha_k y^{(k)}(t) \right) \right) - \epsilon(8(T + 1))^{-1}$$

$$\geq \sum_{k=k_0}^{k_1} \alpha_k w\big(by^{(k)}(t)\big) - \epsilon(8(T + 1))^{-1}. \tag{11.83}$$

Equations (11.73), (11.75), (11.79), (11.82), and (11.83) imply that

$$\int_0^T w(by(t))dt \geq \int_E w(by(t))dt - \epsilon_0^2 |w(0)|$$

$$\geq -\epsilon_0^2 w(0) + \sum_{k=k_0}^{k_1} \alpha_k \int_E w\big(by^{(k)}(t)\big)dt - \epsilon/8$$

$$\geq -\epsilon_0^2 w(0) - \epsilon/8 + \sum_{k=k_0}^{k_1} \alpha_k \int_0^T w\big(by^{(k)}(t)\big)dt - \epsilon_0^2 w(m_1be)$$

$$\geq -2\epsilon_0^2 w(m_1be) - \epsilon/8 + \lim_{i \to \infty} \int_0^T w\big(by^{(i)}(t)\big)dt - \epsilon_0$$

$$\geq \lim_{i \to \infty} \int_0^T w\big(by^{(i)}(t)\big)dt - \epsilon. \tag{11.84}$$

Since ϵ is an arbitrary number from the interval $(0, 1)$, this completes the proof of Proposition 11.9.

Let $x_0 \in R_+^n$ be given. Set

$$\Delta(x_0) = \inf \left\{ \int_0^\infty \delta(x(t), y(t), x'(t))dt \; : \; (x(t), y(t))_{t=0}^\infty \right.$$

$$\text{is program and } x(0) = x_0 \left. \right\}. \tag{11.85}$$

Propositions 11.3 and 11.7 imply that $\Delta(x_0)$ is well-defined and finite.

Proposition 11.10 *Let $x_0 \in R_+^n$. Then there exists a program $(x(t), y(t))_{t=0}^\infty$ which satisfies $x(0) = x_0$ and*

$$\int_0^\infty \delta(x(t), y(t), x'(t))dt = \Delta(x_0).$$

Proof For every integer $k \geq 1$, there exists a program $(x^{(k)}(t), y^{(k)}(t))_{t=0}^\infty$ such that

$$x^{(k)}(0) = x_0, \quad \int_0^\infty \delta\left(x^{(k)}(t), y^{(k)}(t), \left(x^{(k)}\right)'(t)\right)dt \leq \Delta(x_0) + 1/k. \tag{11.86}$$

In view of Proposition 11.2, extracting a subsequence and re-indexing and using a diagonalization process, we may assume without loss of generality that there exists a program $(x(t), y(t))_{t=0}^\infty$ such that for every integer $T \geq 1$, we have

$$x^{(k)}(t) \to x(t) \text{ as } k \to \infty \text{ uniformly on } [0, T], \tag{11.87}$$

$$\left(x^{(k)}\right)' \to x' \text{ as } k \to \infty \text{ weakly in } L^2([0, T]; R^n), \tag{11.88}$$

$$y^{(k)} \to y \text{ as } k \to \infty \text{ weakly in } L^2([0, T]; R^n). \tag{11.89}$$

Let $T \geq 1$ be an integer. Proposition 11.9 and (11.87)–(11.89) imply that

$$\int_0^T w(by(t))dt \geq \limsup_{k\to\infty} \int_0^T w\left(by^{(k)}(t)\right)dt. \tag{11.90}$$

Lemma 11.6, (11.86), (11.87), and (11.90) imply that

$$\int_0^T \delta(x(t), y(t), x'(t))dt$$

$$= -\int_0^T (w(by(t)) - w(b\hat{y}))dt + \hat{p}(x(0) - x(T))$$

$$\leq -\limsup_{k\to\infty} \int_0^T w\big(by^{(k)}(t)\big)dt + Tw(b\widehat{y}) + \lim_{k\to\infty} \widehat{p}\big(x^{(k)}(0) - x^{(k)}(T)\big)$$

$$= \liminf_{k\to\infty} \left[\int_0^T \Big(w(b\widehat{y}) - w\big(by^{(k)}(t)\big)\Big)dt + \widehat{p}\big(x^{(k)}(0) - x^{(k)}(T)\big) \right]$$

$$= \liminf_{k\to\infty} \int_0^T \delta\Big(x^{(k)}(t), y^{(k)}(t), \big(x^{(k)}\big)'(t)\Big)dt$$

$$\leq \lim_{k\to\infty} \int_0^\infty \delta\Big(x^{(k)}(t), y^{(k)}(t), \big(x^{(k)}\big)'(t)\Big)dt = \Delta(x_0).$$

Since the relation above is valid for every integer $T \geq 1$, we conclude that

$$\int_0^\infty \delta(x(t), y(t), x'(t))dt \leq \Delta(x_0).$$

This completes the Proof of Proposition 11.10.

11.4 Proofs of Theorems 11.4 and 11.5

The next auxiliary result easily follows from Lemma 2.13.

Lemma 11.11 *The von Neumann facet*

$$\{(x, z) \in \Omega : \text{ there is } y \in \Lambda(x, z) \text{ such that } \delta(x, y, z) = 0\}$$

is a subset of

$$\{(x, z) \in \Omega : x_i = z_i = 0 \text{ for all } i \in \{1, \ldots, n\} \setminus \{\sigma\},$$

$$z_\sigma = (1/a_\sigma) + (\xi_\sigma - 1)x_\sigma\}$$

with the equality if the function w is linear. If the function w is strictly concave, then the facet is the singleton $\{(\widehat{x}, 0)\}$.

Lemma 11.12 *Let $\{(x(t), y(t))\}_{t=0}^\infty$ be a good program and $T_i > 2i$ for all natural numbers i. Let*

$$\big(x^{(i)}(t), y^{(i)}(t)\big) = (x(t + T_i), y(t + T_i)), \ t \in [-i, i] \qquad (11.91)$$

for every integer $i \geq 1$.

    Then there exist a strictly increasing sequence of natural numbers $\{i_k\}_{k=1}^\infty$, a locally a.c. function $\tilde{x} : R^1 \to R^n$, and a Lebesgue measurable function $\tilde{y} : R^1 \to R^n$ such that for each natural number j,

$$x^{(i_k)}(t) \to \tilde{x}(t) \text{ as } i \to \infty \text{ uniformly on } [-j, j], \tag{11.92}$$

$$\left(x^{(i_k)}\right)' \to \tilde{x}' \text{ as } k \to \infty \text{ weakly in } L^2([-j, j]; R^n), \tag{11.93}$$

$$y^{(i_k)} \to \tilde{y} \text{ as } k \to \infty \text{ weakly in } L^2([-j, j]; R^n). \tag{11.94}$$

Moreover,

$$0 \le \tilde{y}(t) \le \tilde{x}(t) \text{ for a.e. } t \in R^1, \tag{11.95}$$

$$\tilde{x}'(t) + d\tilde{x}(t) \ge 0 \text{ for a.e. } t \in R^1, \tag{11.96}$$

$$e\tilde{y}(t) + a\left(\tilde{x}'(t) + d\tilde{x}(t)\right) \le 1 \text{ for a.e. } t \in R^1 \tag{11.97}$$

and

$$\delta\left(\tilde{x}(t), \tilde{y}(t), \tilde{x}'(t)\right) = 0 \text{ for a.e. } t \in R^1. \tag{11.98}$$

Proof By Propositions 11.1 and 11.8 and a diagonalization process, there exist a strictly increasing sequence of natural numbers $\{i_k\}_{k=1}^\infty$, a locally a.c. function $\tilde{x} : R^1 \to R^n$, and a Lebesgue measurable function $\tilde{y} : R^1 \to R^n$ such that (11.92)–(11.97) are valid for all integers $j \ge 1$.

By (11.19), (11.92)–(11.97), Proposition 11.9, Lemma 11.6, (11.91), and Proposition 11.7, for every integer $j \ge 1$, we have

$$\int_{-j}^{j} \delta(\tilde{x}(t), \tilde{y}(t), \tilde{x}'(t))dt$$

$$= \int_{-j}^{j} \left[w(b\hat{y}) - w(b\tilde{y}(t)) - \hat{p}(\tilde{x}'(t))\right]dt$$

$$= -\int_{-j}^{j} w(b\tilde{y}(t))dt + 2jw(b\hat{y}) + \hat{p}\left(\tilde{x}(j) - \tilde{x}(-j)\right)$$

$$\le -\limsup_{k\to\infty} \int_{-j}^{j} w\left(by^{(i_k)}(t)\right)dt + 2jw(b\hat{y}) - \lim_{k\to\infty} \hat{p}\left(x^{(i_k)}(j) - x^{(i_k)}(-j)\right)$$

$$\le \liminf_{k\to\infty} \int_{-j}^{j} \left[w(b\hat{y}) - w(by^{(i_k)}(t)) - \hat{p}(x^{(i_k)})'(t)\right]dt$$

$$= \liminf_{k\to\infty} \int_{-j}^{j} \delta\left(x^{(i_k)}(t), y^{(i_k)}(t), \left(x^{(i_k)}\right)'(t)\right)dt$$

$$= \lim_{k\to\infty} \int_{T_k-j}^{T_k+j} \delta(x(t), y(t), x'(t))dt = 0.$$

Since this relation holds for every integer $j \ge 1$, we conclude that

$$\delta(x(t), y(t), x'(t)) = 0 \text{ for a.e. } t \in R^1.$$

This completes the proof of Lemma 11.12.

Lemma 11.13 *Let $x : R^1 \to R^n$ be a locally a.c. function and $y : R^1 \to R^n$ be a Lebesgue measurable function such that*

$$(x(t), x'(t)) \in \Omega \text{ for a.e. } t \in R^1, \tag{11.99}$$

$$y(t) \in \Lambda(x(t), x'(t)) \text{ for a.e. } t \in R^1, \tag{11.100}$$

$$\sup\{\|x(t)\| : t \in R^1\} < \infty, \tag{11.101}$$

$$\delta(x(t), y(t), x'(t)) = 0 \text{ for a.e. } t \in R^1. \tag{11.102}$$

Then $x(t) = \widehat{x}, t \in R^1, y(t) = \widehat{x}$ for almost every $t \in R^1$.

Proof Lemma 11.11, (11.99), (11.100), and (11.102) imply that for almost every $t \in R^1$, we have

$$x_i(t) = x'_i(t) = 0 \text{ for all } i \in \{1, \ldots, n\} \setminus \{\sigma\}, \tag{11.103}$$

$$x'_\sigma(t) = a_\sigma^{-1} + (\xi_\sigma - 1)x_\sigma(t).$$

In view of (11.103),

$$x_\sigma(t) = ce^{(\xi_\sigma - 1)t} - (a_\sigma(\xi_\sigma - 1))^{-1}, \ t \in R^1 \tag{11.104}$$

where c is a constant. Note that in view of (11.9),

$$\xi_\sigma \neq 1.$$

By (11.101) and (11.104), $c = 0$. Combined with (11.9) and (11.104), this implies that

$$x(t) = \widehat{x} \text{ for all } t \in R^1. \tag{11.105}$$

It follows from (11.19), (11.102), and (11.105) that for almost every $t \in R^1$,

$$w(by(t)) = w(b\widehat{x}).$$

Since w is strictly increasing, it follows from this equality, (11.100), and (11.105) that

$$y(t) = \widehat{x} \text{ for almost every } t \in R^1.$$

This completes the proof of Lemma 11.13.

Proof of Theorem 11.4 By Proposition 11.1, there exists a positive number m_1 such that

$$x(t) \leq m_1 e \text{ for all } t \geq 0.$$

First we show that

$$x(t) \to \widehat{x} \text{ as } t \to \infty. \tag{11.106}$$

Assume that $\{T_i\}_{i=1}^\infty$ is a strictly increasing sequence of positive numbers and that

$$h = \lim_{i \to \infty} x(T_i). \tag{11.107}$$

In order to prove (11.106), it is sufficient to show that $h = \widehat{x}$. We may assume without loss of generality that $T_i > 2i$ for all integers $i \geq 1$.

Let $x^{(i)}$, $y^{(i)}$ be defined by (11.91) for all natural numbers $i \geq 1$. Lemma 11.12 implies that there exist a strictly increasing sequence of natural numbers $\{i_k\}_{k=1}^\infty$, a locally a.c. function $\tilde{x} : R^1 \to R^n$, and a Lebesgue measurable function $\tilde{y} : R^1 \to R^n$ such that (11.92)–(11.98) are valid. In view of (11.91), (11.92), and (11.107), we have

$$h = \lim_{k \to \infty} x(T_{i_k}) = \lim_{k \to \infty} x^{(i_k)}(0) = \tilde{x}(0). \tag{11.108}$$

By (11.92)–(11.98), the bondedness of $\{x(t) : t \in [0, \infty)\}$, (11.92)

$$\tilde{x}(t) = \widehat{x} \text{ for all } t \in R^1 \text{ and in view of (11.108) } h = \tilde{x}(0) = \widehat{x}.$$

Therefore we have shown that $\lim_{t \to \infty} x(t) = \widehat{x}$.

Now we prove (ii). We may assume that

$$\epsilon < (8(da_\sigma + 1))^{-1}.$$

Fix a positive number γ such that

$$w(b\widehat{x}) > w\big(b\big(\widehat{x} - n^{-1}\epsilon e(\sigma)\big)\big) + 4\gamma \tag{11.109}$$

and a positive number ϵ_0 such that

$$\epsilon_0 < \epsilon \text{ and } \gamma^{-1} L^2 \epsilon_0 < \epsilon. \tag{11.110}$$

There exists a number $\epsilon_1 \in (0, \epsilon)$ such that

$$\epsilon_1 n < \gamma \text{ and } 2\epsilon_1 < \epsilon_0/n, \tag{11.111}$$

$$w(b(\widehat{x} + \epsilon_1 e)) - w(b\widehat{x}) \le \epsilon_0, \tag{11.112}$$

$$|w\big(bz^{(1)}\big) - w\big(bz^{(2)}\big)| \le \gamma \text{ for each } z^{(1)}, z^{(2)} \in R_+^n \tag{11.113}$$

such that

$$z^{(1)}, z^{(2)} \le 4(\widehat{x}_\sigma + 1)e, \ \|z^{(1)} - z^{(2)}\| \le 4\epsilon_1 n.$$

Since the program $(x(t), y(t))_{t=0}^\infty$ is good, we conclude that

$$\int_0^\infty \delta(x(t), y(t), x'(t))dt < \infty.$$

Combined with assertion (i), the inequality above implies that there exists a positive number T_0 such that

$$\int_{T_0}^\infty \delta(x(t), y(t), x'(t))dt < \epsilon_1, \tag{11.114}$$

$$2(1 + \|\widehat{p}\|)\|x(t) - \widehat{x}\| \le \epsilon_1 \text{ for all } t \ge T_0. \tag{11.115}$$

Assume that $T \ge T_0$. Then by (11.19), (11.20), and (11.114), we have

$$\epsilon_1 > \int_T^{T+L} \delta(x(t), y(t), x'(t))dt$$

$$= \int_T^{T+L} [w(b\widehat{y}) - w(by(t))]dt - \widehat{p}(x(T+L) - x(T)).$$

Combined with (11.115) this implies that

$$\int_T^{T+L} [w(b\widehat{y}) - w(by(t))]dt < 2\epsilon_1. \tag{11.116}$$

Define

$$E_1 = \{t \in [T, T+L] : \|\widehat{x} - y(t)\| \le \epsilon\}, \ E_2 = [T, T+L] \setminus E_1. \tag{11.117}$$

In view of (11.112) and (11.115), for a.e. $t \in [T, T+L]$, we have

$$0 \le y(t) \le x(t) \le \widehat{x} + \epsilon_1 e, \tag{11.118}$$

$$w(by(t)) \le w(b(\widehat{x} + \epsilon_1 e)) \le w(b\widehat{y}) + \epsilon_0. \tag{11.119}$$

Let $t \in E_2$ be given. Equation (11.117) implies that there exists $j \in \{1, \ldots, n\}$ such that

$$|\widehat{x}_j - y_j| > \epsilon/n.$$

If $i \in \{1, \ldots, n\} \setminus \{\sigma\}$, then in view of (11.111) and (11.118), we have

$$|\widehat{x}_i - y_i(t)| = |0 - y_i(t)| \leq \epsilon_1 < \epsilon/n.$$

Therefore $j = \sigma$ and

$$|\widehat{x}_\sigma - y_\sigma(t)| > \epsilon/n.$$

Combined with (11.118) this implies that

$$y_\sigma(t) < \widehat{x}_\sigma - \epsilon/n \text{ for a.e. } t \in E_2. \tag{11.120}$$

By (11.118) and (11.120), for a.e. $t \in E_2$,

$$\|y(t) - y_\sigma(t)e(\sigma)\| \leq \epsilon_1 n,$$

and in view of the choice of ϵ_1 (see (11.113)), (11.109), (11.118), and (11.120), we have

$$w(by(t)) \leq \gamma + w(by_\sigma(t)e(\sigma)) < w(b(\widehat{x}_\sigma - \epsilon/n)e(\sigma)) + \gamma$$
$$< \gamma + w(b\widehat{y}) - 4\gamma < w(b\widehat{y}) - 3\gamma. \tag{11.121}$$

It follows from (11.116), (11.117), (11.119), and (11.121) that

$$2\epsilon_1 > \int_T^{T+L} [w(b\widehat{y}) - w(by(t))]dt$$

$$= \int_{E_1} [w(b\widehat{y}) - w(by(t))]dt + \int_{E_2} [w(b\widehat{y}) - w(by(t))]dt$$

$$\geq -\epsilon_0 L + 3\gamma \text{mes}(E_2).$$

Combined with (11.110) and (11.111), this relation implies that

$$\text{mes}(E_2) \leq (3\gamma)^{-1}(2\epsilon_1 + L\epsilon_0) \leq (3\gamma)^{-1}(2L^2\epsilon_0) \leq \gamma^{-1}L^2\epsilon_0 < \epsilon.$$

Assertion (ii) is proved. This completes the Proof of Theorem 11.4.

Proof of Theorem 11.5 Let $x_0 \in R_+^n$ be given. Proposition 11.10 implies that there exists a program $(\tilde{x}(t), \tilde{y}(t))_{t=0}^\infty$ such that

$$\tilde{x}(0) = x_0, \quad \int_0^\infty \delta(\tilde{x}(t), \tilde{y}(t), \tilde{x}'(t))dt = \Delta(x_0). \tag{11.122}$$

Proposition 11.7 and (11.122) imply that $(\tilde{x}(t), \tilde{y}(t))_{t=0}^{\infty}$ is a good program. We claim that $(\tilde{x}(t), \tilde{y}(t))_{t=0}^{\infty}$ is an overtaking optimal program.

Let $(x(t), y(t))_{t=0}^{\infty}$ be a program such that

$$x(0) = x_0.$$

If $(x(t), y(t))_{t=0}^{\infty}$ is bad, then

$$\int_0^T [w(b\tilde{y}(t)) - w(by(t))]dt$$

$$= \int_0^T [w(b\tilde{y}(t)) - w(b\hat{y})]dt + \int_0^T [w(b\hat{y}) - w(by(t))]dt \to \infty \text{ as } T \to \infty.$$

Assume that the program $(x(t), y(t))_{t=0}^{\infty}$ is good. Theorem 11.4 implies that $x(t), \tilde{x}(t) \to \hat{x}$ as $t \to \infty$. Together with (11.19), Proposition 11.7, and (11.122), this implies that

$$\limsup_{T\to\infty} \int_0^T [w(by(t)) - w(b\tilde{y}(t))]dt$$

$$= \limsup_{T\to\infty} \int_0^T [\delta(\tilde{x}(t), \tilde{y}(t), \tilde{x}'(t)) - \delta(x(t), y(t), x'(t)) - \hat{p}(x'(t) - \tilde{x}'(t))]dt$$

$$= \int_0^{\infty} \delta(\tilde{x}(t), \tilde{y}(t), \tilde{x}'(t))dt - \int_0^{\infty} \delta(x(t), y(t), x'(t))dt$$

$$- \hat{p} \lim_{T\to\infty} (x(T) - x(0) - \tilde{x}(T) + \tilde{x}(0))$$

$$= \int_0^{\infty} \delta(\tilde{x}(t), \tilde{y}(t), \tilde{x}'(t))dt - \int_0^{\infty} \delta(x(t), y(t), x'(t))dt \leq 0.$$

Theorem 11.5 is proved.

11.5 Turnpike Results

Let $z \in R_+^n$ and $T > 0$ be given. Define

$$U(z, T) = \sup \left\{ \int_0^T w(by(t))dt : \right.$$

$$\left. (x(t), y(t))_{t=0}^T \text{ is a program such that } x(0) = z \right\}. \tag{11.123}$$

Proposition 11.1, Theorem 11.5, (11.3), (11.5), and (11.123) imply that $U(z, T)$ is a finite number.

Let $x_0, x_1 \in R_+^n$ and let $0 \leq T_1 < T_2$. Define

$$
U(x_0, x_1, T_1, T_2) = \sup \left\{ \int_{T_1}^{T_2} w(by(t))dt : \right.
$$

$$
(x(t), y(t))_{t=T_1}^{T_2} \text{ is a program such that} \qquad (11.124)
$$

$$
\left. x(T_1) = x_0, \ x(T_2) \geq x_1 \right\}.
$$

Here we assume that supremum over empty set is $-\infty$. In view of Proposition 11.1, (11.3), (11.5), and (11.124), $U(x_0, x_1, T_1, T_2) < \infty$. It is not difficult to see that for every point $z \in R_+^n$ and every positive number T, $U(z, T) = U(z, 0, 0, T)$.

We prove the following two theorems, obtained in [110], which describe the structure of approximate optimal solutions of optimal control problems on sufficiently large intervals.

Theorem 11.14 *Let M, ϵ, L be positive numbers and let $\Gamma \in (0, 1)$. Then there exist $T_* > 0$ and a positive number γ such that for each $T > 2T_*$, each $z_0, z_1 \in R_+^n$ satisfying $z_0 \leq Me$ and $az_1 \leq \Gamma d^{-1}$, and each program $(x(t), y(t))_{t=0}^T$ which satisfies*

$$
x(0) = z_0, \ x(T) \geq z_1, \quad \int_0^T w(by(t))dt \geq U(z_0, z_1, 0, T) - \gamma
$$

there are numbers τ_1, τ_2 such that $\tau_1 \in [0, T_*]$, $\tau_2 \in [T - T_*, T]$,

$$
\|x(t) - \widehat{x}\| \leq \epsilon \text{ for all } t \in [\tau_1, \tau_2]
$$

and that for each number S satisfying $\tau_1 \leq S \leq \tau_2 - L$,

$$
mes(\{t \in [S, S + L] : \|y(t) - \widehat{x}\| > \epsilon\}) \leq \epsilon.
$$

Moreover, if $\|x(0) - \widehat{x}\| \leq \gamma$, then $\tau_1 = 0$, and if $\|x(T) - \widehat{x}\| \leq \gamma$, then $\tau_2 = T$.

Theorem 11.15 *Let M_0, M_1, ϵ be positive numbers, $L > 1$, and let $\Gamma \in (0, 1)$. Then there exist $T_* > L$, a natural number Q, and $l > 0$ such that for each $T > T_*$, each $z_0, z_1 \in R_+^n$ satisfying $z_0 \leq Me$ and $az_1 \leq \Gamma d^{-1}$, and each program $(x(t), y(t))_{t=0}^T$ which satisfies*

$$
x(0) = z_0, \ x(T) \geq z_1, \quad \int_0^T w(by(t))dt \geq U(z_0, z_1, 0, T) - M_1
$$

there exists a finite sequence of closed intervals $[S_i, S_i']$, $i = 1, \ldots, q$ such that $q \leq Q$, $S_i' - S_i \leq l$, $i = 1, \ldots, q$, $S_i' \leq S_{i+1}$ for each integer i satisfying $1 \leq i \leq q - 1$,

$$\|x(t) - \widehat{x}\| \le \epsilon, \ t \in [0, T] \setminus \cup_{i=1}^{q} [S_i, S_i']$$

and if $S \in [0, T - L]$ satisfies

$$[S, S + L] \subset [S_i', S_{i+1}] \text{ with } 1 \le i < q,$$

then

$$mes(\{t \in [S, S + L] : \ \|y(t) - \widehat{x}\| > \epsilon\}) \le \epsilon.$$

11.6 Auxiliary Results

Lemma 11.16 *Let $\Gamma \in (0, 1)$. Then there exists a number $k(\Gamma) > 0$ such that for each $z_0 \in R_+^n$ and each $z_1 \in R_+^n$ satisfying $az_1 \le \Gamma d^{-1}$, there is a program $(x(t), y(t))_{t=0}^{\infty}$ such that $x(0) = z_0$ and $x(t) \ge z_1$ for all $t \ge k(\Gamma)$.*

Proof There exists a positive number $k(\Gamma)$ such that

$$1 - e^{-dk(\Gamma)} > \Gamma. \tag{11.125}$$

Assume that $z_0 \in R_+^n$ and $z_1 \in R_+^n$ satisfies

$$az_1 \le \Gamma d^{-1}. \tag{11.126}$$

Set

$$z_2 = \Gamma^{-1} z_1, \tag{11.127}$$

$$x(t) = e^{-dt}(z_0 - z_2) + z_2, \ y(t) = 0, \ t \in [0, \infty), \tag{11.128}$$

In view of (11.128), for all $t \ge 0$, we have

$$x(t) \ge 0, \tag{11.129}$$

$$x'(t) + dx(t) = -de^{-dt}(z_0 - z_2) + de^{-dt}(z_0 - z_2) + dz_2 = dz_2. \tag{11.130}$$

By (11.126), (11.127), and (11.130), for all $t \ge 0$,

$$x'(t) + dx(t) \ge 0,$$

$$a(x'(t) + dx(t)) = adz_2 = ad(\Gamma^{-1} z_1) \le 1.$$

Combined with (11.2), (11.5), (11.128), and (11.129), these inequalities imply that $(x(t), y(t))_{t=0}^{\infty}$ is a program. In view (11.128), we have

$$x(0) = z_0. \tag{11.131}$$

By (11.125)–(11.128), for all $t \geq k(\Gamma)$,

$$x(t) \geq \left(1 - e^{-dt}\right)z_2 \geq \left(1 - e^{-dk(\Gamma)}\right)z_2 \geq \Gamma^{-1}\left(1 - e^{-dk(\gamma)}\right)z_1 \geq z_1.$$

This completes the proof of Lemma 11.16.

In the sequel with each $\Gamma \in (0, 1)$, we associate a number $k(\Gamma) > 0$ for which the assertion of Lemma 11.16 holds.

Lemma 11.17 *There exists a positive number m such that for every $z \in R_+^n$ and every positive number T,*

$$U(z, T) \geq Tw(b\widehat{x}) - m. \tag{11.132}$$

Proof In view of (11.6), we have

$$a\widehat{x} = a_\sigma \widehat{x}_\sigma = a_\sigma(1 + da_\sigma)^{-1} < d^{-1}. \tag{11.133}$$

It follows from (11.133) that there exists a number $\Gamma \in (0, 1)$ such that

$$a\widehat{x} \leq \Gamma d^{-1}. \tag{11.134}$$

Fix a number

$$m > k(\Gamma)[|w(0)| + |w(k(\Gamma))| + |w(b\widehat{x})|].$$

Let $z \in R_+^n$ be given. In view of (11.134), the choice of $k(\Gamma)$, and Lemma 11.17, there exists a program $(x(t), y(t))_{t=0}^{k(\Gamma)}$ such that

$$x(0) = z, \ x(k(\Gamma)) \geq \widehat{x}. \tag{11.135}$$

Let $T > 0$ be given. We claim that (11.132) is valid. There are cases

$$T \leq k(\Gamma) \tag{11.136}$$

and

$$T > k(\Gamma). \tag{11.137}$$

Assume that (11.136) is valid. Then by (11.135), (11.136), and the choice of m, we have

$$U(z, T) \geq \int_0^T w(by(t))dt \geq Tw(0) \geq T(-|w(0)|) \geq -k(\Gamma)|w(0)|$$

$$= Tw(b\widehat{x}) + [-k(\Gamma)|w(0)| - Tw(b\widehat{x})]$$

$$\geq Tw(b\widehat{x}) - k(\Gamma)|w(0)| - k(\Gamma)|w(b\widehat{x})|$$

$$\geq Tw(b\widehat{x}) - m$$

and (11.132) is true.

Assume that (11.137) is valid. For all numbers $t > k(\Gamma)$, define

$$x(t) = \widehat{x} + e^{-d(t-k(\Gamma))}(x(k(\Gamma)) - \widehat{x}), \ y(t) = \widehat{x}. \tag{11.138}$$

In view of (11.135) and (11.138), for all $t \in (k(\Gamma), \infty)$, we have

$$0 \leq y(t) \leq x(t). \tag{11.139}$$

By (11.1), (11.138), for all $t \in (k(\Gamma), \infty)$,

$$x'(t) + dx(t) = d\widehat{x} = d(1 + da_\sigma)^{-1}e_\sigma, \tag{11.140}$$

$$a(x'(t) + dx(t)) \leq 1, \ (x(t), x'(t)) \in \Omega. \tag{11.141}$$

It follows from (11.138), (11.140), for all $t \in (k(\Gamma), \infty)$

$$a(x'(t) + dx(t)) + ey(t) = ad\widehat{x} + (1 + da_\sigma)^{-1}$$

$$= a_\sigma d(1 + da_\sigma)^{-1} + (1 + da_\sigma)^{-1} = 1$$

and together with (11.5), (11.139), and (11.141), this implies that

$$y(t) \in \Lambda(x(t), x'(t)).$$

Hence we have shown that $(x(t), y(t))_{t=0}^{\infty}$ is a program. In view of (11.135), (11.137), (11.138), and the choice of m,

$$U(z, T) \geq \int_0^T w(by(t))dt = \int_0^{k(\Gamma)} w(by(t))dt + (T - k(\Gamma))w(b\widehat{x})$$

$$\geq Tw(b\widehat{x}) + k(\Gamma)(w(0) - w(b\widehat{x})) \geq Tw(b\widehat{x}) - m.$$

Hence (11.132) is valid. This completes the proof of Lemma 11.17.

Lemma 11.18 *Let $\Gamma \in (0, 1)$. Then there exists a positive number m such that for every $z_0 \in R_+^n$, every $z_1 \in R_+^n$ satisfying $az_1 \leq \Gamma d^{-1}$, and every $T > k(\Gamma)$,*

$$U(z_0, z_1, 0, T) \geq Tw(b\widehat{x}) - m.$$

Proof Lemma 11.17 implies that there exists a positive number m_0 such that

$$U(z, T) \geq T w(b\widehat{x}) - m_0 \text{ for every } z \in R^n_+ \text{ and every positive number } T. \tag{11.142}$$

Set

$$m = m_0 + 1 + k(\Gamma)(w(b\widehat{x}) - w(0)). \tag{11.143}$$

Assume that

$$z_0, z_1 \in R^n_+, \ az_1 \leq \Gamma d^{-1}, \ T > k(\Gamma). \tag{11.144}$$

In view of the choice of m_0 (see (11.142)) and (11.144), there exists a program $(x(t), y(t))_{t=0}^{T-k(\Gamma)}$ such that

$$x(0) = z_0,$$

$$\int_0^{T-k(\Gamma)} w(by(t))dt \geq U(z_0, T - k(\Gamma)) - 1 \geq (T - k(\Gamma))w(b\widehat{x}) - m_0 - 1. \tag{11.145}$$

It follows from the choice of $k(\Gamma)$, Lemma 11.16, and (11.144) that there exists a program $(x(t), y(t))_{t=T-k(\Gamma)}^{T}$ such that

$$x(T) \geq z_1. \tag{11.146}$$

It is clear that $(x(t), y(t))_{t=0}^{T}$ is a program. By (11.143), (11.145), and (11.146), we have

$$U(z_0, z_1, 0, T) \geq \int_0^T w(by(t))dt = \int_0^{T-k(\Gamma)} w(by(t))dt + \int_{T-k(\Gamma)}^T w(by(t))dt$$

$$\geq (T - k(\Gamma))w(b\widehat{x}) - m_0 - 1 + k(\Gamma)w(0)$$

$$= T w(b\widehat{x}) - k(\Gamma)(w(b\widehat{x}) - w(0)) - m_0 - 1 = T w(b\widehat{x}) - m.$$

This completes the proof of Lemma 11.18.

Lemma 11.19 *Let m_0 be a positive number. Then there exists a positive number m_2 such that for every positive number T and every program $(x(t), y(t))_{t=0}^{T}$ which satisfies $x(0) \leq m_0 e$, the inequality*

$$\int_0^T [w(by(t)) - w(b\widehat{x})]dt \leq m_2$$

is valid.

Proof Proposition 11.1 implies that there exists a positive number m_1 such that for every positive number T and every program $(x(t), y(t))_{t=0}^{T}$ satisfying $x(0) \leq m_0 e$, the inequality

$$x(t) \leq m_1 e \text{ for all } t \in [0, T]. \tag{11.147}$$

Fix

$$m_2 > 2\|\widehat{p}\|m_1 n. \tag{11.148}$$

Assume that $T > 0$ and that a program $(x(t), y(t))_{t=0}^{T}$ satisfies $x(0) \leq m_0 e$. Then inequality (11.147) is valid. Lemma 11.6, (11.20), (11.147), and (11.148) imply that

$$\int_0^T (w(by(t)) - w(b\widehat{x}))dt \leq -\widehat{p}(x(T) - x(0)) \leq 2\|\widehat{p}\|nm_1 \leq m_2.$$

This completes the proof of Lemma 11.19.

It is not difficult to see that the following auxiliary result holds.

Lemma 11.20 *Assume that nonnegative numbers T_1, T_2 satisfy $T_1 < T_2$,*

$$(x(t), y(t))_{t=T_1}^{T_2}$$

is a program and that $u \in R_+^n$. Then $(x(t) + e^{-d(t-T_1)}u, y(t))_{t=T_1}^{T_2})$ is also a program.

In order to prove Lemma 11.20, it is sufficient to note that for a.e. $t \in [T_1, T_2]$,

$$\left(x(t) + e^{-d(t-T_1)}u\right)' + d\left(x(t) + e^{-d(t-T_1)}u\right) = x'(t) + dx(t).$$

Lemma 11.20 implies the following result.

Lemma 11.21 *Let $0 \leq T_1 < T_2$, $M > 0$, $x_0, x_1 \in R_+^n$, and let*

$$(x(t), y(t))_{t=T_1}^{T_2}$$

be a program such that

$$x(T_1) = x_0, \ x(T_2) \geq x_1, \ \int_{T_1}^{T_2} w(by(t))dt \geq U(x_0, x_1, T_1, T_2) - M.$$

Then for each pair of numbers S_1, S_2 satisfying

$$T_1 \leq S_1 < S_2 \leq T_2$$

the following inequality holds:

$$\int_{S_1}^{S_2} w(by(t))dt \geq U(x(S_1), x(S_2), S_1, S_2) - M.$$

Lemma 11.22 *Let ϵ be a positive number. Then there exists a positive number δ such that for every pair of points $z, z' \in R_+^n$ satisfying*

$$\|z - \widehat{x}\|, \ \|z' - \widehat{x}\| \leq \delta \tag{11.149}$$

and every $T \in [2^{-1}, 2]$, there exists a program $(x(t), y(t))_{t=0}^T$ such that

$$x(0) = z, \ x(T) \geq z',$$

$$\|x(t) - \widehat{x}\|, \ |y(t) - \widehat{x}\| \leq \epsilon, \ t \in [0, T], \ \|x'(t)\| \leq \epsilon, \ t \in [0, T].$$

Proof We may assume without loss of generality that

$$\epsilon < (1 + da_\sigma)^{-1}. \tag{11.150}$$

Fix $\delta > 0$ such that

$$\delta \sum_{i=1}^n a_i < (\epsilon/16), \ 16\delta n < \epsilon. \tag{11.151}$$

Assume that $T \in [2^{-1}, 2]$ and that a pair of points $z, z' \in R_+^n$ satisfies (11.149). For all numbers $t \in [0, T]$, set

$$y(t) = \left((1 + da_\sigma)^{-1} - \epsilon\right)e(\sigma). \tag{11.152}$$

Evidently,

$$\|y(t) - \widehat{x}\| \leq \epsilon, \ t \in [0, T]. \tag{11.153}$$

Set

$$\xi = 4\delta e$$

and define

$$x(t) = z + t\xi, \ t \in [0, T]. \tag{11.154}$$

In view of (11.149)–(11.154) and the choice of ξ,

$$0 \leq y(t) \leq z \leq x(t) \text{ for all } t \in [0, T]. \tag{11.155}$$

By (11.154) and the choice of ξ, for all $t \in [0, T]$, we have

$$x'(t) + dx(t) = \xi + d(z + t\xi) \geq 0. \tag{11.156}$$

It follows from (11.149) and (11.156) that for all $t \in [0, T]$,

$$a(x'(t) + dx(t)) = adz + (1 + dt)a\xi \leq ad\widehat{x} + \delta d \sum_{i=1}^{n} a_i + (1 + 2d)a\xi.$$

Combined with (11.6) and (11.151)–(11.153), this implies that for all $t \in [0, T]$,

$$ey(t) + a(x'(t) + dx(t)) \leq (1 + da_\sigma)^{-1} - \epsilon + a_\sigma d(1 + da_\sigma)^{-1} + \delta \sum_{i=1}^{n} a_i + 3a\xi$$

$$= 1 - \epsilon + \delta \sum_{i=1}^{n} a_i + 3a\xi \leq 1.$$

By the relation above, (11.115), (11.152) (11.154), and (11.156), $(x(t), y(t))_{t=0}^{T}$ is a program. By (11.149), (11.153), (11.154), and the choice of ξ,

$$x(0) = z, \ x(T) = z + T\xi \geq \widehat{x} - \delta e + 2^{-1}\xi \geq \widehat{x} + \delta e \geq z.$$

In view of (11.149), (11.151), and (11.154), for all $t \in [0, T]$,

$$\|x(t) - \widehat{x}\| \leq \|z - \widehat{x}\| + T\|\xi\| \leq \delta + 2\|\xi\| \leq \delta + 8\delta n \leq 16\delta n < \epsilon.$$

Lemma 11.22 is proved.

Lemma 11.23 *Let m_0, m_1, ϵ be positive numbers. Then there exists an integer $\tau \geq 1$ such that for every program $(x(t), y(t))_{t=0}^{\tau}$ satisfying*

$$x(0) \leq m_0 e, \ \int_0^{\tau} w(by(t))dt \geq \tau w(b\widehat{x}) - m_1$$

there exists a number $t \in [0, \tau]$ such that

$$\|x(t) - \widehat{x}\| \leq \epsilon.$$

Proof Assume the contrary. Then for every integer $k \geq 1$, there exists a program $(x^{(k)}(t), y^{(k)}(t))_{t=0}^{k}$ such that

$$x^{(k)}(0) \leq m_0 e, \quad \int_0^k w(by^{(k)}(t))dt \geq kw(b\widehat{x}) - m_1,$$

$$\|x^{(k)}(t) - \widehat{x}\| > \epsilon \text{ for all numbers } t \in [0, k]. \tag{11.157}$$

By (11.157) and Proposition 11.1, there exists a number $m_2 > m_0$ such that for every integer $k \geq 1$,

$$x^{(k)}(t) \leq m_2 e, \ t \in [0, k]. \tag{11.158}$$

Lemma 11.19 implies that there exists a positive number m_3 such that for every positive number T and every program $(x(t), y(t))_{t=0}^{T}$ satisfying

$$x(0) \leq m_2 e$$

we have

$$\int_0^T (w(by(t)) - w(b\widehat{x}))dt \leq m_3. \tag{11.159}$$

Let $k \geq 1$ be an integer and let a number s satisfy $0 < s < k$. By (11.158) and the choice of m_3 (see (11.159)),

$$\int_s^k \left[w(by^{(k)}(t)) - w(b\widehat{x}) \right] dt \leq m_3.$$

Together with (11.158) the relation above implies that

$$\int_0^s \left[w(by^{(k)}(t)) - w(b\widehat{x}) \right] dt = \int_0^k \left[w(by^{(k)}(t)) - w(b\widehat{x}) \right] dt$$

$$- \int_s^k \left[w(by^{(k)}(t)) - w(b\widehat{x}) \right] dt \geq -m_1 - m_3.$$

Therefore for every integer $k \geq 1$ and every $s \in (0, k)$,

$$\int_0^s \left[w(by^{(k)}(t)) - w(b\widehat{x}) \right] dt \geq -m_1 - m_3. \tag{11.160}$$

By extracting a subsequence and using (11.158), Proposition 11.8, and diagonalization process, we obtain that there exist a strictly increasing sequence of natural numbers $\{k_j\}_{j=1}^{\infty}$ and a program $(x^*(t), y^*(t))_{t=0}^{\infty}$ such that for every integer $q \geq 1$, we have

$$x^{(k_j)}(t) \to x^*(t) \text{ as } j \to \infty \text{ uniformly on } [0, q], \tag{11.161}$$

$$\left(x^{(k_j)}\right)'(t) \to (x^*)'(t) \text{ as } j \to \infty \text{ weakly in } L^2([0, q]; R^n), \tag{11.162}$$

$$y^{(k_j)} \to y^* \text{ as } j \to \infty \text{ weakly in } L^2([0, q]; R^n). \tag{11.163}$$

By (11.158) and (11.161),

$$x^*(t) \leq m_2 e \text{ for all } t \geq 0. \tag{11.164}$$

It follows from (11.158), (11.160)–(11.163), and Proposition 11.9 that for all integers $q \geq 1$,

$$\int_0^q w(by^*(t))dt \geq \limsup_{j \to \infty} \int_0^q w\big(by^{k_j}(t)\big)dt \geq qw(b\widehat{x})) - m_3 - m_1.$$

Combined with Proposition 11.3, the relation above implies that

$$(x^*(t), y^*(t))_{t=0}^{\infty}$$

is a good program. Theorem 11.4 implies that

$$\lim_{t \to \infty} x^*(t) = \widehat{x}. \tag{11.165}$$

On the other hand, it follows from (11.157) and (11.161) that

$$\|x^*(t) - \widehat{x}\| \geq \epsilon, \ t \in [0, \infty).$$

This contradicts (11.165). The contradiction we have reached completes the proof of Lemma 11.23.

Lemma 11.24 *Let ϵ be a positive number. Then there exists a positive number γ such that for every number $T > 2$ and every program $(x(t), y(t))_{t=0}^{T}$ satisfying*

$$\|x(0) - \widehat{x}\| \leq \gamma, \ \|x(T) - \widehat{x}\| \leq \gamma, \tag{11.166}$$

$$\int_0^T w(by(t))dt \geq U(x(0), x(T), 0, T) - \gamma \tag{11.167}$$

the following inequality holds:

$$\int_0^T \delta(x(t), y(t), x'(t))dt \leq \epsilon.$$ (11.168)

Proof Fix a number $\epsilon_0 > 0$ such that

$$\epsilon_0 < (\epsilon/18)(\|\widehat{p}\| + 1)^{-1},$$ (11.169)

if $y \in R_+^n$ and $\|y - \widehat{x}\| \leq \epsilon_0$, then $|w(b\widehat{x}) - w(by)| < \epsilon/16$. (11.170)

Lemma 11.22 implies that there exists a number $\gamma \in (0, \epsilon_0)$ such that the following property holds:

(P1) for every $T \in [2^{-1}, 2]$ and every pair of points $z, z' \in R_+^n$ which satisfy

$$\|z - \widehat{x}\|, \ \|z' - \widehat{x}\| \leq \gamma$$

there exists a program $(u(t), v(t))_{t=0}^T$ such that

$$u(0) = z, \ u(T) \geq z', \ \|u(t) - \widehat{x}\|, \ \|v(t) - \widehat{x}\| \leq \epsilon_0, \ t \in [0, T],$$

$$\|u'(t)\| \leq \epsilon_0, \ t \in [0, T].$$

Assume that $T > 2$ and that a program $(x(t), y(t))_{t=0}^T$ satisfies (11.166) and (11.167). In view of (11.166) and property (P1), there exist programs

$$\left(u^{(1)}(t), v^{(1)}(t)\right)_{t=0}^1 \text{ and } \left(u^{(2)}(t), v^{(2)}(t)\right)_{t=T-1}^T$$

such that

$$u^{(1)}(0) = x(0), \ u^{(1)}(1) \geq \widehat{x}, \ \|u^{(1)}(t) - \widehat{x}\|, \ \|v^{(1)}(t) - \widehat{x}\| \leq \epsilon_0, \ t \in [0, 1],$$ (11.171)

$$\|\left(u^{(1)}\right)'(t)\| \leq \epsilon_0, \ t \in [0, 1],$$

$$u^{(2)}(T - 1) = \widehat{x}, \ u^{(2)}(T) \geq x(T), \ \|u^{(2)}(t) - \widehat{x}\|, \ \|v^{(2)}(t) - \widehat{x}\| \leq \epsilon_0,$$

$$\|\left(u^{(2)}\right)'(t)\| \leq \epsilon_0, \ t \in [T - 1, T].$$ (11.172)

Define a program $(\bar{x}(t), \bar{y}(t))_{t=0}^T$ as follows. Set

$$\bar{x}(t) = u^{(1)}(t), \ \bar{y}(t) = v^{(1)}(t), \ t \in [0, 1],$$ (11.173)

$$\bar{x}(t) = \widehat{x} + e^{-d(t-1)}\left(u^{(1)}(1) - \widehat{x}\right), \ \bar{y}(t) = \widehat{x}, \ t \in (1, T - 1].$$

In view of Lemma 11.20, (11.171), and (11.173), $(\bar{x}(t), \bar{y}(t))_{t=0}^{T-1}$ is a program. By (11.171) and (11.173), we have

$$\bar{x}(T-1) \geq \hat{x} = u^{(2)}(T-1). \tag{11.174}$$

For $t \in (T-1, T]$ set

$$\bar{x}(t) = u^{(2)}(t) + e^{-d(t-(T-1))}(\bar{x}(T-1) - u^{(2)}(T-1)), \ \bar{y}(t) = v^{(2)}(t). \tag{11.175}$$

Lemma 11.20, (11.174), and (11.175) imply that $(\bar{x}(t), \bar{y}(t))_{t=0}^{T}$ is a program. By (11.172)–(11.175) we have

$$\bar{x}(0) = x(0), \ \bar{x}(T) \geq u^{(2)}(T) \geq x(T). \tag{11.176}$$

It follows from (11.167) and (11.176) that

$$-\gamma \leq \int_0^T w(by(t))dt - \int_0^T w(b\bar{y}(t))dt. \tag{11.177}$$

Equations (11.171)–(11.173) and (11.175) imply that

$$\|\bar{x}(T) - \hat{x}\| \leq \|\bar{x}(T) - u^{(2)}(T)\| + \|u^{(2)}(T) - \hat{x}\|$$

$$\leq \|\bar{x}(T-1) - u^{(2)}(T-1)\| + \epsilon_0 = \|\bar{x}(T-1) - \hat{x}\| + \epsilon_0$$

$$\leq \|u^{(1)}(1) - \hat{x}\| + \epsilon_0 \leq 2\epsilon_0.$$

In view of Lemma 11.6 and (11.177),

$$-\gamma \leq -\int_0^T (w(b\bar{y}(t)) - w(b\hat{y}))dt - \int_0^T \delta(x(t), y(t), x'(t))dt - \hat{p}(x(T) - x(0)).$$

Combined with (11.166), (11.169)–(11.173), and (11.175), the relation above implies that

$$\int_0^T \delta(x(t), y(t), x'(t))dt \leq \gamma - \int_0^T (w(b\bar{y}(t)) - w(b\hat{x}))dt + \|\hat{p}\| \|x(T) - x(0)\|$$

$$\leq \gamma - \int_0^1 (w(b\bar{y}(t)) - w(b\hat{x}))dt$$

$$- \int_{T-1}^T (w(b\bar{y}(t)) - w(b\hat{x}))dt + \|\hat{p}\|2\epsilon_0$$

$$\leq \gamma + \epsilon/16 + \epsilon/16 + \epsilon/8 < \epsilon.$$

This completes the proof of Lemma 11.24.

Lemma 11.25 *Let $\epsilon > 0$ and $\tau_0 > 0$. Then there exist $\gamma > 0$ and $T_0 > \tau_0$ such that for every number $T \geq T_0$ and every program $(x(t), y(t))_{t=0}^{T}$ which satisfies*

$$\|x(0) - \widehat{x}\|, \ \|x(T) - \widehat{x}\| \leq \gamma, \quad \int_0^T \delta(x(t), y(t), x'(t))dt \leq \gamma$$

the following properties hold:

$$\|x(t) - \widehat{x}\| \leq \epsilon \ \text{for all } t \in [0, T];$$

for every $S \in [0, T - \tau_0]$,

$$mes(\{t \in [S, S + \tau_0] : \ \|y(t) - \widehat{x}\| > \epsilon\}) \leq \epsilon.$$

Proof We may assume that $\epsilon < 1$. Fix

$$\epsilon_0 \in (0, 16^{-1}\epsilon). \tag{11.178}$$

In view of Lemma 11.22 and the continuity of the function $\delta(\cdot, \cdot, \cdot)$, there exists a sequence of positive numbers $\{\gamma_q\}_{q=1}^{\infty}$ such that

$$\gamma_q \leq 4^{-1}\gamma_{q-1} \ \text{for all natural numbers } q \geq 2, \tag{11.179}$$

$$\gamma_q \leq 4^{-q}\epsilon_0 \ \text{for all natural numbers } q. \tag{11.180}$$

and that for every natural number q, the following property holds:
(P2) for every pair of points $z, z' \in R_+^n$ satisfying

$$\|z - \widehat{x}\|, \ \|z' - \widehat{x}\| \leq \gamma_q$$

there exists a program $(x(t), y(t))_{t=0}^{1}$ such that

$$x(0) = z, \ x(1) \geq z', \ \|x(t) - \widehat{x}\|, \ \|y(t) - \widehat{x}\|$$

$$\leq 4^{-q}\epsilon_0, \ t \in [0, 1],$$

$$\|x'(t)\| \leq 4^{-q}\epsilon_0, \ t \in [0, 1],$$

$$\int_0^1 \delta(x(t), y(t), x'(t))dt \leq 4^{-q}\epsilon_0.$$

Assume that the lemma does not hold. Then for every integer $q \geq 1$, there exist

$$T_q \geq \tau_0 + q \tag{11.181}$$

and a program $(x^{(q)}(t), y^{(q)}(t))_{t=0}^{T_q}$ such that

$$\|x^{(q)}(0) - \widehat{x}\|, \ \|x^{(q)}(T_q) - \widehat{x}\| \le \gamma_q, \tag{11.182}$$

$$\int_0^{T_q} \delta\big(x^{(q)}(t), y^{(q)}(t), (x^{(q)})'(t)\big)dt \le \gamma_q$$

and that at least one of the following properties holds:

$$\sup\big\{\|x^{(q)}(t) - \widehat{x}\| : \ t \in [0, T_q]\big\} > \epsilon; \tag{11.183}$$

(P3) there exists $S \in [0, T_q - \tau_0]$ such that

$$\mathrm{mes}\big(\{t \in [S, S + \tau_0] : \ \|y^{(q)}(t) - \widehat{x}\| > \epsilon\}\big) > \epsilon. \tag{11.184}$$

Extracting a subsequence and re-indexing, we may assume without loss of generality that one of the following cases holds:
  Equation (11.183) is valid for all integers $q \ge 1$; (P3) holds for all integers $q \ge 1$.
  Property (P2), (11.79), and (11.182) imply that for every integer $q \ge 1$, there exists a program $(u^{(q)}(t), v^{(q)}(t))_{t=0}^1$ such that

$$u^{(q)}(0) = x^{(q)}(T_q), \ u^{(q)}(1) \ge x^{(q+1)}(0), \tag{11.185}$$

$$\|u^{(q)}(t) - \widehat{x}\|, \ \|v^{(q)}(t) - \widehat{x}\| \le 4^{-q}\epsilon_0, \ t \in [0, 1], \tag{11.186}$$

$$\|(u^{(q)})'(t)\| \le 4^{-q}\epsilon_0, \ t \in [0, 1],$$

$$\int_0^1 \delta\Big(u^{(q)}(t), v^{(q)}(t), \big(u^{(q)}\big)'(t)\Big)dt \le 4^{-q}\epsilon_0. \tag{11.187}$$

We construct a program $(\bar{x}(t), \bar{y}(t))_{t=0}^\infty$ by induction. Put

$$\bar{x}(t) = x^{(1)}(t), \ \bar{y}(t) = y^{(1)}(t), \ t \in [0, T_1]. \tag{11.188}$$

Assume that q is a natural number and that we have already defined a program

$$(\bar{x}(t), \bar{y}(t))_{t=0}^{\sum_{i=1}^q T_i + q - 1}$$

such that

$$\bar{x}\left(\sum_{i=1}^q T_i + q - 1\right) \ge x^{(q)}(T_q). \tag{11.189}$$

(Evidently, for $q = 1$ our assumption holds.) For $t \in (\sum_{i=1}^q T_i + q - 1, \sum_{i=1}^q T_i + q]$ set

$$\bar{x}(t) = u^{(q)}\left(t - \left(\sum_{i=1}^{q} T_i + q - 1\right)\right) + e^{-d(t-(\sum_{i=1}^{q} T_i + q - 1))}$$

$$\times \left[\bar{x}\left(\sum_{i=1}^{q} T_i + q - 1\right) - x^{(q)}(T_q)\right], \qquad (11.190)$$

$$\bar{y}(t) = v^{(q)}\left(t - \left(\sum_{i=1}^{q} T_i + q - 1\right)\right).$$

It follows from (11.185), (11.189), (11.190), and Lemma 11.20 that

$$(\bar{x}(t), \bar{y}(t))_{t=0}^{\sum_{i=1}^{q} T_i + q}$$

is a program,

$$\bar{x}\left(\sum_{i=1}^{q} T_i + q\right) = u^{(q)}(1) + e^{-d}\left[\bar{x}\left(\sum_{i=1}^{q} T_i + q - 1\right) - x^{(q)}(T_q)\right]$$

$$\geq u^{(q)}(1) \geq x^{(q+1)}(0). \qquad (11.191)$$

For every number $t \in (\sum_{i=1}^{q} T_i + q, \sum_{i=1}^{q+1} T_i + q]$, define

$$\bar{x}(t) = x^{(q+1)}\left(t - \left(\sum_{i=1}^{q} T_i + q\right)\right) + e^{-d(t-(\sum_{i=1}^{q} T_i + q))}$$

$$\times \left[\bar{x}\left(\sum_{i=1}^{q} T_i + q\right) - x^{(q+1)}(0)\right],$$

$$\bar{y}(t) = y^{(q+1)}\left(t - \left(\sum_{i=1}^{q} T_i + q\right)\right). \qquad (11.192)$$

It follows from (11.191), (11.192), and Lemma 11.20 that $(\bar{x}(t), \bar{y}(t))_{0}^{\sum_{i=1}^{q+1} T_i + q}$ is a program and that

$$\bar{x}\left(\sum_{i=0}^{q+1} T_i + q\right) \geq x^{(q+1)}(T_{q+1}). \qquad (11.193)$$

Hence the program $(\bar{x}(t), \bar{y}(t))_{t=0}^{\infty}$ has been constructed by induction.

It follows from (11.190), (11.186) (with $q = 1$), and (11.188) that for every $t \in [T_1, T_1 + 1]$,

$$\|\bar{x}(t) - \widehat{x}\| \leq \|\bar{x}(t) - u^{(1)}(t - T_1)\| + \|u^{(1)}(t - T_1) - \widehat{x}\|$$

$$\leq \|\bar{x}(T_1) - x^{(1)}(T_1)\| + 4^{-1}\epsilon_0 = 4^{-1}\epsilon_0. \tag{11.194}$$

We show by induction that for every integer $q \geq 1$, we have

$$\left\| \bar{x}\left(t + \sum_{i=1}^{q} T_i + q - 1 - T_q\right) - x^{(q)}(t) \right\| \leq 2\left(\sum_{i=1}^{q} 4^{-i}\epsilon_0\right), \ t \in [0, T_q],$$

$$\tag{11.195}$$

$$\|\bar{x}(t) - \widehat{x}\| \leq \sum_{i=1}^{q} 2 \cdot 4^{-i}\epsilon_0, \ t \in \left[\sum_{i=1}^{q} T_i + q - 1, \sum_{i=1}^{q} T_i + q\right]. \tag{11.196}$$

In view of (11.188) and (11.194), Equations (11.195) and (11.196) are valid for $q = 1$.

Assume that $q \geq 1$ is an integer and that (11.195) and (11.196) are true. For every number $t \in [0, T_{q+1}]$, it follows from (11.179), (11.180), (11.182), (11.192), and (11.196) that

$$\left\| \bar{x}\left(t + \sum_{i=1}^{q} T_i + q\right) - x^{(q+1)}(t) \right\| \leq \left\| \bar{x}\left(\sum_{i=1}^{q} T_i + q\right) - x^{(q+1)}(0) \right\|$$

$$\leq \left\| \bar{x}\left(\sum_{i=1}^{q} T_i + q\right) - \widehat{x}\right\| + \|\widehat{x} - x^{(q+1)}(0)\|$$

$$\leq \left\| \bar{x}\left(\sum_{i=1}^{q} T_i + q\right) - \widehat{x}\right\| + 4^{-q-1}\epsilon_0$$

$$\leq \sum_{i=1}^{q} 2 \cdot 4^{-i}\epsilon_0 + 4^{-q-1}\epsilon_0. \tag{11.197}$$

In view of (11.186) and (11.190) (which holds for every integer $q \geq 1$), (11.196), and (11.197), for every number $t \in [\sum_{i=1}^{q+1} T_i + q, \sum_{i=1}^{q+1} T_i + q + 1]$,

$$\|\bar{x}(t) - \widehat{x}\| \leq \left\| \bar{x}(t) - u^{(q+1)}\left(t - \left(\sum_{i=1}^{q+1} T_i + q\right)\right) \right\|$$

$$+ \left\| u^{(q+1)}\left(t - \left(\sum_{i=1}^{q+1} T_i + q\right)\right) - \widehat{x}\right\|$$

$$\leq \|\bar{x}\left(\sum_{i=1}^{q+1} T_i + q\right) - x^{(q+1)}(T_{q+1})\| + 4^{-q-1}\epsilon_0$$

$$\leq \|\bar{x}\left(\sum_{i=1}^{q} T_i + q\right) - \widehat{x}\| + 2\cdot 4^{-q-1}\epsilon_0 \leq \sum_{i=1}^{q+1} 2\cdot 4^{-i}\epsilon_0.$$

Thus we have shown by induction that (11.195) and (11.196) are true for every integer $q \geq 1$.

We show that $(\bar{x}(t), \bar{y}(t))_{t=0}^{\infty}$ is a good program. In view of Proposition 11.2, in order to meet this goal, it is sufficient to show that

$$\int_0^T (w(b\bar{y}(t)) - w(b\widehat{y}))dt$$

does not tend to $-\infty$ as $T \to \infty$.

Lemma 11.6, (11.180), and (11.182) imply that for every integer $q \geq 1$,

$$\int_0^{T_q} \left(w\big(by^{(q)}(t)\big) - w(b\widehat{y})\right)dt \geq -\int_0^{T_q} \delta\big(x^{(q)}(t), y^{(q)}(t), \big(x^{(q)}\big)'(t)\big)dt$$

$$- \|\widehat{p}\|\big(\|x^{(q)}(0) - x^{(q)}(T_q)\|\big)$$

$$\geq -\gamma_q - 2\|\widehat{p}\|\gamma_q = -\gamma_q(1 + 2\|\widehat{p}\|)$$

$$\geq -4^{-q}(1 + 2\|\widehat{p}\|)\epsilon_0 \qquad (11.198)$$

and in view of Lemma 11.6 and (11.186), we have

$$\int_0^1 \left(w\big(bv^{(q)}(t)\big) - w(b\widehat{y})\right)dt \geq -\int_0^1 \delta\big(u^{(q)}(t), v^{(q)}(t), \big(u^{(q)}\big)'(t)\big)dt$$

$$- \|\widehat{p}\|\big(\|u^{(q)}(0) - u^{(q)}(1)\|\big)$$

$$\geq -4^{-q}\epsilon_0 - 2\cdot 4^{-q}(\epsilon_0)\|\widehat{p}\| \geq -4^{-q}\epsilon_0(1 + 2\|\widehat{p}\|).$$

$$(11.199)$$

By (11.198), (11.199), and the construction of the program $(\bar{x}(t), \bar{y}(t))_{t=0}^{\infty}$ (see (11.188)–(11.192)), for every natural number q,

$$\int_0^{\sum_{i=1}^{q} T_i + q - 1} (w(b\bar{y}(t)) - w(b\widehat{x}))dt \geq \sum_{i=1}^{q} -4^{-i}\epsilon_0(2 + 4\|\widehat{p}\|) \geq -(2 + 4\|\widehat{p}\|)2\epsilon_0.$$

Thus $(\bar{x}(t), \bar{y}(t))_{t=0}^{\infty}$ is a good program. By Theorem 11.4,

$$\lim_{t \to \infty} \bar{x}(t) = \widehat{x}$$

and there exists a positive number S_0 such that

$$\|\bar{x}(t) - \widehat{x}\| \le \epsilon_0 \text{ for every } t \ge S_0. \tag{11.200}$$

It follows from (11.195), (11.200), and (11.178), which is true for every natural number q, that

$$\|x^{(q)}(t) - \widehat{x}\| \le \epsilon, \ t \in [0, T_q] \text{ for all sufficiently large natural numbers } q. \tag{11.201}$$

In view of Theorem 11.4, there exists a positive number S_1 such that for every number $T \ge S_1$, we have

$$\mathrm{mes}([T, T + \tau_0] : \ \|\bar{y}(t) - \widehat{x}\| > \epsilon\}) \le \epsilon. \tag{11.202}$$

By (11.202), (11.192) (which is true for every integer $q \ge 1$), and (11.181), for all sufficiently large natural numbers q and for all numbers $S \in [0, T_q - \tau_0]$, we have

$$\mathrm{mes}\{t \in [S, S + \tau_0] : \ \|y^{(q)}(t) - \widehat{x}\| > \epsilon\} \le \epsilon.$$

This contradicts (P3), while (11.201) contradicts (11.183). The contradiction we have reached proves Lemma 11.25.

11.7 Proof of Theorem 11.4

Proposition 11.1 implies that there exists a positive number M_1 such that for every $T > 0$ and every program $(x(t), y(t))_{t=0}^{T}$ which satisfies $x(0) \le Me$, the following inequality is valid:

$$x(t) \le M_1 e \text{ for all } t \in [0, T]. \tag{11.203}$$

Lemma 11.18 implies that there exists a positive number M_2 such that for every $z_0 \in R_+^n$, every $z_1 \in R_+^n$ satisfying $az_1 \le \Gamma d^{-1}$, and every $T > k(\Gamma)$, we have

$$U(z_0, z_1, 0, T) \ge Tw(b\widehat{x}) - M_2. \tag{11.204}$$

Lemma 11.19 implies that there exists a positive number M_3 such that for every positive number T and every program $(x(t), y(t))_{t=0}^{T}$ satisfying $x(0) \le M_1 e$, the following inequality is valid:

$$\int_0^T [w(by(t)) - w(b\widehat{y})]dt \le M_3. \tag{11.205}$$

Lemma 11.25 implies that there exist numbers $\epsilon_1 > 0$, $L_1 > L$ such that for every number $T \ge L_1$ and every program $(x(t), y(t))_{t=0}^T$ which satisfies

$$\|x(0) - \widehat{x}\| \le \epsilon_1, \ \|x(T) - \widehat{x}\| \le \epsilon_1,$$

$$\int_0^T \delta(x(t), y(t), x'(t))dt \le \epsilon_1$$

the following properties hold:

$$\|x(t) - \widehat{x}\| \le \epsilon \ \text{ for all } t \in [0, T]; \tag{11.206}$$

for every $S \in [0, T - L]$,

$$\text{mes}(\{t \in [S, S + L] : \ \|y(t) - \widehat{x}\| > \epsilon\}) \le \epsilon. \tag{11.207}$$

Lemma 11.24 implies that there exists a number

$$\gamma \in (0, \min\{1, \epsilon, \epsilon_1\}) \tag{11.208}$$

such that for every $T > 2$ and every program $(x(t), y(t))_{t=0}^T$ satisfying

$$\|x(0) - \widehat{x}\| \le \gamma, \ \|x(T) - \widehat{x}\| \le \gamma,$$

$$\int_0^T w(by(t))dt \ge U(x(0), x(T), 0, T) - \gamma \tag{11.209}$$

the following inequality is valid:

$$\int_0^T \delta(x(t), y(t), x'(t))dt \le \epsilon_1. \tag{11.210}$$

Lemma 11.23 implies that there exists an integer $L_2 \ge 1$ such that for every program

$$(x(t), y(t))_{t=0}^{L_2}$$

which satisfies

$$x(0) \le M_1 e, \ \int_0^{L_2} w(by(t))dt \ge L_2 w(b\widehat{x}) - M_2 - M_3 - 1 \tag{11.211}$$

there exists a number $t \in [0, L_2]$ for which

$$\|x(t) - \widehat{x}\| \le \gamma. \tag{11.212}$$

Define

$$l = 2L_2 + 2L_1 + L, \quad Q > 4\epsilon_1^{-1}(M_3 + M_2 + 2\|\widehat{p}\|nM_2). \tag{11.213}$$

Set

$$T_* = L_2 + L_1 + k(\Gamma) + 2 + Ql. \tag{11.214}$$

Assume that

$$T > 2T_*, \ z_0, \ z_1 \in R_+^n, \ z_0 \le Me, \ az_1 \le \Gamma d^{-1} \tag{11.215}$$

and that a program $(x(t), y(t))_{t=0}^{T}$ satisfies

$$x(0) = z_0, \ x(T) \ge z_1, \ \int_0^T w(by(t))dt \ge U(z_0, z_1, 0, T) - \gamma. \tag{11.216}$$

By (11.215) and (11.216), relation (11.203) is valid. In view of (11.208), (11.214)–(11.216), and the choice of M_2 (see (11.204)), we have

$$\int_0^T w(by(t))dt \ge U(z_0, z_1, 0, T) - \gamma \ge Tw(b\widehat{x}) - M_2 - 1. \tag{11.217}$$

By the choice of M_3 (see (11.205)) and (11.203),

$$\int_{L_2}^T [w(by(t)) - w(b\widehat{x})]dt \le M_3, \quad \int_0^{T-L_2} [w(by(t)) - w(b\widehat{x})]dt \le M_3. \tag{11.218}$$

Equations (11.217) and (11.218) imply that

$$\int_0^{L_2} [w(by(t)) - w(b\widehat{x})]dt \ge -M_2 - 1 - M_3, \tag{11.219}$$

$$\int_{T-L_2}^T [w(by(t)) - w(b\widehat{x})]dt \ge -M_2 - 1 - M_3. \tag{11.220}$$

By (11.203), (11.219), (11.220), and the choice of L_2 (see (11.211) and (11.212)), there exist

$$\tau_1 \in [0, L_2], \ \tau_2 \in [T - L_2, T] \tag{11.221}$$

such that

$$\|x(\tau_i) - \widehat{x}\| \le \gamma, \ i = 1, 2. \tag{11.222}$$

If $\|x(0) - \widehat{x}\| \le \gamma$, then we set $\tau_1 = 0$, and if $\|x(T) - \widehat{x}\| \le \gamma$, then we set $\tau_2 = T$. In view of (11.216) and Lemma 11.21, we have

$$\int_{\tau_1}^{\tau_2} w(by(t))dt \ge U(x(\tau_1), x(\tau_2), \tau_1, \tau_2) - \gamma. \tag{11.223}$$

It follows from (11.222), (11.223), and the choice of γ (see (11.208)–(11.210)) that

$$\int_{\tau_1}^{\tau_2} \delta(x(t), y(t), x'(t))dt \le \epsilon_1. \tag{11.224}$$

By (11.215), (11.221), (11.222), (11.224), and the choice of ϵ_1, L_2 (see (11.206)–(11.208)),

$$\|x(t) - \widehat{x}\| \le \epsilon, \ t \in [\tau_1, \tau_2]$$

and if a number S satisfies $\tau_1 \le S \le \tau_2 - L$, then

$$\mathrm{mes}(\{t \in [S, S + L] : \ \|y(t) - \widehat{x}\| > \epsilon\}) \le \epsilon.$$

This completes the proof of Theorem 11.4.

11.8 Proof of Theorem 11.5

We may assume that $\epsilon < 1/4$. Proposition 11.1 implies that there exists a positive number M_2 such that for every $T > 0$ and every program $(x(t), y(t))_{t=0}^{T}$ satisfying $x(0) \le M_0 e$, the following inequality is valid:

$$x(t) \le M_2 e \text{ for all } t \in [0, T]. \tag{11.225}$$

Lemma 11.18 implies that there exists a positive number M_3 such that for every $z_0 \in R_+^n$, every $z_1 \in R_+^n$ satisfying $az_1 \le \Gamma d^{-1}$, and every number $T > k(\Gamma)$, we have

$$U(z_0, z_1, 0, T) \ge Tw(b\widehat{x}) - M_3. \tag{11.226}$$

In view of Lemma 11.25, there exist $\epsilon_1 \in (0, \epsilon)$, $L_1 > L$ such that for every number $T \ge L_1$ and every program $(x(t), y(t))_{t=0}^{T}$ satisfying

$$\|x(0) - \widehat{x}\| \le \epsilon_1, \ \|x(T) - \widehat{x}\| \le \epsilon_1, \tag{11.227}$$

$$\int_0^T \delta(x(t), y(t), x'(t))dt \le 2\epsilon_1,$$

the inequality

$$\|x(t) - \widehat{x}\| \le \epsilon, \ t \in [0, T] \tag{11.228}$$

is true and for every $S \in [0, T - L]$, we have

$$\mathrm{mes}(\{t \in [S, S + L]: \ \|y(t) - \widehat{x}\| > \epsilon\}) \le \epsilon. \tag{11.229}$$

Lemma 11.23 implies that there exists an integer $L_2 \ge 1$ such that for every program

$$(x(t), y(t))_{t=0}^{L_2}$$

satisfying

$$x(0) \le M_2 e, \ \int_0^{L_2} w(by(t))dt \ge L_2 w(b\widehat{x}) - M_1 - M_3 - 2 - 4\|\widehat{p}\|nM_2 \tag{11.230}$$

there exists $t \in [0, L_2]$ such that

$$\|x(t) - \widehat{x}\| \le \epsilon_1. \tag{11.231}$$

Fix

$$l = 2L_2 + 2L_1 + 8, \ \text{a natural number } Q > 4\epsilon_1^{-1}(M_3 + M_1 + M_2 + 2\|\widehat{p}\|nM_2), \tag{11.232}$$

$$T_* > 8L + 8L_1 + 8L_2 + k(\Gamma).$$

Assume that

$$T > T_*, \ z_0, \ z_1 \in R_+^n, \ z_0 \le Me, \ az_1 \le \Gamma d^{-1} \tag{11.233}$$

and that a program $(x(t), y(t))_{t=0}^T$ satisfies

$$x(0) = z_0, \ x(T) \ge z_1, \ \int_0^T w(by(t))dt \ge U(z_0, z_1, 0, T) - M_1. \tag{11.234}$$

By (11.233), (11.234), and the choice of M_2, relation (11.225) is valid. In view of (11.232)–(11.234) and the choice of M_3 (see (11.226)),

$$\int_0^T w(by(t))dt \geq Tw(b\widehat{x}) - M_3 - M_1.$$ (11.235)

Lemma 11.6 and (11.235) imply that

$$\int_0^T \delta(x(t), y(t), x'(t))dt$$

$$= \int_0^T (w(b\widehat{x}) - w(by(t))dt + \widehat{p}(x(0) - x(T)) \leq M_3 + M_1 + 2\|\widehat{p}\|nM_2.$$
(11.236)

It is easy to see that there exists a finite sequence of numbers $\{T_i\}_{i=0}^q$ such that $T_0 = 0$, $T_i < T_{i+1}$ for each integer i satisfying $0 \leq i < q$, $T_q = T$, for every integer i satisfying $0 \leq i < q$

$$\int_{T_i}^{T_{i+1}} \delta(x(t), y(t), x'(t))dt = \epsilon_1,$$ (11.237)

$$\int_{T_{q-1}}^{T_q} \delta(x(t), y(t), x'(t))dt \leq \epsilon_1.$$ (11.238)

In view of (11.236)–(11.238),

$$q\epsilon_1 \leq M_3 + M_1 + 2\|\widehat{p}\|nM_2$$

and

$$q \leq \epsilon_1^{-1}(M_3 + M_1 + 2\|\widehat{p}\|nM_2).$$ (11.239)

Lemma 11.6, (11.225), (11.237), and (11.238) imply that for every $i \in \{0, \ldots, q - 1\}$ and every pair of numbers $S_1, S_2 \in [T_i, T_{i+1}]$ satisfying $S_1 < S_2$,

$$\int_{S_1}^{S_2} (w(by(t)) - w(b\widehat{x}))dt$$

$$= -\int_{S_1}^{S_2} \delta(x(t), y(t), x'(t))dt + \widehat{p}(x(S_1) - x(S_2)) \geq -1 - 2\|\widehat{p}\|nM_2.$$
(11.240)

Define

$$J = \{i \in \{0, \ldots, q - 1\} : T_{i+1} - T_i \geq 2L_2 + 2L_1\}.$$ (11.241)

Let $i \in J$. In view of (11.241), the choice of L_2 (see (11.230) and (11.231)), (11.225), and (11.240), there exist numbers t_{i1}, t_{i2} such that

$$t_{i1} \in [T_i, L_2 + T_i], \ t_{i2} \in [T_{i+1} - L_2, T_{i+1}], \ \|x(t_{ij}) - \widehat{x}\| \le \epsilon_1, \ j = 1, 2.$$
(11.242)

It follows from (11.237), (11.238), (11.241), (11.242), and the choice of ϵ_1, L_1 (see (11.227)–(11.229)) that

$$\|x(t) - \widehat{x}\| \le \epsilon, \ t \in [t_{i1}, t_{i2}]$$
(11.243)

and if $S \in [t_{i1}, t_{i2} - L]$, then

$$\text{mes}(\{t \in [S, S + L]: \ \|y(t) - \widehat{x}\| > \epsilon\}) \le \epsilon.$$
(11.244)

Set

$$\mathcal{A} = \{[T_i, T_{i+1}]: \ i \in \{0, \ldots, q - 1\} \setminus J\} \cup \{[T_i, t_{i1}], \ [t_{i2}, T_{i+1}]: \ i \in J\}.$$
(11.245)

Evidently, the length of all the intervals belonging to $\mathcal{A}$ does not exceed $2L_2 + 2L_1 < l$.

In view of (11.232), (11.239), and (11.245), the number of elements of $\mathcal{A}$ does not exceed

$$4q \le 4\epsilon_1^{-1}(M_3 + M_1 + 2\|\widehat{p}\| n M_2) \le Q.$$

The inequalities above and (11.243) imply the validity of Theorem 11.5.

11.9 Stability of the Turnpike Phenomenon

In this chapter we prove the following turnpike result obtained in [116].

Theorem 11.26 *Let M, ϵ, L_1 be positive numbers and $\Gamma \in (0, 1)$. Then there exist a positive number L and a positive number γ such that for each $T > 2L$, each $z_0, z_1 \in R_+^n$ satisfying $z_0 \le Me$ and $az_1 \le \Gamma d^{-1}$, and each program $(x(t), y(t))_{t=0}^T$ which satisfies*

$$x(0) = z_0, \ x(T) \ge z_1,$$

$$\int_\tau^{\tau+L} w(by(t))dt \ge U(x(\tau), x(\tau + L), 0, L) - \gamma \text{ for all } \tau \in [0, T - L]$$

and

$$\int_{T-L}^{T} w(by(t))dt \geq U(x(T-L), z_1, 0, L) - \gamma$$

there are real numbers τ_1, τ_2 such that

$$\tau_1 \in [0, L], \ \tau_2 \in [T - L, T],$$

$$\|x(t) - \widehat{x}\| \leq \epsilon \text{ for all } t \in [\tau_1, \tau_2]$$

and that for each number S satisfying $\tau_1 \leq S \leq \tau_2 - L_1$,

$$mes(\{t \in [S, S + L_1] : \|y(t) - \widehat{x}\| > \epsilon\}) \leq \epsilon.$$

Moreover, if $\|x(0) - \widehat{x}\| \leq \gamma$, then $\tau_1 = 0$, and if $\|x(T) - \widehat{x}\| \leq \gamma$, then $\tau_2 = T$.

Let $-\infty < T_1 < T_2 < \infty$. A function $\phi : [T_1, T_2] \times R_+^n \rightarrow R^1$ is called $\mathcal{L} \times \mathcal{B}$-measurable if it is measurable with respect to the σ-algebra generated by products of Lebesgue subsets of $[T_1, T_2]$ and Borel subsets of R^n.

For each $M > 0$ and each function $\phi : R_+^n \rightarrow R^1$, set

$$\|\phi\|_M = \sup\{|\phi(z)| : z \in R_+^n \text{ and } z \leq Me\}. \tag{11.246}$$

Let numbers T_1, T_2 satisfy $0 \leq T_1 < T_2$, and let $W : [T_1, T_2] \times R_+^n \rightarrow R^1$ be an $\mathcal{L} \times \mathcal{B}$-measurable function which is bounded on bounded subsets of $[T_1, T_2] \times R_+^n$. For each $z_0, z_1 \in R_+^n$ set

$$U(z_0, z_1, T_1, T_2, W) = \sup \left\{ \int_{T_1}^{T_2} W(t, y(t))dt : \right.$$

$$\left. (x(t), y(t))_{t=T_1}^{T_2} \text{ is a program such that } x(T_1) = z_0, \ x(T_2) \geq z_1 \right\}, \tag{11.247}$$

$$U(z_0, T_1, T_2, W) = \sup \left\{ \int_{T_1}^{T_2} W(t, y(t))dt : \right.$$

$$\left. (x(t), y(t))_{t=T_1}^{T_2} \text{ is a program such that } x(T_1) = z_0 \right\}. \tag{11.248}$$

(Here we assume that supremum over the empty set is $-\infty$.)

Theorem 11.27 *Let M, L_1, ϵ be positive numbers and $\Gamma \in (0, 1)$. Then there exist $M_0 > M, L > 0, \delta > 0$ such that for each $T_1 \geq 0$, each $T_2 > T_1 + 2L$, each $z_0, z_1 \in R_+^n$ which satisfy $z_0 \leq Me$ and $az_1 \leq \Gamma d^{-1}$, each $\mathcal{L} \times \mathcal{B}$-measurable function $W : [T_1, T_2] \times R_+^n \rightarrow R^1$ which is bounded on bounded subsets of $[T_1, T_2] \times R_+^n$ and such that for almost every $t \in [T_1, T_2]$,*

$$\|W(t, \cdot) - w(b(\cdot))\|_{M_0} \leq \delta$$

and each program $(x(t), y(t))_{t=T_1}^{T_2}$ *which satisfies*

$$x(T_1) = z_0, \ x(T_2) \geq z_1,$$

$$\int_\tau^{\tau+L} W(t, y(t))dt \geq U(x(\tau), x(\tau + L), \tau, \tau + L, W) - \delta$$

for all $\tau \in [T_1, T_2 - L]$ *and*

$$\int_{T_2-L}^{T_2} W(t, y(t))dt \geq U(x(T_2 - L), z_1, T_2 - L, T_2, W) - \delta$$

there are real numbers τ_1, τ_2 *such that*

$$\tau_1 \in [T_1, T_1 + L], \ \tau_2 \in [T_2 - L, T_2],$$

$$\|x(t) - \widehat{x}\| \leq \epsilon \ \text{for all } t \in [\tau_1, \tau_2]$$

and that for each number S *satisfying* $\tau_1 \leq S \leq \tau_2 - L_1$,

$$mes(\{t \in [S, S + L_1] : \|y(t) - \widehat{x}\| > \epsilon\}) \leq \epsilon.$$

Moreover, if $\|x(T_1) - \widehat{x}\| \leq \delta$, *then* $\tau_1 = T_1$, *and if* $\|x(T_2) - \widehat{x}\| \leq \delta$, *then* $\tau_2 = T_2$.

Proof Theorem 11.27 follows easily from Theorem 11.26. Namely, let $L > 0$ and $\gamma > 0$ be as guaranteed by Theorem 11.26. Proposition 11.1 implies that there exists a positive number $M_0 > 0$ such that for every positive number T and every program $(x(t), y(t))_{t=0}^T$ satisfying $x(0) \leq Me$, the inequality $x(t) \leq M_0 e$ is true for all numbers $t \in [0, T]$. Set

$$\delta = \gamma \big(4^{-1}(2L + 1)\big)^{-1}.$$

Now it is easy to see that the assertion of Theorem 11.27 holds.

Theorem 11.27 and Lemma 11.21 imply the following result.

Theorem 11.28 *Let* M, L_1, ϵ *be positive numbers and* $\Gamma \in (0, 1)$. *Then there exist* $M_0 > M, L > 0, \delta > 0$ *such that for each* $T_1 \geq 0$, *each* $T_2 > T_1 + 2L$, *each* $z_0, z_1 \in R_+^n$ *which satisfy* $z_0 \leq Me$ *and* $az_1 \leq \Gamma d^{-1}$, *each* $\mathcal{L} \times \mathcal{B}$- *measurable function* $W : [T_1, T_2] \times R_+^n \to R^1$ *which is bounded on bounded subsets of* $[T_1, T_2] \times R_+^n$ *and such that for almost every* $t \in [T_1, T_2]$,

$$\|W(t, \cdot) - w(b(\cdot))\|_{M_0} \leq \delta$$

and each program $(x(t), y(t))_{t=T_1}^{T_2}$ *which satisfies*

$$x(T_1) = z_0, \ x(T_2) \geq z_1,$$

$$\int_{T_1}^{T_2} W(t, y(t))dt \geq U(z_0, z_1, T_1, T_2, W) - \delta$$

there are real numbers τ_1, τ_2 such that

$$\tau_1 \in [T_1, T_1 + L], \ \tau_2 \in [T_2 - L, T_2],$$

$$\|x(t) - \widehat{x}\| \leq \epsilon \ \text{for all } t \in [\tau_1, \tau_2]$$

and that for each number S satisfying $\tau_1 \leq S \leq \tau_2 - L_1$,

$$mes(\{t \in [S, S + L_1] : \|y(t) - \widehat{x}\| > \epsilon\}) \leq \epsilon.$$

Moreover, if $\|x(T_1) - \widehat{x}\| \leq \delta$, then $\tau_1 = T_1$, and if $\|x(T_2) - \widehat{x}\| \leq \delta$, then $\tau_2 = T_2$.

The following theorem is proved in Section 11.12. It was obtained in [116].

Theorem 11.29 *Let M, M_1, L_1, ϵ be positive numbers and $\Gamma \in (0, 1)$. Then there exist $M_* > 0, l > 0, \delta > 0$, a natural number Q such that for each $T_1 \geq 0$, each $T_2 > T_1 + Ql$, each $z_0, z_1 \in R_+^n$ which satisfy $z_0 \leq Me$ and $az_1 \leq \Gamma d^{-1}$, each $\mathcal{L} \times \mathcal{B}$-measurable function $W : [T_1, T_2] \times R_+^n \to R^1$ which is bounded on bounded subsets of $[T_1, T_2] \times R_+^n$ and such that for almost every $t \in [T_1, T_2]$,*

$$\|W(t, \cdot) - w(b(\cdot))\|_{M_*} \leq \delta$$

and each program $(x(t), y(t))_{t=T_1}^{T_2}$ which satisfies

$$x(T_1) = z_0, \ x(T_2) \geq z_1,$$

$$\int_{T_1}^{T_2} W(t, y(t))dt \geq U(z_0, z_1, T_1, T_2, W) - M_1$$

there exist a natural number $q \leq Q$ and monotone increasing finite sequences

$$\{a_i\}_{i=1}^q, \ \{b_i\}_{i=1}^q \subset [T_1, T_2]$$

such that

$$0 \leq b_i - a_i \leq l, \ i = 1, \ldots, q, \tag{11.249}$$

$$b_i \leq a_{i+1} \text{ for each integer } i \text{ satisfying } 1 \leq i \leq q - 1, \tag{11.250}$$

$$\|x(t) - \widehat{x}\| \leq \epsilon, \ t \in [T_1, T_2] \setminus \cup_{i=1}^q [a_i, b_i] \tag{11.251}$$

and if a number S satisfies

$$[S, S + L_1] \subset [T_1, T_2] \setminus \cup_{i=1}^{q}[a_i, b_i], \tag{11.252}$$

then

$$mes(\{t \in [S, S + L_1] : \|y(t) - \widehat{x}\| > \epsilon\}) < \epsilon. \tag{11.253}$$

11.10 Discount Case

Theorem 11.27 and Proposition 11.1 imply the following result.

Theorem 11.30 *Let M, L_1, ϵ be positive numbers and $\Gamma \in (0, 1)$. Then there exist $M_0 > 0, L > 0, \gamma > 0, \lambda > 1$ such that for each $T_1 \geq 0$, each $T_2 > T_1 + 2L$, each $z_0, z_1 \in R_+^n$ which satisfy $z_0 \leq Me$ and $az_1 \leq \Gamma d^{-1}$, each $\mathcal{L} \times \mathcal{B}$-measurable function $W : [T_1, T_2] \times R_+^n \to R^1$ which is bounded on bounded subsets of $[T_1, T_2] \times R_+^n$ and such that for almost every $t \in [T_1, T_2]$,*

$$\|W(t, \cdot) - w(b(\cdot))\|_{M_0} \leq \gamma,$$

each Lebesgue measurable function $\alpha : [T_1, T_2] \to (0, 1]$ such that for each $t_1, t_2 \in [T_1, T_2]$ satisfying $|t_2 - t_1| \leq L$ the inequality $\alpha(t_1)\alpha(t_2)^{-1} \leq \lambda$ holds and each program $(x(t), y(t))_{t=T_1}^{T_2}$ which satisfies

$$x(T_1) = z_0, \ x(T_2) \geq z_1,$$

$$\int_{\tau}^{\tau+L} \alpha(t)W(t, y(t))dt \geq U(x(\tau), x(\tau + L), \tau, \tau + L, \alpha W) - \gamma\alpha(\tau)$$

for all $\tau \in [T_1, T_2 - L]$ and

$$\int_{T_2-L}^{T_2} \alpha(t)W(t, y(t))dt \geq U(x(T_2 - L), z_1, T_2 - L, T_2, \alpha W) - \gamma\alpha(T_2 - L)$$

there are real numbers τ_1, τ_2 such that

$$\tau_1 \in [T_1, T_1 + L], \ \tau_2 \in [T_2 - L, T_2],$$

$$\|x(t) - \widehat{x}\| \leq \epsilon \text{ for all } t \in [\tau_1, \tau_2]$$

and that for each number S satisfying $\tau_1 \leq S \leq \tau_2 - L_1$,

$$mes(\{t \in [S, S + L_1] : \|y(t) - \widehat{x}\| > \epsilon\}) \leq \epsilon.$$

Moreover, if $\|x(T_1) - \widehat{x}\| \leq \delta$, then $\tau_1 = T_1$, and if $\|x(T_2) - \widehat{x}\| \leq \delta$, then $\tau_2 = T_2$.

11.11   Proof of Theorem 11.26

Proposition 11.1 implies that there exists a positive number M_0 such that the following property holds:

(P1) for every positive number T and every program $(x(t), y(t))_{t=0}^{T}$ satisfying $x(0) \leq Me$, the inequality $x(t) \leq M_0 e$ is valid for all $t \in [0, T]$.

Note that

$$a\widehat{x} = a_\sigma \left(1 + da_\sigma^{-1}\right) < d^{-1}.$$

Therefore we may assume without loss of generality that

$$a\widehat{x} < \Gamma d^{-1}, \tag{11.254}$$

$$\sup\{ay : y \in R^n \text{ and } \|y - \widehat{x}\| \leq \epsilon\} < \Gamma d^{-1}. \tag{11.255}$$

Theorem 11.4 implies that there exist $L_0 > L_1$ and a positive number γ such that the following property holds:

(P2) for every $T > 2L_0$, every pair of points $z_0, z_1 \in R_+^n$ satisfying $z_0 \leq M_0 e$, $az_1 \leq \Gamma d^{-1}$, and every program $(x(t), y(t))_{t=0}^{T}$ which satisfies

$$x(0) = z_0, \ x(T) \geq z_1, \tag{11.256}$$

$$\int_0^T w(by(t))dt \geq U(z_0, z_1, 0, T) - \gamma \tag{11.257}$$

there exist τ_1, τ_2 such that

$$\tau_1 \in [0, L_0], \ \tau_2 \in [T - L_0, T], \tag{11.258}$$

$$\|x(t) - \widehat{x}\| \leq \epsilon \text{ for all } t \in [\tau_1, \tau_2], \tag{11.259}$$

for every number S satisfying $\tau_1 \leq S \leq \tau_2 - L_1$,

$$\text{mes}(\{t \in [S, S + L_1] : \|y(t) - \widehat{x}\| > 4^{-1}\epsilon\}) \leq 4^{-1}\epsilon \tag{11.260}$$

and that

$$\text{if } \|x(0) - \widehat{x}\| \leq \gamma, \text{ then } \tau_1 = 0 \text{ and if } \|x(T) - \widehat{x}\| \leq \gamma, \text{ then } \tau_2 = T. \tag{11.261}$$

Set

$$L = 8L_0 + 4. \tag{11.262}$$

Let $z_0, z_1 \in R_+^n$ satisfy

$$z_0 \leq Me, \; az_1 \leq \Gamma d^{-1} \tag{11.263}$$

and let a number $T > 2L$. Assume that a program $(x(t), y(t))_{t=0}^{T}$ satisfies (11.251), for every $\tau \in [0, T - L]$, we have

$$\int_{\tau}^{\tau+L} w(by(t))dt \geq U(x(\tau), x(\tau + L), 0, L) - \gamma \tag{11.264}$$

and that

$$\int_{T-L}^{T} w(by(t))dt \geq U(x(T - L), z_1, 0, L) - \gamma. \tag{11.265}$$

We claim that there exist numbers τ_1, τ_2 such that

$$\tau_1 \in [0, L], \; \tau_2 \in [T - L, T], \tag{11.266}$$

Equation (11.259) is true, for every number S satisfying $\tau_1 \leq S \leq \tau_2 - L_1$, we have

$$\mathrm{mes}(\{t \in [S, S + L_1] : \|y(t) - \widehat{x}\| > \epsilon\}) \leq \epsilon$$

and that (11.261) is valid.
  In view of (11.256), (11.263), and property (P1), we have

$$x(t) \leq M_0 e, \; t \in [0, T]. \tag{11.267}$$

Consider the program $(x(t), y(t))_{t=T-L}^{T}$. In view of property (P2), applied to this program, (11.256), (11.262), (11.263), (11.265), and (11.267), there exist $\widetilde{\tau}_1, \widetilde{\tau}_2$ such that

$$\widetilde{\tau}_1 \in [T - L, T - L + L_0], \; \widetilde{\tau}_2 \in [T - L_0, T], \tag{11.268}$$

$$\|x(t) - \widehat{x}\| \leq \epsilon, \; t \in [\widetilde{\tau}_1, \widetilde{\tau}_2], \tag{11.269}$$

for every number S satisfying $\widetilde{\tau}_1 \leq S \leq \widetilde{\tau}_2 - L_1$, (11.260) is valid and that

if $\|x(T - L) - \widehat{x}\| \leq \gamma$, then $\widetilde{\tau}_1 = T - L$ and if $\|x(T) - \widehat{x}\| \leq \gamma$, then $\widetilde{\tau}_2 = T$.
$$\tag{11.270}$$

Suppose that

$$\{t \in [0, \widetilde{\tau}_1] : \|x(t) - \widehat{x}\| > \epsilon\} \neq \emptyset. \tag{11.271}$$

Define

$$t_0 = \sup\{t \in [0, \tilde{\tau}_1] : \|x(t) - \widehat{x}\| > \epsilon\}. \tag{11.272}$$

It is not difficult to see that

$$t_0 > 0, \ t_0 \leq \tilde{\tau}_1, \ \|x(t_0) - \widehat{x}\| = \epsilon. \tag{11.273}$$

We claim that

$$t_0 \leq 2L_0. \tag{11.274}$$

Assume the contrary. Then

$$t_0 > 2L_0$$

and there exists a number t_1 such that

$$0 \leq t_1 < t_0, \ t_1 > t_0 - L_0 > L_0, \ \|x(t_1) - \widehat{x}\| > \epsilon. \tag{11.275}$$

Consider the program $(x(t), y(t))_{t=t_1-L_0}^{t_0+L_0}$. In view of (11.275),

$$(t_0 + L_0) - (t_1 - L_0) = t_0 - t_1 + 2L_0 > 2L_0. \tag{11.276}$$

By (11.267),

$$x(t_1 - L_0) \leq M_0 e. \tag{11.277}$$

Equations (11.262), (11.268), and (11.272) imply that

$$t_0 < t_0 + L_0 \leq \tilde{\tau}_1 + L_0 \leq T - L + 2L_0 \leq T - L_0 \leq \tilde{\tau}_2. \tag{11.278}$$

It follows from (11.269), (11.272), and (11.278) that

$$\|x(t_0 + L_0) - \widehat{x}\| \leq \epsilon. \tag{11.279}$$

By (11.265) and (11.279),

$$ax(t_0 + L_0) < \Gamma d^{-1}. \tag{11.280}$$

By (11.262), (11.264), (11.267), (11.269), (11.275), (11.276), (11.280), Lemma 11.21, and (P2) applied to the program $(x(t), y(t))_{t=t_1-L_0}^{t_0+L_0}$, we have

$$\|x(t) - \widehat{x}\| \leq \epsilon \text{ for all } t \in [t_1, t_0]$$

and that

$$\|x(t_1) - \widehat{x}\| \le \epsilon.$$

This contradicts (11.275). The contradiction we have reached proves that $t_0 \le 2L_0$. Then in view of (11.268)–(11.270), (11.272), and (11.273),

$$0 \le t_0 \le 2L_0, \quad \tilde{\tau}_2 \in [T - L_0, T], \tag{11.281}$$

$$\|x(t) - \widehat{x}\| \le \epsilon \text{ for all } t \in [t_0, \tilde{\tau}_2], \tag{11.282}$$

$$\text{if } \|x(T) - \widehat{x}\| \le \gamma, \text{ then } \tilde{\tau}_2 = T. \tag{11.283}$$

There are two cases:

$$\|x(0) - \widehat{x}\| \le \gamma; \tag{11.284}$$

$$\|x(0) - \widehat{x}\| > \gamma. \tag{11.285}$$

If (11.285) is valid, then we put

$$\tau_1 = 2L_0, \quad \tau_2 = \tilde{\tau}_2. \tag{11.286}$$

Assume that (11.284) is true and consider the program $(x(t), y(t))_{t=0}^{3L_0}$. In view of (11.262), (11.267), (11.281), and (11.282),

$$x(0) \le M_0 e, \quad \|x(3L_0) - \widehat{x}\| \le \epsilon. \tag{11.287}$$

By (11.255) and (11.287),

$$ax(3L_0) < \Gamma d^{-1}. \tag{11.288}$$

Equations (11.264), (11.284), (11.287), (11.288), Lemma 11.21, and property (P2) applied to the program $(x(t), y(t))_{t=0}^{3L_0}$ imply that

$$\|x(t) - \widehat{x}\| \le \epsilon \text{ for all } t \in [0, 2L_0].$$

Combined with (11.281) and (11.282), the inequality above implies that

$$\|x(t) - \widehat{x}\| \le \epsilon, \quad t \in [0, \tilde{\tau}_2]. \tag{11.289}$$

Put

$$\tau_1 = 0, \quad \tau_2 = \tilde{\tau}_2. \tag{11.290}$$

In view of (11.289) and (11.290), we have

$$\|x(t) - \widehat{x}\| \le \epsilon \text{ for all } t \in [\tau_1, \tau_2]. \tag{11.291}$$

It follows from (11.181)–(11.283), (11.286), (11.290), and (11.291) that in both cases, we have defined $\tau_1, \tau_2 \in [0, T]$ such that:
   in the case of (11.284)

$$\tau_1 = 0 \tag{11.292}$$

and in the case of (11.285)

$$\tau_1 = 2L_0; \tag{11.293}$$

$$\tau_2 = \tilde{\tau}_2 \ge T - L_0, \tag{11.294}$$

$$\text{if } \|x(T) - \widehat{x}\| \le \gamma, \text{ then } \tau_2 = T, \tag{11.295}$$

$$\|x(t) - \widehat{x}\| \le \epsilon, \ t \in [\tau_1, \tau_2]. \tag{11.296}$$

Assume that a real number S satisfies

$$2L_0 \le S \le \tau_2 - L_1. \tag{11.297}$$

We claim that

$$\text{mes}(\{t \in [S, S + L_1] : \|y(t) - \widehat{x}\| > \epsilon\}) \le \epsilon. \tag{11.298}$$

If $S \ge \tilde{\tau}_1$, then (11.298) follows from (11.260), (11.270), (11.294), and the choice of $\tilde{\tau}_1, \tilde{\tau}_2$. Consider the case with

$$S < \tilde{\tau}_1 \le T - L + L_0 \tag{11.299}$$

(see (11.268)). By (11.262), (11.295), and (11.299),

$$S + 2L_0 \le T - L + 3L_0 < T - L_0 \le \tilde{\tau}_2 = \tau_2. \tag{11.300}$$

Consider the program $(x(t), y(t))_{t=S-2L_0}^{S+2L_0}$. By (11.264), (11.297), (11.298), and Lemma 11.21,

$$\int_{S-2L_0}^{S+2L_0} w(by(t))dt \ge U(x(S - 2L_0), x(S + 2L_0), 0, 4L_0) - \gamma. \tag{11.301}$$

By (11.292)–(11.294), (11.296), and (11.300),

$$\|x(S + 2L_0) - \widehat{x}\| \le \epsilon.$$

Combined with (11.255) this implies that

$$ax(S + 2L_0) < \Gamma d^{-1}. \tag{11.302}$$

Equations (11.267), (11.301), (11.302), and (P2) imply that (11.298) holds. Thus for every number S satisfying (11.297), relation (11.298) is valid.

Assume that (11.284) is valid,

$$S \in [0, 2L_0] \tag{11.303}$$

and consider the program $(x(t), y(t))_{t=0}^{4L_0}$. By (11.262), (11.284), (11.292), (11.295), and (11.296),

$$\|x(0) - \widehat{x}\| \leq \gamma, \tag{11.304}$$

$$\|x(4L_0) - \widehat{x}\| \leq \epsilon. \tag{11.305}$$

In view of (11.255) and (11.305),

$$ax(4L_0) < \Gamma d^{-1}. \tag{11.306}$$

It follows from (11.262), (11.264), and Lemma 11.21 that

$$\int_0^{4L_0} w(by(t))dt \geq U(x(0), x(4L_0), 0, 4L_0) - \gamma. \tag{11.307}$$

In view of (11.303)–(11.307) and (P2), inequality (11.298) is valid. This completes the proof of Theorem 11.26.

11.12   Proof of Theorem 11.29

We suppose that the sum over the empty set is zero.

In view of Proposition 11.1, there exists a positive number $M_0 > M$ such that the following property holds:

(P6) for every positive number T and every program $(x(t), y(t))_{t=0}^T$ satisfying $x(0) \leq Me$, the inequality $x(t) \leq M_0 e$ is true for all $t \in [0, T]$, and if $S \in [0, T]$, $\tau > 0$ and $(u(t), v(t))_{t=S}^{S+\tau}$ is a program with $u(S) = x(S)$, then $u(t) \leq M_0 e$ for all $t \in [S, S + \tau]$.

We may assume without loss of generality that

$$a\widehat{x} < \Gamma d^{-1}, \tag{11.308}$$

$$\sup \{ay : y \in R_+^n \text{ and } \|y - \widehat{x}\| \leq \epsilon\} < \Gamma d^{-1}. \tag{11.309}$$

Lemma 11.19 implies that there exists a positive number m_0 such that the following property holds:

(P7) for every positive number T and every program $(x(t), y(t))_{t=0}^T$ satisfying $x(0) \leq M_0 e$, we have

$$\int_0^T [w(by(t)) - w(b\widehat{x})]dt \leq m_0.$$

Lemma 11.17 implies that there exists a number $m_1 > m_0$ such that for every $z \in R_+^n$ and every positive number T, we have

$$U(z, T) \geq Tw(b\widehat{x}) - m_1. \tag{11.310}$$

Lemma 11.18 implies that there exists $m_2 > m_1$ such that the following property holds:

(P8) for every $z_0 \in R_+^n$, every $z_1 \in R_+^n$ satisfying $az_1 \leq \Gamma d^{-1}$, and every number $T > m_2$, we have

$$U(z_0, z_1, 0, T) \geq Tw(b\widehat{x}) - m_2.$$

Theorem 11.28 and (11.309) imply that there exist $M_* > M_0, L_0 > 0, \delta_1 \in (0, \epsilon)$ such that the following property holds:

(P9) for each $T_1 \geq 0$, each $T_2 > T_1 + 2L_0$, each $\mathcal{L} \times \mathcal{B}$-measurable function $W : [T_1, T_2] \times R_+^n \to R^1$ which is bounded on bounded subsets of $[T_1, T_2] \times R_+^n$ and such that for almost every $t \in [T_1, T_2]$,

$$\|W(t, \cdot) - w(b(\cdot))\|_{M_*} \leq \delta_1$$

and each program $(x(t), y(t))_{t=T_1}^{T_2}$ which satisfies

$$x(T_1) \leq M_0 e,$$

$$\|x(T_i) - \widehat{x}\| \leq \delta_1, \ i = 1, 2,$$

$$\int_{T_1}^{T_2} W(t, y(t))dt \geq U(x(T_1), x(T_2), T_1, T_2, W) - \delta_1$$

we have

$$\|x(t) - \widehat{x}\| \leq \epsilon \text{ for all } t \in [T_1, T_2]$$

and for each number S satisfying $T_1 \leq S \leq T_2 - L_1$,

$$\text{mes}(\{t \in [S, S + L_1] : \|y(t) - \widehat{x}\| > \epsilon\}) \leq \epsilon.$$

In view of Lemma 11.23, there exists an integer $L_2 \geq 1$ such that the following property holds:

(P10) for every program $(x(t), y(t))_{t=0}^{L_2}$ which satisfies

$$x(0) \leq M_0 e, \quad \int_0^{L_2} w(by(t))dt \geq L_2 w(b\widehat{x}) - M_1 - m_0 - m_2 - 1$$

there exists $t \in [0, L_2]$ such that

$$\|x(t) - \widehat{x}\| \leq \delta_1.$$

Choose

$$l > 4L_1 + 4L_2 + 4m_2 + 4L_0, \tag{11.311}$$

a number δ such that

$$0 < \delta < 8^{-1}\delta_1, \quad 4\delta m_2 + 4\delta L_2 < 1, \tag{11.312}$$

and a natural number

$$Q > 4 + 4\delta_1^{-1} M_1. \tag{11.313}$$

Assume that

$$T_1 \geq 0, \ T_2 > T_1 + Ql, \tag{11.314}$$

$$z_0 \in R_+^n, \ z_0 \leq Me, \tag{11.315}$$

$W : [T_1, T_2] \times R_+^n \to R^1$ is an $\mathcal{L} \times \mathcal{B}$-measurable function which is bounded on bounded subsets of $[T_1, T_2] \times R_+^n$ and such that for almost every $t \in [T_1, T_2]$, we have

$$\|W(t, \cdot) - w(b(\cdot))\|_{M_*} \leq \delta \tag{11.316}$$

and that a program $(x(t), y(t))_{t=T_1}^{T_2}$ and $z_1 \in R_+^n$ satisfy

$$x(T_1) = z_0, \tag{11.317}$$

$$az_1 \leq \Gamma d^{-1}, x(T_2) \geq z_1, \tag{11.318}$$

$$\int_{T_1}^{T_2} W(t, y(t))dt \geq U(z_0, z_1, T_1, T_2, W) - M_1. \tag{11.319}$$

Property (P6), (11.315), and (11.317) imply that

$$x(t) \leq M_0 e \text{ for all } t \in [0, T].\tag{11.320}$$

We show that the following property holds:

(P11) Let a number S satisfy $T_1 \leq S \leq T_2 - 2L_2 - m_2$, and let there exists $\tilde{x}(S + 2L_2 + m_2) \in R_+^n$ such that

$$\tilde{x}(S + 2L_2 + m_2) \leq x(S + 2L_2 + m_2),\tag{11.321}$$

$$a\tilde{x}(S + 2L_2 + m_2) \leq \Gamma d^{-1},\tag{11.322}$$

$$\int_S^{S+2L_2+m_2} W(t, y(t))dt \geq U(x(S), \tilde{x}(S + 2L_2 + m_2), S, S$$

$$+2L_2 + m_2, W) - M_1.\tag{11.323}$$

Then there exists $\tau \in [S, S + L_2]$ such that

$$\|x(\tau) - \hat{x}\| \leq \delta_1.$$

Let a number S satisfy $T_1 \leq S \leq T_2 - 2L_2 - m_2$, and let (11.321)–(11.323) be valid with $\tilde{x}(S+2L_2+m_2) \in R_+^n$. Property (P6), (11.312), (11.316), and (11.320)–(11.323) imply that

$$\int_S^{S+2L_2+m_2} w(by(t))dt$$

$$\geq \int_S^{S+2L_2+m_2} W(t, y(t))dt - \delta(2L_2 + m_2)$$

$$\geq U(x(S), \tilde{x}(S + 2L_2 + m_2), S, S + 2L_2 + m_2, W) - M_1 - \delta(2L_2 + m_2)$$

$$\geq U(x(S), \tilde{x}(S + 2L_2 + m_2), 0, 2L_2 + m_2) - M_1 - 2\delta(2L_2 + m_2)$$

$$\geq U(x(S), \tilde{x}(S + 2L_2 + m_2), 0, 2L_2 + m_2) - M_1 - 1.\tag{11.324}$$

Combined with (11.310), (11.321)–(11.323), and property (P8), this implies that

$$\int_S^{S+2L_2+m_2} w(by(t))dt \geq (2L_2 + m_2)w(b\hat{x}) - m_2 - M_1 - 1.\tag{11.325}$$

In view of (11.325) and property (P7), we have

$$\int_S^{S+L_2} w(by(t))dt = \int_S^{S+2L_2+m_2} w(by(t))dt - \int_{S+L_2}^{S+2L_2+m_2} w(by(t))dt$$

$$\geq (2L_2 + m_2)w(b\hat{x}) - m_2 - M_1 - 1$$

$$- (L_2 + m_2)w(b\hat{x}) - m_0$$

$$= L_2 w(b\widehat{x}) - m_2 - M_1 - m_0 - 1. \tag{11.326}$$

In view of (11.320), (11.326), and (P10) there exists a number $\tau \in [S, S + L_2]$ such that $\|x(\tau) - \widehat{x}\| \leq \delta_1$. Hence property (P11) holds.

By property (P11), Lemma 11.21, (11.309), (11.318), and (11.319), the following property holds:

(P12) Let a number S satisfy $T_1 \leq S \leq T_2 - 2L_2 - m_2$ and

$$\|x(S + 2L_2 + m_2) - \widehat{x}\| < \epsilon.$$

Then there exists a number $\tau \in [S, S + L_2]$ for which $\|x(\tau) - \widehat{x}\| \leq \delta_1$.

Property (P11), (11.318), and (11.319) imply that there exists a number S_1 such that

$$S_1 \in [T_2 - 2L_2 - m_2, T_2 - L_2 - m_2], \quad \|x(S_1) - \widehat{x}\| \leq \delta_1. \tag{11.327}$$

By induction applying (P12), we construct a finite strictly decreasing sequence of real numbers $S_1, \ldots, S_p$ where p is a natural number such that

$$\|x(S_i) - \widehat{x}\| \leq \delta_1, \quad i = 1, \ldots, p, \tag{11.328}$$

$$S_p \in [0, 2L_2 + m_2),$$

Equation (11.327) is valid, and for all integers i satisfying $1 \leq i < p$, we have

$$S_i - S_{i+1} \in [L_2, 2L_2 + m_2]. \tag{11.329}$$

By induction we construct a finite strictly increasing sequence of real numbers

$$\tau_1, \ldots, \tau_q \in \{S_i : i = 1, \ldots, p\} \tag{11.330}$$

such that

$$\tau_1 = S_p,$$

for each integer $i \in \{1, \ldots, q\}$ satisfying $i < q$ we have: if

$$\int_{\tau_i}^{S_1} W(t, y(t))dt \geq U(x(\tau_i), x(S_1), \tau_i, S_1, W) - \delta_1, \tag{11.331}$$

then

$$q = i + 1, \quad \tau_q = S_1; \tag{11.332}$$

otherwise

$$\tau_{i+1} = \min\{\tau \in \{S_1, \ldots, S_p\} : \ \tau > \tau_i \text{ and}$$

$$\int_{\tau_i}^{\tau} W(t, y(t))dt < U(x(\tau_i), x(\tau), \tau_i, \tau, W) - \delta_1\}. \tag{11.333}$$

It follows from (11.318), (11.319), (11.330)–(11.333), and Lemma 11.21 that

$$M_1 \geq U(x(T_1), x(T_2), T_1, T_2, W) - \int_{T_1}^{T_2} W(t, y(t))dt$$

$$\geq \sum \{U(x(\tau_i), x(\tau_{i+1}), \tau_i, \tau_{i+1}, W) - \int_{\tau_i}^{\tau_{i+1}} W(t, y(t))dt :$$

$$i \in \{1, \ldots, q\} \text{ and } i < q - 1\} \geq (q - 2)\delta_1,$$

$$q \leq 2 + \delta_1^{-1} M_1. \tag{11.334}$$

Define

$$A = \{i \in \{1, \ldots, q\} : \ i < q, \ \tau_{i+1} - \tau_i \geq l\}. \tag{11.335}$$

Let $i \in A$ be given. In view of (11.311), (11.329)–(11.333), and (11.335), there exists a natural number $j \in \{1, \ldots, p\}$ such that

$$\tau_{i+1} - 2L_2 - m_2 \leq S_{j(i)} \leq \tau_{i+1}, \tag{11.336}$$

$$\int_{\tau_i}^{S_{j(i)}} W(t, y(t))dt \geq U(x(\tau_i), x(S_{j(i)}), \tau_i, S_{j(i)}, W) - \delta_1. \tag{11.337}$$

It follows from (11.312), (11.316), (11.328), (11.330), (11.336), (11.337), and (P9) that

$$\|x(t) - \widehat{x}\| \leq \epsilon, \ t \in [\tau_i, S_{j(i)}]$$

and that for every number S satisfying

$$\tau_i \leq S \leq S_{j(i)} - L_1$$

we have

$$\text{mes}(\{t \in [S, S + L_1] : \ \|y(t) - \widehat{x}\| \geq \epsilon\}) < \epsilon.$$

In order to complete the proof, it is sufficient to note that

$$[T_1, T_2] \setminus \cup\{[\tau_i, S_{j(i)}] : \ i \in A\}$$

is a finite number of intervals, their maximal length does not exceed l, and in view of (11.313) and (11.334), their number does not exceed

$$2(q+2) \leq 4 + 4\delta_1^{-1} M_1 < Q.$$

This completes the proof of Theorem 11.29.

11.13 Optimal Programs over Infinite Horizon

A program $(x(t), y(t))_{t=0}^{\infty}$ is called weakly maximal if for every positive number T,

$$\int_0^T w(by(t))dt = U(x(0), x(T), 0, T).$$

We prove the following result.

Theorem 11.31 *A program $(x(t), y(t))_{t=0}^{\infty}$ is overtaking optimal if and only if it is good and weakly maximal.*

Proof Assume that a program $(x(t), y(t))_{t=0}^{\infty}$ is overtaking optimal. Let us show that it is good. Assume the contrary. Then in view of Proposition 11.2, it is bad and

$$\lim_{T \to \infty} \int_0^T (w(by(t)) - w(b\widehat{y}))dt = -\infty. \tag{11.338}$$

Proposition 11.3 implies that there exists a good program $(\bar{x}(t), \bar{y}(t))_{t=0}^{\infty}$ which satisfies $\bar{x}(0) = x(0)$. Then there exists a number $M \in R^1$ such that

$$\int_0^T (w(b\bar{y}(t)) - w(b\widehat{y}))dt \geq M \text{ for all positive numbers } T.$$

In view of (11.338) and the inequality above, we have

$$\lim_{T \to \infty} \left[\int_0^T w(b\bar{y}(t))dt - \int_0^T w(by(t))dt \right] = \infty,$$

a contradiction. Hence the program $(x(t), y(t))_{t=0}^{\infty}$ is good.

Let us show that the program $(x(t), y(t))_{t=0}^{\infty}$ is weakly maximal. Assume the contrary. Then there exists a positive number T_0 such that

$$\int_0^{T_0} w(by(t))dt < U(x(0), x(T_0), 0, T_0).$$

This implies that there exists a program $(\bar{x}(t), \bar{y}(t))_{t=0}^{T_0}$ such that

$$\bar{x}(0) = x(0), \quad \bar{x}(T_0) \geq x(T_0),$$

$$\int_0^{T_0} w(b\bar{y}(t))dt > \int_0^{T_0} w(by(t))dt. \tag{11.339}$$

Lemma 11.20 and (11.339) imply that there exists a program $(\tilde{x}(t), \tilde{y}(t))_{t=T_0}^{\infty}$ such that

$$\tilde{x}(T_0) = \bar{x}(T_0), \quad \tilde{y}(t) = y(t), \quad t \in [T_0, \infty). \tag{11.340}$$

Set

$$x_1(t) = \bar{x}(t), \ t \in [0, T_0], \ x_1(t) = \tilde{x}(t), \ t \in (T_0, \infty),$$

$$y_1(t) = \bar{y}(t), \ t \in [0, T_0], \ y_1(t) = \tilde{y}(t) = y(t), \ t \in (T_0, \infty). \tag{11.341}$$

In view of (11.339)–(11.341), $(x_1(t), y_1(t))_{t=0}^{\infty}$ is a program. By (11.339)–(11.341), for all numbers $T > T_0$, we have

$$\int_0^T w(by_1(t))dt - \int_0^T w(by(t))dt$$

$$= \int_0^{T_0} w(b\bar{y}(t))dt - \int_0^{T_0} w(by(t))dt > 0$$

and

$$\limsup_{T \to \infty} \left[\int_0^T w(by_1(t))dt - \int_0^T w(by(t))dt \right] > 0,$$

a contradiction. The contradiction we have reached proves that the program $(x(t), y(t))_{t=0}^{\infty}$ is weakly maximal.

Assume that a program $(x(t), y(t))_{t=0}^{\infty}$ is good and weakly maximal. We show that it is overtaking optimal. In view of Theorem 11.5, there exists an overtaking optimal program $(\tilde{x}(t), \tilde{y}(t))_{t=0}^{\infty}$ such that

$$\tilde{x}(0) = x(0). \tag{11.342}$$

Evidently,

$$\limsup_{T \to \infty} \left[\int_0^T w(by(t))dt - \int_0^T w(b\tilde{y}(t))dt \right] \leq 0.$$

In order to complete the proof, it is sufficient to show that

$$\limsup_{T\to\infty} \left[\int_0^T w(b\tilde{y}(t))dt - \int_0^T w(by(t))dt \right] \leq 0.$$

Assume the contrary. Then there exist a positive number ϵ and a strictly increasing sequence of positive numbers $\{T_k\}_{k=1}^\infty$ such that

$$\lim_{k\to\infty} T_k = \infty$$

and that for all integers $k \geq 1$, we have

$$\int_0^{T_k} w(b\tilde{y}(t))dt - \int_0^{T_k} w(by(t))dt > \epsilon. \tag{11.343}$$

Fix $\Delta > 0$ for which

$$w(b(\widehat{x} + \Delta e)) - w(b(\widehat{x} - \Delta e)) < \epsilon/2. \tag{11.344}$$

In view of Lemma 11.22, there exists $\delta \in (0, \Delta)$ such that the following property holds:

(P) for every pair of points $z, z' \in R_+^n$ which satisfy

$$\|z - \widehat{x}\|, \ \|z' - \widehat{x}\| \leq \delta$$

and every $\tau \in [2^{-1}, 2]$, there exists a program $(u(t), v(t))_{t=0}^\tau$ such that

$$u(0) = z, \ u(\tau) \geq z',$$

$$\|u(t) - \widehat{x}\|, \ \|v(t) - \widehat{x}\| \leq \Delta, \ t \in [0, \tau],$$

$$\|u'(t)\| \leq \Delta, \ t \in [0, \tau].$$

Since the programs $(x(t), y(t))_{t=0}^\infty$ and $(\tilde{x}(t), \tilde{y}(t))_{t=0}^\infty$ are good, it follows from Theorem 11.4 that

$$\lim_{t\to\infty} x(t) = \widehat{x}, \ \lim_{t\to\infty} \tilde{x}(t) = \widehat{x},$$

and there exists a positive number S such that

$$\|\tilde{x}(t) - \widehat{x}\| \leq \delta, \ \|x(t) - \widehat{x}\| \leq \delta \text{ for all } t \geq S. \tag{11.345}$$

Fix an integer $k \geq 1$ such that

$$T_k > S. \tag{11.346}$$

In view of (11.345) and (11.346), we have

$$\|x(T_k) - \widehat{x}\| \le \delta, \ \|x(T_k + 1) - \widehat{x}\| \le \delta,$$
$$\|\tilde{x}(T_k) - \widehat{x}\| \le \delta, \ \|\tilde{x}(T_k + 1) - \widehat{x}\| \le \delta. \tag{11.347}$$

It follows from (11.347) and property (P) that there exists a program

$$(x_0(t), y_0(t))_{t=0}^{T_k+1}$$

such that

$$x_0(t) = \tilde{x}(t), \ y_0(t) = \tilde{y}(t), \ t \in [0, T_k], \tag{11.348}$$

$$x_0(T_k + 1) \ge x(T_k + 1), \tag{11.349}$$

$$\|x_0(t) - \widehat{x}\|, \ \|y_0(t) - \widehat{x}\|, \ \|x_0'(t)\| \le \Delta, \ t \in [T_k, T_k + 1]. \tag{11.350}$$

In view of (11.342) and (11.348), we have

$$x_0(0) = x(0). \tag{11.351}$$

It follows from (11.343) and (11.348) that

$$\int_0^{T_k+1} w(by_0(t))dt - \int_0^{T_k+1} w(by(t))dt$$

$$= \int_0^{T_k} w(b\tilde{y}(t))dt - \int_0^{T_k} w(by(t))dt$$

$$+ \int_{T_k}^{T_k+1} w(by_0(t))dt - \int_{T_k}^{T_k+1} w(by(t))dt$$

$$> \epsilon + \int_{T_k}^{T_k+1} w(by_0(t))dt - \int_{T_k}^{T_k+1} w(by(t))dt. \tag{11.352}$$

By (11.345) and (11.346), for for almost every $t \in [T_k, T_{k+1}]$, we have

$$y(t) \le x(t) \le \widehat{x} + \delta e. \tag{11.353}$$

In view of (11.350), for almost every $t \in [T_k, T_{k+1}]$,

$$y_0(t) \ge \widehat{x} - \Delta e(\sigma). \tag{11.354}$$

It follows from (11.344), (11.353), and (11.354) that for almost every $t \in [T_k, T_{k+1}]$,

$$w(by_0(t)) - w(by(t)) \ge w(b(\widehat{x} - \Delta e(\sigma))) - w(b(\widehat{x} + \delta e))$$
$$\ge w(b(\widehat{x} - \Delta e)) - w(b(\widehat{x} + \Delta e)) > -\epsilon/2.$$

Combined with (11.352) the relation above implies that

$$\int_0^{T_k+1} w(by_0(t))dt - \int_0^{T_k+1} w(by(t))dt \geq \epsilon - \epsilon/2.$$

This contradicts (11.349) and (11.351). The contradiction we have reached proves that the program $(x(t), y(t))_{t=0}^{\infty}$ is overtaking optimal. Theorem 11.31 is proved.

References

1. Artstein Z, Leizarowitz A (1985) Tracking periodic signals with the overtaking criterion. IEEE Trans Autom Control AC-30:1123–1126
2. Aseev SM, Kryazhimskiy AV (2004) The Pontryagin Maximum principle and transversality conditions for a class of optimal control problems with infinite time horizons. SIAM J Control Optim 43:1094–1119
3. Aseev SM, Veliov VM (2012) Maximum principle for infinite-horizon optimal control problems with dominating discount. Dyn Contin. Discrete Impuls Syst B 19:43–63
4. Aseev SM, Veliov VM (2012) Necessary optimality conditions for improper infinite-horizon control problems. Oper Res Proc 2011:21–26
5. Aseev SM, Krastanov MI, Veliov VM (2017) Optimality conditions for discrete-time optimal control on infinite horizon. Pure Appl Funct Anal 2:395–409
6. Atsumi H (1965) Neoclassical growth and the efficient program of capital accumulation. Rev Econ Stud 3:127–136
7. Aubry S, Le Daeron PY (1983) The discrete Frenkel-Kontorova model and its extensions I. Phys D 8:381–422
8. Bachir M, Blot J (2015) Infinite dimensional infinite-horizon Pontryagin principles for discrete-time problems. Set-Valued Variational Anal 23:43–54
9. Bachir M, Blot J (2017) Infinite dimensional multipliers and Pontryagin principles for discrete-time problems. Pure Appl Funct Anal 2:411–426
10. Basu K, Mitra T (2007) Utilitarianism for infinite utility streams: a new welfare criterion and its axiomatic characterization. J. Econ. Theory 133:350–373
11. Baumeister J, Leitao A, Silva GN (2007) On the value function for nonautonomous optimal control problem with infinite horizon. Syst Control Lett 56:188–196
12. Blot J (2009) Infinite-horizon Pontryagin principles without invertibility. J Nonlinear Convex Anal 10:177–189
13. Blot J, Cartigny P (2000) Optimality in infinite-horizon variational problems under sign conditions. J Optim Theory Appl 106:411–419
14. Blot J, Hayek N (2000) Sufficient conditions for infinite-horizon calculus of variations problems. ESAIM Control Optim Calc Var 5:279–292
15. Blot J, Hayek N (2014) Infinite-horizon optimal control in the discrete-time framework. Springer briefs in optimization. Springer, New York
16. Bright I (2012) A reduction of topological infinite-horizon optimization to periodic optimization in a class of compact 2-manifolds. J Math Anal Appl 394:84–101

A. J. Zaslavski, *Turnpike Theory for the Robinson–Solow–Srinivasan Model*,
Springer Optimization and Its Applications 166,
https://doi.org/10.1007/978-3-030-60307-6

17. Brock WA (1970) On existence of weakly maximal programmes in a multi-sector economy. Rev Econ Stud 37:275–280
18. Carlson DA (1990) The existence of catching-up optimal solutions for a class of infinite horizon optimal control problems with time delay. SIAM J Control Optim 28:402–422
19. Carlson DA, Haurie A, Leizarowitz A (1991) Infinite horizon optimal control. Springer, Berlin
20. Cartigny P, Michel P (2003) On a sufficient transversality condition for infinite horizon optimal control problems. Autom J IFAC 39:1007–1010
21. Coleman BD, Marcus M, Mizel VJ (1992) On the thermodynamics of periodic phases. Arch. Rational Mech Anal 117:321–347
22. Damm T, Grune L, Stieler M, Worthmann K (2014) An exponential turnpike theorem for dissipative discrete time optimal control problems. SIAM J Control Optim 52:1935–1957
23. De Oliveira VA, Silva GN (2009) Optimality conditions for infinite horizon control problems with state constraints. Nonlinear Anal 71: 1788–1795
24. Gaitsgory V, Rossomakhine S, Thatcher N (2012) Approximate solution of the HJB inequality related to the infinite horizon optimal control problem with discounting. Dyn Contin Discrete Impuls Syst B 19:65–92
25. Gaitsgory V, Grune L, Thatcher N (2015) Stabilization with discounted optimal control. Syst Control Lett 82:91–98
26. Gaitsgory V, Mammadov M, Manic L (2017) On stability under perturbations of long-run average optimal control problems. Pure Appl Funct Anal 2:461–476
27. Gaitsgory V, Parkinson A, Shvartsman I (2017) Linear programming formulations of deterministic infinite horizon optimal control problems in discrete time. Discrete Contin Dyn Syst B 22:3821–3838
28. Gale D (1967) On optimal development in a multi-sector economy. Rev Econ Stud 34:1–18
29. Glizer VY, Kelis O (2017) Upper value of a singular infinite horizon zero-sum linear-quadratic differential game. Pure Appl Funct Anal 2:511–534
30. Gugat M, Trelat E, Zuazua, E (2016) Optimal Neumann control for the 1D wave equation: finite horizon, infinite horizon, boundary tracking terms and the turnpike property. Syst Control Lett 90:61–70
31. Guo X, Hernandez-Lerma O (2005) Zero-sum continuous-time Markov games with unbounded transition and discounted payoff rates. Bernoulli 11:1009–1029
32. Hammond PJ (1974) Consistent planning and intertemporal welfare economics. University of Cambridge, Cambridge
33. Hammond PJ (1975) Agreeable plans with many capital goods. Rev Econ Stud 42:1–14
34. Hammond PJ, Mirrlees JA (1973) Agreeable plans, models of economic growth. Mirrlees J, Stern N.H. (eds). Wiley, New York, pp 283–299
35. Hayek N (2011) Infinite horizon multiobjective optimal control problems in the discrete time case. Optimization 60:509–529
36. Jasso-Fuentes H, Hernandez-Lerma O (2008) Characterizations of overtaking optimality for controlled diffusion processes. Appl Math Optim 57:349–369
37. Khan MA, Mitra T (2005) On choice of technique in the Robinson-Solow-Srinivasan model. Int J Econ Theory 1:83–110
38. Khan MA, Mitra T (2006) Optimal growth in the two-sector RSS model: a continuous time analysis. In: Proceedings of the seventh portugese conference on automatic control, Electronic publication
39. Khan MA, Mitra T (2006) Undiscounted optimal growth in the two-sector Robinson-Solow-Srinivasan model: a synthesis of the value-loss approach and dynamic programming. Econ Theory 29:341–362
40. Khan MA, Mitra T(2006) Discounted optimal growth in the two-sector RSS model: a geometric investigation. Adv Math Econ 8:349–381
41. Khan MA, Mitra T (2007) Optimal growth in a two-sector RSS model without discounting: a geometric investigation. Jpn Econ Rev 58:191–225
42. Khan MA, Mitra T (2007) Optimal growth under discounting in the two-sector Robinson-Solow-Srinivasan model: a dynamic programming approach. J Differ Equ Appl 13:151–168

43. Khan MA, Mitra T (2008) Growth in the Robinson-Solow-Srinivasan model: undiscounted optimal policy with a strictly concave welfare function. J. Math. Econ. 44:707–732
44. Khan MA, Mitra T (2012) Impatience and dynamic optimal behavior: a bifurcation analysis of the Robinson-Solow-Srinivasan model. Nonlinear Anal. 75, 1400–1418
45. Khan MA, Mitra T (2013) Discounted optimal growth in a two-sector RSS model: a further geometric investigation. Adv Math Econ 17:39–70
46. Khan MA, Piazza A (2010) On the non-existence of optimal programs in the Robinson-Solow-Srinivasan (RSS) model. Econ. Lett. 109:94–98
47. Khan MA, Piazza A (2011) The economics of forestry and a set-valued turnpike of the classical type. Nonlinear Anal 74:171–181
48. Khan MA, Piazza A (2011) The concavity assumption on felicities and asymptotic dynamics in the RSS model. Set-Valued Var Anal 19:135–156
49. Khan MA, Piazza A (2011) An overview of turnpike theory: towards the discounted deterministic case. Adv Math Econ 14:39–67
50. Khan MA, Piazza A (2012) On the Mitra-Wan forestry model: a unified analysis. J Econ Theory 147:230–260
51. Khan MA, Zaslavski AJ (2007) On a uniform turnpike of the third kind in the Robinson-Solow-Srinivasan model. J Econ 92:137–166
52. Khan MA, Zaslavski AJ (2009) On existence of optimal programs: the RSS model without concavity assumptions on felicities. J. Math Econ 45:624–633
53. Khan MA, Zaslavski AJ (2010) On two classical turnpike results for the Robinson-Solow-Srinivisan (RSS) model. Adv Math Econ 13:47–97
54. Khan MA, Zaslavski AJ (2010) On locally optimal programs in the RSS (Robinson-Solow-Srinivasan) model. J Econ 99:65–92
55. Khlopin DV (2017) On Lipschitz continuity of value functions for infinite horizon problem. Pure Appl Funct Anal 2:535–552
56. Kolokoltsov V, Yang W (2012) The turnpike theorems for Markov games. Dyn Games Appl 2:294–312
57. Leizarowitz A (1985) Infinite horizon autonomous systems with unbounded cost. Appl. Math. Opt. 13:19–43
58. Leizarowitz A (1986) Tracking nonperiodic trajectories with the overtaking criterion. Appl. Math. Opt. 14:155–171
59. Leizarowitz A, Mizel VJ (1989) One dimensional infinite horizon variational problems arising in continuum mechanics. Arch Rational Mech Anal 106:161–194
60. Lykina V, Pickenhain S, Wagner M (2008) Different interpretations of the improper integral objective in an infinite horizon control problem. J Math Anal Appl 340:498–510
61. Makarov VL, Rubinov AM (1977) Mathematical theory of economic dynamics and equilibria. Springer, New York
62. Malinowska AB, Martins N, Torres DFM (2011) Transversality conditions for infinite horizon variational problems on time scales. Optim Lett 5:41–53
63. Mammadov M (2014) Turnpike theorem for an infinite horizon optimal control problem with time delay. SIAM J Control Optim 52:420–438
64. Marcus M, Zaslavski AJ (1999) On a class of second order variational problems with constraints. Isr J Math 111:1–28
65. Marcus M, Zaslavski AJ (1999) The structure of extremals of a class of second order variational problems. Ann Inst H Poincaré Anal Non Linéaire 16:593–629
66. Marcus M, Zaslavski AJ (2002) The structure and limiting behavior of locally optimal minimizers. Ann Inst H Poincaré Anal Non Linéaire 19:343–370
67. McKenzie LW (1976) Turnpike theory. Econometrica 44:841–866
68. Mordukhovich BS (1990) Minimax design for a class of distributed parameter systems. Autom. Remote Control 50:1333–1340
69. Mordukhovich BS (2011) Optimal control and feedback design of state-constrained parabolic systems in uncertainly conditions. Appl Anal 90:1075–1109

70. Mordukhovich BS, Nam NM (2014) An easy path to convex analysis and applications. Morgan Claypool Publishers, San Rafael
71. Mordukhovich BS, Shvartsman I (2004) Optimization and feedback control of constrained parabolic systems under uncertain perturbations, Optimal control, stabilization and nonsmooth analysis. Lecture notes control inform. sci., Springer, Berlin, 121–132
72. Ocana Anaya E, Cartigny P, Loisel P (2009) Singular infinite horizon calculus of variations. Applications to fisheries management. J. Nonlinear Convex Anal. 10:157–176
73. Okishio N (1966) Technical choice under full employment in a socialist economy. Econ J 76:585–592
74. Pickenhain S, Lykina V, Wagner M (2008) On the lower semicontinuity of functionals involving Lebesgue or improper Riemann integrals in infinite horizon optimal control problems. Control Cybern. 37:451–468
75. Porretta A, Zuazua E (2013) Long time versus steady state optimal control. SIAM J Control Optim 51:4242–4273
76. Radner R (1961) Paths of economics growth that are optimal with regard only to final states: a turnpike theorem. Rev Econ Stud 28:98–104
77. Robinson J (1960) Exercises in economic analysis. MacMillan, London
78. Robinson J (1969) A model for accumulation proposed by J.E. Stiglitz. Econ J 79:412–413
79. Rockafellar RT (1970) Convex analysis. Princeton University Press, Princeton
80. Rubinov AM (1984) Economic dynamics. J. Soviet Math. 26:1975–2012
81. Sagara N (2018) Recursive variational problems in nonreflexive Banach spaces with an infinite horizon: an existence result. Discrete Contin Dyn Syst S 11:1219–1232
82. Samuelson PA (1965) A catenary turnpike theorem involving consumption and the golden rule. Am Econ Rev 55:486–496
83. Solow RM (1962) Substitution and fixed proportions in the theory of capital. Rev Econ Stud 29:207–218
84. Srinivasan TN (1962) Investment criteria and choice of techniques of production. Yale Econ Essays 1:58–115
85. Stiglitz JE (1968) A note on technical choice under full employment in a socialist economy. Econ J 78:603–609
86. Stiglitz JE (1970) Reply to Mrs. Robinson on the choice of technique. Econ J 80:420–422
87. Stiglitz JE (1973) Recurrence of techniques in a dynamic economy. Models of economic growth, Mirrlees J, Stern NH (eds.) Wiley, New York, 283–299
88. Trelat E, Zhang C, Zuazua E (2018) Optimal shape design for 2D heat equations in large time. Pure Appl Funct Anal 3:255–269
89. von Weizsacker CC (1965) Existence of optimal programs of accumulation for an infinite horizon. Rev Econ Stud 32:85–104
90. Zaslavski AJ (1987) Ground states in Frenkel-Kontorova model. Math. USSR Izvestiya 29:323–354
91. Zaslavski AJ (1999) Turnpike property for dynamic discrete time zero-sum games. Abstract Appl Anal 4:21–48
92. Zaslavski AJ (2005) Optimal programs in the RSS model. Int J Econ Theory 1:151–165
93. Zaslavski AJ (2006) Turnpike properties in the calculus of variations and optimal control. Springer, New York
94. Zaslavski AJ (2006) Good programs in the RSS model with a nonconcave utility function. J. Ind Manag Optim 2:399–423
95. Zaslavski AJ (2007) Turnpike results for a discrete-time optimal control systems arising in economic dynamics. Nonlinear Anal. 67:2024–2049
96. Zaslavski AJ (2008) A turnpike result for a class of problems of the calculus of variations with extended-valued integrands. J Convex Anal 15:869–890
97. Zaslavski AJ (2009) Structure of approximate solutions of variational problems with extended-valued convex integrands. ESAIM Control Optim Calculus Var 15:872–894

98. Zaslavski AJ (2009) The structure of good programs in the RSS Model. In: Proceedings of annual meeting at the Kyoto University, RIMS Kokyuroku:1654, 166–178, Departmental Bulletin Paper, repository.kulib.kyoto-u.ac.jp

99. Zaslavski AJ (2009) Good solutions for a class of infinite horizon discrete-time optimal control problems. Taiwanese J Math 13:1637–1669

100. Zaslavski AJ (2009) Turnpike results for a class of infinite horizon discrete-time optimal control problems arising in economic dynamics. Set-Valued Variational Anal 17:285–318

101. Zaslavski AJ (2009) Existence and structure of solutions for an infinite horizon optimal control problem arising in economic dynamics. Adv Differ Equ 14:477–496

102. Zaslavski AJ (2010) Optimal solutions for a class of infinite horizon variational problems with extended-valued integrands. Optimization 59:181–197

103. Zaslavski AJ (2010) Good locally maximal programs for the Robinson-Solow-Srinivasan (RSS) model. Adv Math Econ 13:161–176

104. Zaslavski AJ (2010) Overtaking optimal solutions for a class of infinite horizon discrete-time optimal control problems. Dyn Contin Discrete Impulsive Syst B 17:607–620

105. Zaslavski AJ (2010) Locally maximal solutions of control systems arising in economic dynamics. Commun Appl Nonlinear Anal 17:61–68

106. Zaslavski AJ (2010) Structure of approximate solutions for discrete-time control systems arising in economic dynamics. Nonlinear Anal 73:952–970

107. Zaslavski AJ (2011) Two turnpike results for a continuous-time optimal control systems. In: Proceedings of an international conference, complex analysis and dynamical systems IV: function theory and optimization, vol 553, pp 305–317

108. Zaslavski AJ (2011) Stability of a turnpike phenomenon for the Robinson-Solow-Srinivasan model. Dyn Syst Appl 20:25–44

109. Zaslavski AJ (2011) The existence and structure of approximate solutions of dynamic discrete time zero-sum games. J Nonlinear Convex Anal 12:49–68

110. Zaslavski AJ (2011) A turnpike property of approximate solutions of an optimal control problem arising in economic dynamics. Dyn Syst Appl 20:395–422

111. Zaslavski AJ (2012) Weakly agreeable programs for the Robinson-Solow-Srinivasan (RSS) model. In: Optimization theory and related topics. Contemporary Mathematics, vol 568, American Mathematical Society, Providence, pp 259–271

112. Zaslavski AJ (2013) Structure of solutions of variational problems. SpringerBriefs in optimization. Springer, New York

113. Zaslavski AJ (2013) Structure of approximate solutions of optimal control problems. SpringerBriefs in Optimization. Springer, New York

114. Zaslavski AJ (2013) On turnpike properties of approximate optimal programs of the discrete-time Robinson-Solow-Srinivasan model. Commun Appl Anal 17:129–145

115. Zaslavski AJ (2013) On a class of discrete-time optimal control problems arising in economic dynamics. Pan-Am Math J 23:1–12

116. Zaslavski AJ (2013) Stability of the turnpike phenomenon for a convex optimal control problem arising in economic dynamics. Commun Appl Anal 17:271–288

117. Zaslavski AJ (2014) Turnpike phenomenon and infinite horizon optimal control. Springer optimization and its applications. Springer, Cham

118. Zaslavski AJ (2014) Structure of approximate solutions of dynamic continuous-time zero-sum games. J Dyn Games 1:153–179

119. Zaslavski AJ (2014) Structure of solutions of variational problems with extended-valued integrands in the regions close to the endpoints. Set-Valued Variational Anal 22:809–842

120. Zaslavski AJ (2016) Structure of solutions of optimal control problems on large intervals: a survey of recent results. Pure Appl Funct Anal 1:123–158

121. Zaslavski AJ (2017) Discrete-time optimal control and games on large intervals. Springer optimization and its applications. Springer, Cham

122. Zaslavski AJ (2018) Equivalence of optimality criterions for discrete time optimal control problems. Pure Appl Funct Anal 3:505–517

123. Zaslavski AJ (2019) Turnpike conditions in infinite dimensional optimal control. Springer optimization and its applications. Springer, Cham
124. Zaslavski AJ (2019) Optimal control problems arising in the forest management. Springer-Briefs in optimization. Springer, Cham
125. Zaslavski AJ, Leizarowitz A (1997) Optimal solutions of linear control systems with nonperiodic integrands. Math Oper Res 22:726–746
126. Zaslavski AJ, Leizarowitz A (1998) Optimal solutions of linear periodic control systems with convex integrands. Appl Math Optim 37:127–150

Index

Printed in the United States
by Baker & Taylor Publisher Services